M. Horbatsch Quantum Mechanics Using **Maple**®

Springer
Berlin
Heidelberg
New York
Barcelona
Budapest
Hong Kong
London
Milan
Paris
Tokyo

Marko Horbatsch

Quantum Mechanics Using Maple®

With 75 Exercises
and Cross Platform Diskette
Containing 39 Guided Maple Sesssions

Professor Dr. Marko Horbatsch
Department of Physics and Astronomy
York University
4700 Keele St
Toronto, Ontario, M3J 1P3
Canada

Cover graphic explanation:
The cover displays a color-coded contour diagram of a two-dimensional harmonic oscillator wave-function for a state with two excitation quanta in the horizontal direction and three in the vertical. Such graphs and equivalent surface plots are easily created in Maple V. Explicit instructions for their generation on a computer screen or with a color inkjet printer are included in a separate worksheet on the diskette.

ISBN 3-540-58875-2 Springer-Verlag Berlin Heidelberg New York

CIP data applied for

Typesetting: Camera-ready copy from the authors using a Springer TEX macro package
SPIN 10480684 56/3144 - 5 4 3 2 1 0 - Printed on acid-free paper

Preface

The need to introduce computers into the university curriculum in the sciences has been stressed for many years. For a long time this meant training students in problem solving using numerical analysis and traditional programming languages coupled with graphing. Training in this area has been emphasized as it has been, and probably will remain, one of the more practical aspects of the students' university education. The lack of a universal platform that combines the numerical computing and graphing capabilities of a computer, as well as the complexity of numerical analysis and programming, presents a hindrance to introducing these tools in courses that would benefit from the presentation of solutions that are not (easily) accessible by analytic means.

With the advent of powerful, yet inexpensive personal computers and computer algebra systems (CASs) that exist on a variety of platforms it is possible to rectify the situation. It is also possible to mollify the lecturer who is critical of numerical techniques and results (usually for aesthetic reasons) that exact computations are feasible in this environment. The main advantage, however, of these CASs is that they encourage intuitive understanding by making graphing available literally at the press of a button. Also, it is possible to get away from the drudgery of calculations for a time, while concentrating on the ideas.

Alas, the critical voices will counter this argument, *'Will these students know how to calculate anything?'*; *'Will they be able to survive without the notebook computer on their lap?'* The complete answer to these questions is not obvious and will have to be evaluated in the future. Is it possible to work in CASs and actually not know how the calculation is done? The answer is 'yes' and 'no'. Another important question is *'How much overhead is involved in working with these CASs?'* The latter issue of spending time to understand what went wrong in a particular CAS calculation, and how to be critical of its results, is coupled strongly with knowing how the calculational goal can be achieved and how a CAS can go about it. Presumably all CASs that are currently popular have their weaknesses, and developers are working on how to remove these.

This book is concerned with the use of an increasingly popular CAS, namely Maple, for the exploration of quantum mechanics. One may ask why the physicist picks this subject in order to gain familiarity with a modern CAS. Currently there is a great push for more application-oriented studies in the sciences. With this text I offer a CAS-based collection of milestone problems in quantum mechanics (following in a number of problems the well-known classic text by Siegfried Flügge). I hope to convince the pragmatists that learning quantum mechanics can be accompanied by very useful training for future engineering-type physicists that will outwit their competitors in problem-solving skills. At the same time I hope to provide an enjoyable environment for learning the practical aspects of quantum mechanics for those who missed this opportunity at university and wish to pursue the subject in self-study. I do emphasize, however, that a beginner of the subject will also need to consult a standard text on quantum mechanics or modern physics.

As a course instructor, does one have to know Maple to recommend the use of the worksheets to students? Frankly, I don't know. It is worth trying and referring the students to the Maple literature. At York university I managed to introduce the problems in the form of assignments in a quantum mechanics course for students who had not been exposed to Maple previously. The students did learn very quickly, but also fell into traps occasionally (as I did when I learned Maple, as well as when I figured out how to do some of the problems). It helps if one can get assistance before getting frustrated too much. I do hope that the book provides much of it. In addition, it is possible to consult the growing Maple literature and an electronic bulletin board with specific questions. If a CAS is adopted as a learning tool at a university, it makes sense to offer help through a helpdesk that could be operated by graduate students.

The game plan for the present textbook is as follows: there are some sections (worksheets) that present the mathematical development of materials covered traditionally in class, either in an introductory course (often called Modern Physics) or in a regular quantum mechanics course. Simple modifications of the relevant section worksheet allow one to perform the exercises presented usually at the end of the section. Sections that fall into this category are 1.1, 1.2, 1.3, 1.5, 2.3, 3.1, 3.2, 3.4, 4.1, 4.2, 4.4, 4.5, 5.1, 5.3, 6.1, 6.2, 6.3, 7.1, 7.2, 7.3, 7.4. Then, there are sections describing important topics or methods that can be left to independent additional study, not necessarily covered in depth in class. They could form the basis of small independently conducted studies. Sections in this category are 1.4, 1.6, 2.1, 2.2, 3.3, 4.3, 5.2, 5.4, 6.4. More specialized techniques for the future researcher are found in sections 2.4, 2.5, 2.6, 3.5, 7.5. Whether the numerical approach to the Schrödinger equation that is presented in these sections is considered to be essential in a course on quantum mechanics depends, of course, on the taste of the lecturer. Finally, Chapter 8 deals with special functions and orthogonal polynomials that can be dealt with either in a separate course on advanced calculus or as part of the quantum mechanics course.

Most sections or worksheets can be used in stand-alone mode, i.e., there is only a minimum of interdependence (a counterexample is Section 7.5 which cannot be understood without dealing with 2.4). This causes some repetition in Chapter 8, but permits the lecturer to assign worksheet problems without following the strict order. Sections 1.1 to 1.3 can be considered a crash-course in Maple with examples from introductory quantum mechanics.

The text is restricted to nonrelativistic quantum mechanics. The worksheets contain minimal text explanations in order to be useful even if the book is not close by. The graphs are not included as they would take up much disk space, but they can be produced easily in a Maple session. The naming of the files is consistent with the chapters in the text. An additional worksheet that covers elements of relativistic quantum mechanics, namely the Dirac equation, has been included on the disk. Also included on the disk is the Maple worksheet that produced the graphics for the cover. While I do not make too much use of 3D graphics in the text, I stress that the user should creatively explore Maple's capabilities in this respect to learn through visualization.

Finally, I wish to express my sincere thanks to all my teachers and colleagues who helped me to gain an understanding of quantum mechanics and who supported me in my quest to use Maple for pedagogical purposes. Some research papers with pedagogical value for beginning graduate students in atomic or few-particle physics found their way into the bibliography to encourage students to begin to read the journal literature. I wish to thank Dr. H.J. Kölsch from the Springer-Verlag in Heidelberg for his encouragement to produce the book. Dr. Victoria Wicks and the production team assisted me to put the camera-ready TEX-manuscript into final form. The responsibility for errors (of which I hope there aren't too many) is, of course, only mine. I will appreciate comments by mail (postal or electronic). Please send them to

Marko Horbatsch
Department of Physics and Astronomy
York University
4700 Keele St
Toronto, Ontario, M3J 1P3
CANADA

marko @ theory2.phys.yorku.ca

How to use this book

Given the special format of the text and the fact that electronic worksheets are included with it, some explanations are in order. A complete integration of the electronic medium with the text was attempted. Each section in the eight chapters corresponds to one worksheet on the diskette, and thus represents a self-contained entity. All Maple commands required in the given section are printed in the text together with their output. The output is sometimes truncated in the book to save space. The Maple commands together with their mathematics and plot output are embedded in the explanatory text. Some text information is also provided in the Maple worksheets to make them more useful even if the book is not at hand. Only standard Maple is used, i.e., no special external Maple packages have to be pre-loaded.

The Maple worksheets as included on the diskette can be used on any computer running Maple in the windowing environment. It is possible to convert them to worksessions without the windowing interface (e.g., for the straight DOS-Maple user). For details of how to read the worksheets the reader is referred to the README file on the diskette. A prerequisite to make direct use of the worksheets is Maple V Release 2 or Release 3 (or higher) with a windowing interface (the worksheets are `.ms` and not `.m` files). Both the full research or student versions can run all of the worksheets. In the PC environment I suggest you use them on a 486-based (or equivalent) machine as a minimum platform.

The text deviates in some respects from standard typesetting standarts. Each section represents a stand-alone unit and, therefore, begins on a separate page. The end of each section provides exercises that range in scope from minor modifications of the worksheet to small projects for independent study. The graphs are sometimes so small that the legends on the axes are hard to read. This I consider a small penalty for maintaining a continuous flow in the book, and particularly for allowing it to be read without running the worksheet simultaneously on a computer. It is straightforward for the non-expert of Maple to prepare oneself using the book, and to proceed at the workstation using the text on the side.

At last some explanation for the different typefaces employed in the text. Maple input commands are typeset using a `typewriter` style font. The Maple prompt (not to be entered) is displayed as a $>$ sign. Maple output is provided in math style and as such cannot be distinguished, a priori, from typeset formulae. It can be recognized in that it always follows a Maple input command and has no punctuation at the end. Also no use is made of symbols such as $\hbar$ in it. Instead a variable name entered in Maple as `hbar` is printed as hbar. The Maple output for the text could have been modified to show an $\hbar$ symbol, but then the text would not reflect the screen output during a session. For readability the printed Maple output was changed to the norm when it came to matters such as *italics* vs. roman display of long variable names.

Table of Contents

1. Problems in One Dimension

1.1 Introduction

While I do not assume the student has much familiarity with Maple, some previous experience based on introductory texts [CG+93, EJ+92, He93, Ba94] or reference books [BM94, Re93] is extremely helpful. In this introductory chapter I discuss the one-dimensional Schrödinger equation in the coordinate space representation. I begin with this topic to permit the non-expert of Maple to make some first steps.

I recommend you browse through as many texts on the subject of quantum mechanics (QM) as possible to obtain an overview of the different tastes that exist as far as the style of presentation is concerned. I provide an overview of reference texts sorted according to category:

- (a) introductory texts that emphasize atomic physics applications, usually used in modern physics courses [Ga79, An71, McG71, Mo90, Li92];
- (b) intermediate-level QM texts that emphasize applications or practical problems [Ga74, BM89, Gr93, Fl71, BD91, ER74];
- (c) QM texts that emphasize the theoretical development [Sch68, Da69, Bl64, Sa94, Mes70, CD+77, Mer70, BS57, LL77, Ba90, To92];
- (d) computer-algebra related texts [Fe94, Th94].

No use is made in the text of the path integral formulation of QM[FH65], even though it is very useful for a further understanding of QM and relativistic quantum field theory[Ka93]. I also refer to texts that explain the mathematics relevant to quantum mechanics [AS65, Ar85, GR80, CH89]. Furthermore, an undergraduate education in physics is considered incomplete these days without ample use of numerical computing. Some references to the literature in this field are [PF+92, Ko86, Th92, Ac90, Zw92, LP82, VK81].

We begin with the case of a free particle of mass m and linear momentum p. The classical Hamiltonian

$$H = \frac{p^2}{2m}$$

is quantized according to the prescription

$$p \to p_{\mathrm{op}} = -\mathrm{i}\hbar \frac{\mathrm{d}}{\mathrm{d}x} .$$

The Schrödinger Hamiltonian for a free particle moving in one dimension (1D) therefore becomes

$$H_{\text{op}} = -\frac{\hbar^2}{2m}\frac{\mathrm{d}^2}{\mathrm{d}x^2}\,.$$

How do we implement this in Maple? There are two ways to define derivatives: one is through the function **diff**, which permits us to differentiate n times, but requires Maple expressions as an argument. Thus we define a function to execute $H\psi$ that requires the wavefunction ψ as an argument. We assume that ψ is an expression (not a function) that depends on x.

```
> Ham:=psi->-hbar^2/(2*m)*diff(psi,x$2);
```

$$\mathrm{Ham} := \psi \to -\frac{1}{2}\frac{\mathrm{hbar}^2\,\mathrm{diff}(\psi, x\,\$\,2)}{m}$$

```
> Ham(exp(I*k*x));
```

$$\frac{1}{2}\frac{\mathrm{hbar}^2\,k^2\,\mathrm{e}^{(I\,k\,x)}}{m}$$

We observe that the chosen wavefunction $\exp(ikx)$ apparently is an eigenfunction of the free Hamiltonian: the function $\phi = H\psi$ equals a constant (eigenvalue) times the original function ψ. In analogy with linear algebra (or analytic geometry) we can say that the action of H_{op} on this particular ψ is such that the resulting vector in function space $|\phi\rangle$ is not rotated, but just stretched with respect to $|\psi\rangle$.

The energy eigenvalue corresponding to the plane wave $\exp(ikx)$ with wave number k is

$$E = \frac{(\hbar k)^2}{2m}\,.$$

This can also be expressed in terms of the particle's momentum $p = \hbar k$. While acting with H_{op} on the eigenfunction in this form we obtain:

```
> Ham(exp(I/hbar*p*x));
```

$$\frac{1}{2}\frac{p^2\,\mathrm{e}^{\left(\frac{I\,p\,x}{\mathrm{hbar}}\right)}}{m}$$

We would like to demonstrate that the eigenvalue $p^2/(2m)$ is doubly degenerate. The plane wave describes a flux of particles with momentum p in the positive x direction. We can also have a flux in the opposite direction with the same energy:

```
> Ham(exp(-I/hbar*p*x));
```

$$\frac{1}{2}\frac{p^2\,\mathrm{e}^{\left(-\frac{I\,p\,x}{\mathrm{hbar}}\right)}}{m}$$

Furthermore, we see that the momentum value is known precisely at the same time that the energy is sharp. To demonstrate this property we define the momentum operator itself:

```
> Pop:=psi->-I*hbar*diff(psi,x);
```

$$Pop := \psi \to -I\,\text{hbar}\left(\frac{\partial}{\partial x}\,\psi\right)$$

```
> Pop(exp(I/hbar*p*x));
```

$$p\,e^{\left(\frac{I\,p\,x}{\text{hbar}}\right)}$$

```
> Pop(exp(-I/hbar*p*x));
```

$$-p\,e^{\left(-\frac{I\,p\,x}{\text{hbar}}\right)}$$

We can ask ourselves the question: *what is the probability distribution for the 'free particle' to be found in coordinate space?* We declare a function that calculates the magnitude squared for a complex expression. The **evalc** function forces the complex calculation to display real and imaginary parts.

```
> rho:=psi->evalc(conjugate(psi)*psi);
```

$$\rho := \psi \to \text{evalc}(\,\text{conjugate}(\,\psi\,)\,\psi\,)$$

```
> Pofx:=rho(exp(I/hbar*p*x));
```

$$Pofx := \cos\left(\frac{p\,x}{\text{hbar}}\right)^2 + \sin\left(\frac{p\,x}{\text{hbar}}\right)^2$$

Here we run into one of Maple's quirks. It is not trimmed to make use automatically of well-known trigonometric relations. It is obvious to us that the answer is 1. Maple has to be told that the trigonometric relation should be employed to simplify the expression, which is done below. Also note that in Maple's math display $\sin^2 x$ is shown in a way that suggests $\sin(x^2)$. This is to avoid confusion with repeated function application (functional composition), i.e., $\sin^2$ is reserved in Maple for $\sin(\sin x)$. Therefore, at present, the onus is on the Maple user to realize where to imagine the brackets in math display in order to understand clearly what the result means.

```
> Pofx:=combine(Pofx,trig);
```

$$Pofx := 1$$

We show the graph for the Maple novice to learn the command, and also to emphasize that there is a normalization problem with this wavefunction. The squared magnitude of the wavefunction $\psi^*(x)\psi(x)$ cannot be interpreted as a probability density in this case as it sums to infinity if all x are allowed for $-\infty < x < \infty$.

```
> plot(Pofx,x=-10..10,title=`Probability density for
> plane wave`);
```

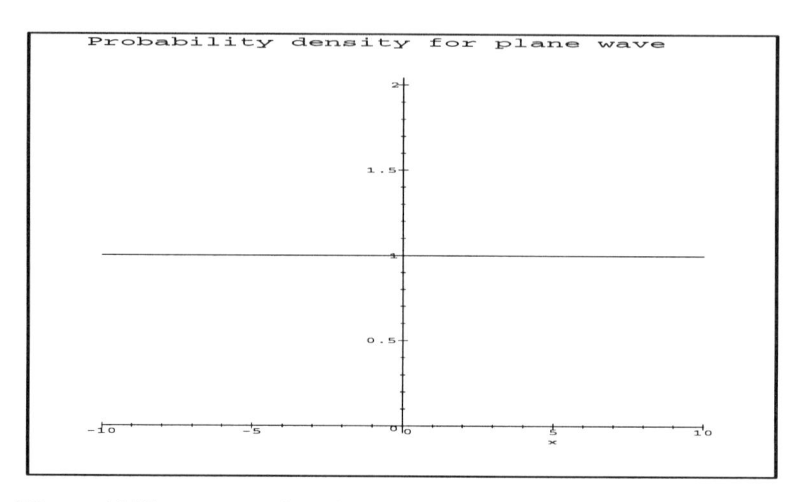

We would like to normalize this state such that the sum over all probabilities to find the particle at any location x gives unity (otherwise $\rho(x)$ is not a probability density). We introduce the dot product in Hilbert space and allow space to be truncated at $\pm L$:

```
> dotpro:=(psi,phi,L)->int(evalc(conjugate(psi)*phi),x=-L..L);
```

$$\text{dotpro} := (\,\psi, \phi, L\,) \to \int_{-L}^{L} \text{evalc}(\,\text{conjugate}(\,\psi\,)\,\phi\,)\,dx$$

```
> dotpro(exp(I*k*x),exp(I*k*x),infinity);
```

$$\infty$$

```
> dotpro(exp(I*k*x),exp(I*k*x),L);
```

$$2\,L$$

```
> dotpro(exp(I*k*x)/sqrt(L),exp(I*k*x)/sqrt(L),L);
```

$$2L\left(\cos\left(\frac{1}{4}(\,-1+\text{signum}(\,L\,)\,)\,\pi\right)^2 + \sin\left(\frac{1}{4}(\,-1+\text{signum}(\,L\,)\,)\,\pi\right)^2\right)/(\,|L|\,)$$

What's gone wrong here? Maple is concerned about the sign of L.

```
> assume(L>0);
> dotpro(exp(I*k*x)/sqrt(2*L),exp(I*k*x)/sqrt(2*L),L);
```

$$1$$

We see that the plane wave state can be normalized such that it permits a probabilistic interpretation, but that the normalization constant depends on the choice of volume of space.

For states describing the flux of free (or interacting) particles one has two choices:

- (a) Enclose the system in a box of size $2L$. The wavefunction vanishes outside the box, i.e., $\psi(x) = 0$ for $x \leq -L$ and for $x \geq L$. The wavefunction is normalized in the usual sense. The energy eigenvalues for the scattering states become discrete (see below).
- (b) Consider the system in the limit $L \to \infty$. The energy eigenvalues for the scattering states become continuous. Impose delta-function normalization.

The inner product in the function space can be used to demonstrate the orthonormality (orthogonality) of the scattering states for different momenta p and q.

```
> dotpro(exp(I/hbar*p*x)/sqrt(2*L),exp(I/hbar*q*x)/sqrt(2*L),L);
```

$$\frac{1}{2}\,\frac{\left(\sin\left(\frac{L\tilde{}\,(-p+q)}{\mathrm{hbar}}\right) - I\cos\left(\frac{L\tilde{}\,(-p+q)}{\mathrm{hbar}}\right)\right)\mathrm{hbar}}{L\tilde{}\,(-p+q)} + \frac{1}{2}\,\frac{\left(\sin\left(\frac{L\tilde{}\,(-p+q)}{\mathrm{hbar}}\right) + I\cos\left(\frac{L\tilde{}\,(-p+q)}{\mathrm{hbar}}\right)\right)\mathrm{hbar}}{L\tilde{}\,(-p+q)}$$

```
> limit(",L=infinity);
```

$$0$$

Let us emphasize that Maple, in order to save CPU cycles, performs by default only a minimum of simplifications. Remedy comes from applying `simplify`:

```
> simplify(dotpro(exp(I/hbar*p*x)/sqrt(2*L),exp(I/hbar*q*x)
> /sqrt(2*L),L));
```

$$\frac{\mathrm{hbar}\sin\left(\frac{L\tilde{}\,(-p+q)}{hbar}\right)}{L\tilde{}\,(-p+q)}$$

We may wonder why it is that the states for finite L are not orthogonal. The problem is buried in the fact that p and q are still arbitrary. To see that they, in fact, cannot be chosen arbitrarily, if box normalization is imposed, one has to solve the 'bound-state' problem of a particle enclosed in a box. The boundary condition that ψ vanishes at $x = +L$ and $-L$ is equivalent to the demand that the potential becomes infinite at these points. This justifies the notion of a particle that cannot escape from the box.

Suppose the box is quite finite, $L = 2$. We can see quickly that the criterion of a vanishing wave function at $x = -L$, L can only be satisfied for particular momenta. Also, the plane-wave solutions are not quite suitable as $\exp(ikx) = \cos(kx) + i\sin(kx)$ never actually vanishes. We see that we can have real symmetric solutions of pure cosine-type and also anti-symmetric solutions of pure sine-type.

```
> Ham(cos(k*x));
```

$$\frac{1}{2}\frac{\mathrm{hbar}^2 \cos(k\,x)\,k^2}{m}$$

```
> Ham(sin(k*x));
```

$$\frac{1}{2}\frac{\mathrm{hbar}^2 \sin(k\,x)\,k^2}{m}$$

Now observe that the demand for a vanishing wavefunction at $x = -2, 2$ selects k

```
> plot({cos(1*x),cos(2*x),cos(3*x)},x=-2..2);
```

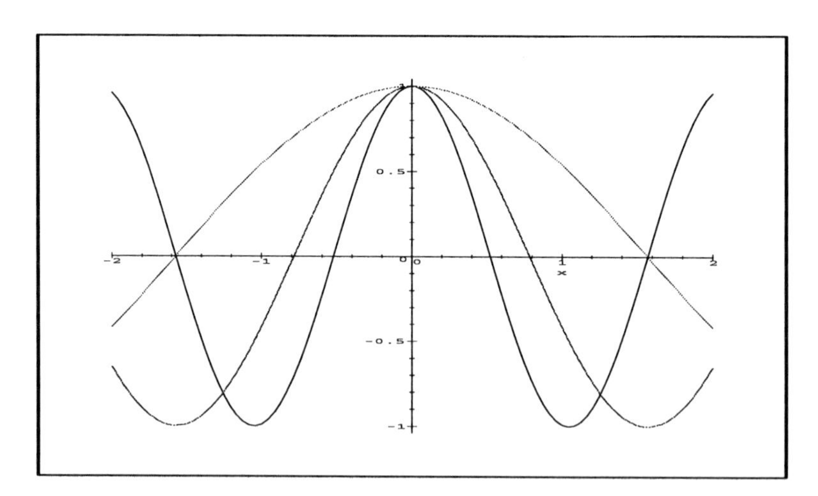

None of the above choices of $k(= 1, 2, 3)$ is suitable for $L = 2$. We have to impose the boundary condition. Due to the symmetry (anti-symmetry) of the eigenfunctions we have one condition, namely $\psi(x = L) = 0$.

```
> bcS:=L->cos(k*L)=0;
```

$$bcS := L \rightarrow \cos(k\,L) = 0$$

Next we wish to solve the condition for the symmetric eigenfunctions with the choice $L = 2$:

```
> solve(bcS(2),k);
```

$$\frac{1}{4}\,\pi$$

Maple returns just one value of k. This can't be everything, as is evident from a graphical solution, i.e., a graph of the left-hand side of bcS with roots given by those k values at which the graph crosses the abscissa, i.e., the k axis:

```
> plot(lhs(bcS(2)),k=0..10);
```

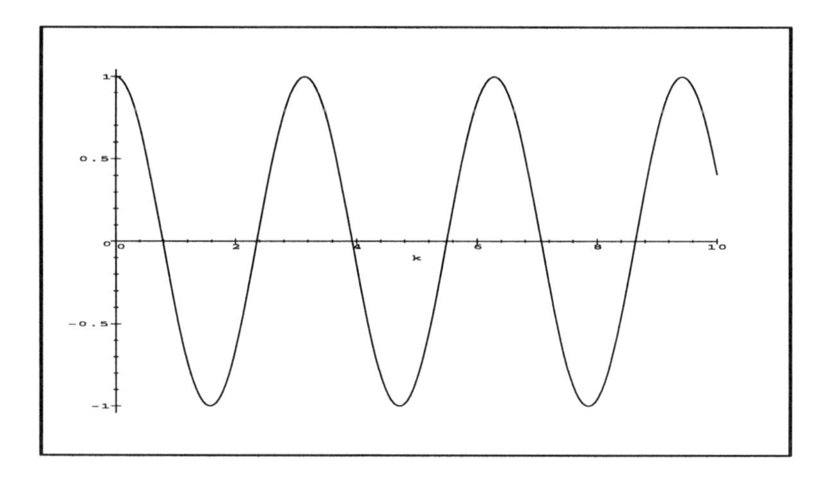

Maple returned only the first solution. At present solve does not make use of the periodicity of the trig-functions to return multiple roots. One remedy is a numerical approach by which we can obtain the roots one by one. This can be done to arbitrary precision by changing the variable Digits to some other value from the default of 10. Observe from the graph that the values $k = 2$ and $k = 3$ bracket the next root.

For a future release of Maple V (possibly release 4) there is a promise that the software will recognize that for integer n, $\sin(n\pi) = 0$, and other seemingly trivial results for the trig functions. It is possible to replace the internal sine and cosine functions by routines placed on the Maple User's Group bulletin board by their author M. Monagan. Whether solve will be clever enough to recognize these properties (resulting from the periodicity) remains to be seen. This is a problem for most CASs, not just Maple.

```
> fsolve(bcS(2),k=2..3);
```

$$2.356194490$$

It is easy to find that the eigenvalue condition reads:

$$kL = \frac{(2n+1)}{2\pi} \quad ; \qquad n = 0, 1, 2, \ldots .$$

We define a procedure that contains this relation and evaluate it for some choices of n and L. To save space we pool three commands on a single line and display the results side-by-side.

```
> kn:=(n,L)->((2*n+1)/(2*L))*Pi;
```

$$kn := (n, L) \to \frac{1}{2} \frac{(2n+1)\pi}{L}$$

```
> kn(0,2);    kn(1,2);    evalf(");
```

$$\frac{1}{4}\pi \qquad \frac{3}{4}\pi \qquad 2.356194491$$

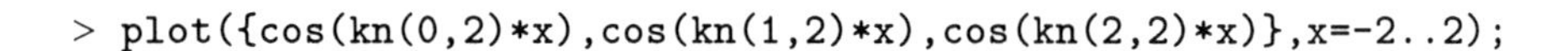

```
> plot({cos(kn(0,2)*x),cos(kn(1,2)*x),cos(kn(2,2)*x)},x=-2..2);
```

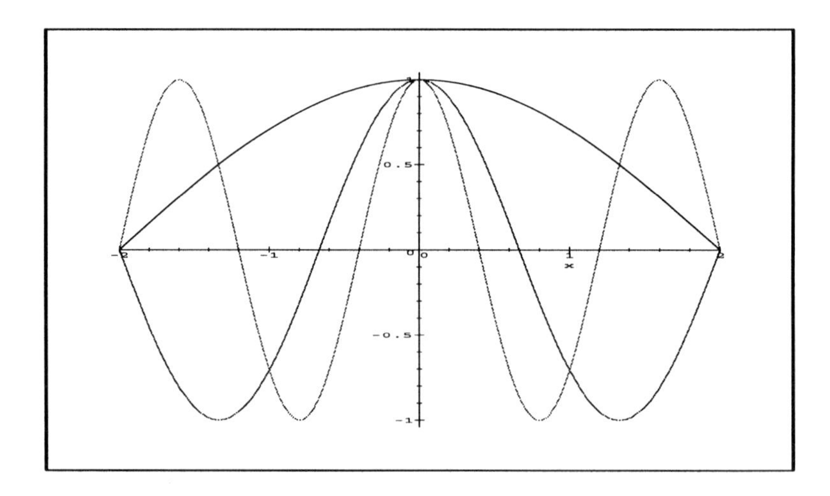

Exercise 1.1.1: Find the corresponding eigenvalue condition for the anti-symmetric solutions.

I close by emphasizing that in the context of scattering (large box size L) the physical interpretation of the different eigenfunctions $\exp(+ikx)$, $\exp(-ikx)$, $\cos(kx)$, $\sin(kx)$ is different. Only the complex functions represent travelling waves. The real ones correspond to standing waves as is obvious from $2\cos(kx) = \exp(+ikx) + \exp(-ikx)$. These standing waves can be very useful for practical calculations (real scattering matrix K), but physical information has to be extracted by converting the information to physical boundary conditions (Heitler's damping equation; cf. texts on quantum scattering theory [GW64, Jo79, Si91, RT67, Ne82, Br70, MM71]).

For the courageous Maple user (novices should at least go through the first 10 following commands) I complete the introduction by an alternative that makes use of functions and operators in Maple. Before showing the use of the D differential operator, I demonstrate how to convert between functions and expressions. We define a wavefunction with argument x that can be changed.

```
> psi:=x->exp(I/hbar*p*x);
```

$$\psi := x \to e^{\left(\frac{I\,p\,x}{\mathrm{hbar}}\right)}$$

We evaluate the wavefunction at fixed argument $x = 0$, and at x, and pool the results side-by-side to save space in the book:

```
> psi(0);       psi(x);
```

$$1 \qquad e^{\left(\frac{I\,p\,x}{\mathrm{hbar}}\right)}$$

Suppose we want $\phi(x) = (\mathrm{d}/\mathrm{d}x)\psi(x)$ and need $\phi(y)$. There are various ways to achieve this in Maple.

```
> res1:=diff(psi(x),x);
```

$$\mathit{res1} := \frac{I\,p\,e^{\left(\frac{I\,p\,x}{\mathrm{hbar}}\right)}}{\mathrm{hbar}}$$

The Maple function **unapply** converts from expressions to mappings in one or more variables.

```
> phi:=unapply(res1,x);
```

$$\phi := x \to \frac{I\,p\,e^{\left(\frac{I\,p\,x}{\mathrm{hbar}}\right)}}{\mathrm{hbar}}$$

```
> phi(y);
```

$$\frac{I\,p\,e^{\left(\frac{I\,p\,y}{\mathrm{hbar}}\right)}}{\mathrm{hbar}}$$

Alternatively we can use the substitute command **subs** in expressions:

```
> subs(x=y,res1);
```

$$\frac{I\,p\,e^{\left(\frac{I\,p\,y}{\mathrm{hbar}}\right)}}{\mathrm{hbar}}$$

The D differential operator acts on functions (not expressions) and returns functions. Note that ϕ and ψ are functions, while $\phi(x)$ and $\psi(y)$ are expressions!

```
> D(psi);
```

$$x \to \frac{I\,p\,e^{\left(\frac{I\,p\,x}{\text{hbar}}\right)}}{\text{hbar}}$$

```
> chi:=D(psi);
```

$$\chi := x \to \frac{I\,p\,e^{\left(\frac{I\,p\,x}{\text{hbar}}\right)}}{\text{hbar}}$$

```
> chi(0);
```

$$\frac{I\,p}{\text{hbar}}$$

Observe that the application of D to an expression can be confusing (for Maple as well as the user):

```
> D(psi(x));
Error, (in D) univariate operand expected
```

```
> D[1](psi(x));
```

$$D_1\left(e^{\left(\frac{I\,p\,x}{\text{hbar}}\right)}\right)$$

We can define a Hamiltonian such that it acts on functions and not on expressions. We keep the potential function as an argument.

```
> c1:=x->-hbar^2/(2*m)*x;
```

$$c1 := x \to -\frac{1}{2}\frac{\text{hbar}^2\,x}{m}$$

The above function generates a factor by functional composition. However, our straightforward attempts at functional compositions didn't give the desired results, i.e., they did not result in final functions $x \to c\psi(x)$. We need a function that allows a factor to be attached to a function:

```
> Attach:=(c1,psi)->unapply(c1(psi(x)),x);
```

$$\text{Attach} := (\,c1, \psi\,) \to \text{unapply}(\,c1(\,\psi(\,x\,)\,), x\,)$$

```
> Attach(c1,psi);
```

$$x \to -\frac{1}{2}\frac{\mathrm{hbar}^2\,\mathrm{e}^{\left(\frac{I\,p\,x}{\mathrm{hbar}}\right)}}{m}$$

```
> Hamiltonian:=arg->Attach(c1,(D@@2)(arg));
```

$$\mathrm{Hamiltonian} := \mathrm{arg} \to \mathrm{Attach}(\,c1, D^{(2)}(\,\mathrm{arg}\,)\,)$$

```
> phi:=Hamiltonian(psi);
```

$$\phi := x \to \frac{1}{2}\frac{p^2\,\mathrm{e}^{\left(\frac{I\,p\,x}{\mathrm{hbar}}\right)}}{m}$$

```
> phi(y);
```

$$\frac{1}{2}\frac{p^2\,\mathrm{e}^{\left(\frac{I\,p\,y}{\mathrm{hbar}}\right)}}{m}$$

We can generalize from the free to an interacting Hamiltonian:

```
> Vpot:=x->1/2*m*omega^2*x^2;
```

$$\mathrm{Vpot} := x \to \frac{1}{2}\,m\,\omega^2\,x^2$$

This is the potential function. We seek, however, a multiplicative operator:

```
> VP:=arg->Vpot(x)*arg;
```

$$VP := \mathrm{arg} \to \mathrm{Vpot}(\,x\,)\,\mathrm{arg}$$

The Hamiltonian for an interacting particle is given as

```
> Hint:=arg->Attach(c1,(D@@2)(arg))+Attach(VP,arg);
```

$$\mathrm{Hint} := \mathrm{arg} \to \mathrm{Attach}(\,c1, D^{(2)}(\,\mathrm{arg}\,)\,) + \mathrm{Attach}(\,VP, \mathrm{arg}\,)$$

```
> chi:=Hint(psi);
```

$$\chi := \left(x \to \frac{1}{2}\frac{p^2\,\mathrm{e}^{\left(\frac{I\,p\,x}{\mathrm{hbar}}\right)}}{m}\right) + \left(x \to \frac{1}{2}\,m\,\omega^2\,x^2\,\mathrm{e}^{\left(\frac{I\,p\,x}{\mathrm{hbar}}\right)}\right)$$

```
> chi(x);
```

$$\frac{1}{2}\frac{p^2\,\mathrm{e}^{\left(\frac{I\,p\,x}{\mathrm{hbar}}\right)}}{m} + \frac{1}{2}\,m\,\omega^2\,x^2\,\mathrm{e}^{\left(\frac{I\,p\,x}{\mathrm{hbar}}\right)}$$

The latter is clearly not an eigenstate as can be seen from

```
> chi(x)/psi(x);
```

$$\frac{\frac{1}{2}\frac{p^2 \, e^{\left(\frac{I\,p\,x}{\text{hbar}}\right)}}{m} + \frac{1}{2}\, m\, \omega^2\, x^2\, e^{\left(\frac{I\,p\,x}{\text{hbar}}\right)}}{e^{\left(\frac{I\,p\,x}{\text{hbar}}\right)}}$$

```
> simplify(");
```

$$\frac{1}{2}\frac{p^2 + m^2\,\omega^2\,x^2}{m}$$

which is a function of x and not a constant (eigenvalue). Thus, with the potential added, the state vector $|\chi\rangle$ 'points in a different direction' in function space than the original vector $|\psi\rangle$.

An eigenfunction for the harmonic-oscillator Hamiltonian is given by a Gaussian:

```
> zeta:=Hint(x->exp(-m*omega/hbar*x^2/2)):
> zeta(x);
```

$$-\frac{1}{2}\,\text{hbar}^2\left(-\frac{m\,\omega\, e^{\left(-1/2\,\frac{m\,\omega\,x^2}{\text{hbar}}\right)}}{\text{hbar}} + \frac{m^2\,\omega^2\,x^2\, e^{\left(-1/2\,\frac{m\,\omega\,x^2}{\text{hbar}}\right)}}{\text{hbar}^2}\right)\Big/m$$
$$+\frac{1}{2}\, m\,\omega^2\,x^2\, e^{\left(-1/2\,\frac{m\,\omega\,x^2}{\text{hbar}}\right)}$$

```
> simplify(");
```

$$\frac{1}{2}\,\text{hbar}\,\omega\, e^{\left(-1/2\,\frac{m\,\omega\,x^2}{\text{hbar}}\right)}$$

This proves that $\exp[-m\omega x^2/(2\hbar)]$ is an eigenfunction of the harmonic-oscillator Hamiltonian with eigenvalue $\hbar\omega/2$.

Exercise 1.1.2: Find the first two excited states of the harmonic oscillator (HO) by trial and error, by multiplying the ground state function with simple polynomials.

1.2 Harmonic Oscillator

The harmonic oscillator (HO) is a classic textbook example in both classical and quantum mechanics. In classical mechanics the motivation for its study is given by the problem of a spring (Hooke's law) and the linearized pendulum as well, and the fact that Newton's equation can be solved easily. In QM the fact that Schrödinger's equation can be solved exactly for this problem, and that the HO represents an important classical problem appears to provide a sufficient motivation. However, the problem appears to be abstract for the beginning student, one may ask about the meaning of a spring or a pendulum for a microscopic particle, such as the electron. The reliance on analytical solutions in the early days of QM has turned the HO problem into the standard bound-state model for most branches of modern physics. Electronic binding is often characterized by spring constants, nuclear shell models were HO based for some time, quark models of baryons and mesons relied on the confining features of this potential. Perhaps the most convincing and justified application of the HO potential is in problems where a bound-state system is in equilibrium. Near a stationary point of a two-particle interaction the potential is dominated by the lowest nontrivial term in a Taylor expansion, which is the quadratic term. We present an application of this idea in the treatment of molecular vibration spectra in section 7.3, and emphasize that an equivalent problem exists in the motion of nuclear quasimolecules.

The 1D stationary Schrödinger equation in coordinate representation simplifies for a particular choice of units. Physicists tend to keep notation simple by working with dimensionless expressions. Usually one sets $\hbar = 1$. High-energy physicists are interested in relativity and set the speed of light $c = 1$. They keep mass (= energy, if $c = 1$ according to $E = mc^2$) as a dimensionful quantity. In a non-relativistic framework, where the rest mass (energy) is not part of the particle energy it is convenient to set the particle's (usually the electron's) mass to unity. A third constant can be set to unity to make everything dimensionless. In Coulomb problems (atomic physics) this is the electron charge $e = 1$, and the result is the atomic unit system (Bohr units).

For the harmonic oscillator the Hamiltonian is given as

$$H = \frac{p^2}{2m} + \frac{1}{2}m\omega^2 x^2$$

with

$$V = \frac{1}{2}kx^2, \qquad \omega = \sqrt{k/m}\,.$$

We can scale the problem using $\hbar = m = 1$ leaving one dimensionful quantity. Why this is desirable will follow from the algebra.

We generate kinetic and potential energy expressions that allow us to pass the wavefunction expressions, as well as their independent variable, as arguments, but to differentiate with respect to x. Note that we have to let

Maple know that ψ is a function of x when we call Tkin, which is defined below.

```
> Tkin:=(psi,x)->-hbar^2/(2*m)*diff(psi(x),x$2);
```

$$\text{Tkin} := (\psi, x) \to -\frac{1}{2}\frac{\text{hbar}^2\,\text{diff}(\psi(x), x\,\$\,2)}{m}$$

```
> Vpot:=(psi,x)->1/2*m*omega^2*x^2*psi(x);
```

$$\text{Vpot} := (\psi, x) \to \frac{1}{2}\,m\,\omega^2\,x^2\,\psi(x)$$

```
> Tkin(psi,x)+Vpot(psi,x);
```

$$-\frac{1}{2}\frac{hbar^2\left(\frac{\partial^2}{\partial x^2}\psi(x)\right)}{m} + \frac{1}{2}\,m\,\omega^2\,x^2\,\psi(x)$$

The stationary Schrödinger equation is an eigenvalue problem for the energy En. It is homogeneous and can be written such that the right-hand side vanishes.

```
> SE:=Tkin(psi,x)+Vpot(psi,x)-En*psi(x)=0;
```

$$\text{SE} := -\frac{1}{2}\frac{\text{hbar}^2\left(\frac{\partial^2}{\partial x^2}\psi(x)\right)}{m} + \frac{1}{2}\,m\,\omega^2\,x^2\,\psi(x) - En\,\psi(x) = 0$$

Ideally we would like to use Maple's differential equation solver dsolve.

Maybe in the future one will be able to type dsolve(SE,psi(x),E) and simply obtain the solutions. What would one learn from that though?

The point is that we have to incorporate the boundary condition that $\psi(x) \to 0$ as $x \to \pm\infty$. This condition (normalizable solutions) should select special, discrete energy values.

We perform some parameter substitutions:

– (1) isolate the second derivative:

```
> SE1:=expand(SE*(-2*m/hbar^2));
```

$$\text{SE1} := \left(\frac{\partial^2}{\partial x^2}\psi(x)\right) - \frac{m^2\,\omega^2\,x^2\,\psi(x)}{\text{hbar}^2} + 2\,\frac{m\,En\,\psi(x)}{\text{hbar}^2} = 0$$

– (2) substitute: $m\omega/\hbar = \lambda$

```
> SE2:=subs(omega=lambda*hbar/m,SE1);
```

$$\text{SE2} := \left(\frac{\partial^2}{\partial x^2}\psi(x)\right) - \lambda^2\,x^2\,\psi(x) + 2\,\frac{m\,En\,\psi(x)}{\text{hbar}^2} = 0$$

– (3) substitute $En = \mu\hbar\omega$, where μ plays the role of a scaled eigenvalue, or $En = \hbar^2k^2/(2m)$ with $\mu = En/(\hbar\omega) = k^2/(2\lambda)$

```
> SE3:=subs(En=hbar^2*k^2/(2*m),SE2);
```

$$\mathrm{SE3} := \left(\frac{\partial^2}{\partial x^2}\psi(x)\right) - \lambda^2 x^2 \psi(x) + k^2 \psi(x) = 0$$

Now we analyze the asymptotic behaviour of the scaled Schrödinger equation. The term involving the eigenenergy $k^2\psi$ is negligible compared to the second, i.e., we seek solutions to

```
> SE3as:=subs(k=0,SE3);
```

$$\mathrm{SE3as} := \left(\frac{\partial^2}{\partial x^2}\psi(x)\right) - \lambda^2 x^2 \psi(x) = 0$$

```
> subs(psi(x)=exp(-lambda*x^2/2),SE3as);
```

$$\left(\frac{\partial^2}{\partial x^2}e^{(-1/2\,\lambda\,x^2)}\right) - \lambda^2 x^2 e^{(-1/2\,\lambda\,x^2)} = 0$$

```
> simplify(");
```

$$-\lambda\, e^{(-1/2\,\lambda\,x^2)} = 0$$

No x dependence is left apart from the attempted eigenfunction. The constant factor can be absorbed by the eigenvalue. Therefore, we can consider the attempt successful and substitute in the full equation:

```
> subs(psi(x)=exp(-lambda*x^2/2),SE3);
```

$$\left(\frac{\partial^2}{\partial x^2}e^{(-1/2\,\lambda\,x^2)}\right) - \lambda^2 x^2 e^{(-1/2\,\lambda\,x^2)} + k^2 e^{(-1/2\,\lambda\,x^2)} = 0$$

```
> simplify(");
```

$$-\lambda\, e^{(-1/2\,\lambda\,x^2)} + k^2 e^{(-1/2\,\lambda\,x^2)} = 0$$

This leads clearly to a solution as the eigenvalue k^2 can be chosen to cancel $-\lambda$:

$$k^2 = \lambda, \qquad \text{or} \qquad E_0 = \frac{\hbar\omega k^2}{2\lambda} = \frac{1}{2}\hbar\omega\,.$$

The question arises as to how to calculate the excited states. We expect that they can be obtained by factoring the function $\exp(-\lambda x^2/2)$ away, i.e.,

```
> SE4:=expand(exp(lambda*x^2/2)*subs(psi(x)=f(x)*
> exp(-lambda*x^2/2),SE3));
```

$$\mathrm{SE4} := \left(\frac{\partial^2}{\partial x^2} f(x)\right) - 2\left(\frac{\partial}{\partial x} f(x)\right)\lambda x - f(x)\lambda + k^2 f(x) = 0$$

We have now absorbed into the exponential the boundary condition that the solution has to vanish as x approaches positive or negative infinity. The boundary condition will be satisfied as long as $f(x)$ represents at most a polynomial of finite degree n. In fact, one solves SE4 by a series ansatz

```
> dsolve(SE4,f(x),series);
```

$$\Big\{ f(x) = f(0) + \mathrm{D}(f)(0)\,x + \left(\frac{1}{2} f(0)\lambda - \frac{1}{2} k^2 f(0)\right) x^2$$
$$+ \left(\frac{1}{2}\mathrm{D}(f)(0)\lambda - \frac{1}{6} k^2\,\mathrm{D}(f)(0)\right) x^3$$
$$+ \frac{1}{24}(-5\lambda + k^2) f(0)(-\lambda + k^2)\,x^4$$
$$+ \frac{1}{120}(-7\lambda + k^2)\,\mathrm{D}(f)(0)(-3\lambda + k^2)\,x^5 + \mathrm{O}(x^6),$$
$$f(x) = f(0) + \mathrm{D}(f)(0)\,x + \left(\frac{1}{2} f(0)\lambda - \frac{1}{2} k^2 f(0)\right) x^2$$
$$+ \left(\frac{1}{2}\mathrm{D}(f)(0)\lambda - \frac{1}{6} k^2\,\mathrm{D}(f)(0)\right) x^3$$
$$+ \left(\frac{5}{24} f(0)\lambda^2 - \frac{1}{4}\lambda k^2 f(0) + \frac{1}{24} k^4\,\mathrm{f}(0)\right) x^4$$
$$+ \left(\frac{7}{40}\mathrm{D}(f)(0)\lambda^2 - \frac{1}{12}\lambda k^2\,\mathrm{D}(f)(0) + \frac{1}{120} k^4\,\mathrm{D}(f)(0)\right) x^5$$
$$+\mathrm{O}(x^6)\Big\}$$

For some reason Maple does not realize that it offers two supposedly independent solutions that, in fact, are identical. The polynomials are fixed on the basis of two initial conditions ($f(0)$ and $f'(0)$) and the series will terminate for particular values of k^2. This is most easily seen from the recursion relation.

One can show that a symmetric potential has symmetric eigenstates f_+ (even parity) as well as anti-symmetric eigenstates f_- (odd parity) that correspond to the eigenvalues $+1$ and -1, respectively, of the parity operator, i.e.,

$$\hat{P} f_+ = +1 f_+ \qquad \text{and} \qquad \hat{P} f_- = -1 f_- \quad ,$$

where the parity operator is defined by $\hat{P} f(x) = f(-x)$. The symmetry translates into two possibilities for the initial conditions at $x = 0$ for $f(x)$.

We do not worry about normalizing the eigensolutions right now. Even states are selected by $f_+(0) = 1$ (arbitrary normalization) and $f'_+(0) = 0$

(symmetry). Odd states are selected by $f_-(0) = 0$ (anti-symmetry) and $f'_-(0) = 1$ (arbitrary normalization).

```
> Even:=dsolve({SE4,f(0)=1,D(f)(0)=0},f(x),series);
```

$$\text{Even} := f(x) = 1+\left(\frac{1}{2}\lambda - \frac{1}{2}k^2\right)x^2+\left(\frac{5}{24}\lambda^2 - \frac{1}{4}\lambda k^2 + \frac{1}{24}k^4\right)x^4+\mathrm{O}(x^6)$$

```
> Odd:=dsolve({SE4,f(0)=0,D(f)(0)=1},f(x),series);
```

$$\text{Odd} := f(x) =$$
$$x + \left(\frac{1}{2}\lambda - \frac{1}{6}k^2\right)x^3 + \left(\frac{7}{40}\lambda^2 - \frac{1}{12}\lambda k^2 + \frac{1}{120}k^4\right)x^5 + \mathrm{O}(x^6)$$

Interestingly the eigenvalue condition for the ground state $k^2 = \lambda$ makes the terms with higher powers in x vanish in the even case. It is obvious for x^2, but let's check the x^4 coefficient. We pool two commands below to a single line and display the result side-by-side to save space in the book.

```
> op(2,Even);
```

$$1 + \left(\frac{1}{2}\lambda - \frac{1}{2}k^2\right)x^2 + \left(\frac{5}{24}\lambda^2 - \frac{1}{4}\lambda k^2 + \frac{1}{24}k^4\right)x^4 + \mathrm{O}(x^6)$$

```
> coeff(",x,4);            factor(");
```

$$\frac{5}{24}\lambda^2 - \frac{1}{4}\lambda k^2 + \frac{1}{24}k^4 \qquad\qquad \frac{1}{24}(-5\lambda + k^2)(-\lambda + k^2)$$

We read off the next eigenvalue in the even-symmetry sector:

$$k^2 = 5\lambda, \qquad \text{or} \qquad E_2 = \frac{5}{2}\hbar\omega\,.$$

Similarly we can investigate the odd series

```
> op(2,Odd);
```

$$x + \left(\frac{1}{2}\lambda - \frac{1}{6}k^2\right)x^3 + \left(\frac{7}{40}\lambda^2 - \frac{1}{12}\lambda k^2 + \frac{1}{120}k^4\right)x^5 + \mathrm{O}(x^6)$$

The condition for $f(x) = x$ being a solution is given by the a truncation of the series before x^3:

$$k^2 = 3\lambda, \qquad \text{or} \qquad E_1 = \frac{3}{2}\hbar\omega\,.$$

Again, we can demonstrate the factorization of the polynomial coefficients:

```
> coeff(",x,5);            factor(");
```

$$\frac{7}{40}\lambda^2 - \frac{1}{12}\lambda k^2 + \frac{1}{120}k^4 \qquad\qquad \frac{1}{120}(-7\lambda + k^2)(-3\lambda + k^2)$$

We can guess that the next eigenvalue is obtained from the other root:

$$k^2 = 7\lambda, \qquad \text{or} \qquad E_3 = \frac{7}{2}\hbar\omega \,.$$

A more systematic approach is to derive the recursion relation that forms the basis for the found series solution. This leads to the Hermite polynomials that are discussed in Sect. 8.1.

We can calculate the lowest few polynomials by substituting the eigenvalue condition. For the $n = 2$ even-parity state we have

```
> subs(k=sqrt(5*lambda),Even);
> simplify(convert(rhs("),polynom));
```

$$f(x) = 1 - 2\lambda x^2 + O(x^6) \qquad 1 - 2\lambda x^2$$

while for the $n = 3$ odd-parity state we obtain

```
> subs(k=sqrt(7*lambda),Odd);
> simplify(convert(rhs("),polynom));
```

$$f(x) = x - \frac{2}{3}\lambda x^3 + O(x^6) \qquad x - \frac{2}{3}\lambda x^3$$

The Hermite polynomials, for reference, are called up by:

```
> with(orthopoly);
```

$$[G, H, L, P, T, U]$$

```
> H(2,sqrt(lambda)*x);      H(3,sqrt(lambda)*x);
```

$$4\lambda x^2 - 2 \qquad 8\lambda^{3/2}x^3 - 12\sqrt{\lambda}\,x$$

For plotting we choose $\lambda = 1$. We will worry about the normalization later.

```
> psiHO:=(x,n)->exp(-x^2/2)*H(n,x);
```

$$\mathrm{psiHO} := (x, n) \to e^{(-1/2\,x^2)}\,\mathrm{H}(n, x)$$

Exercise 1.2.1: Learn to identify the HO eigenfunctions according to their symmetry and number of nodes!

We display the unnormalized lowest four harmonic oscillator eigenstates to show their nodal structure and their symmetry properties under the parity transformation. From the point of view of solving the differential equation – eigenvalue problem the normalization of the solution is arbitrary as the equation is homogeneous. The physicist requires a normalization such that the square of the wavefunction can be interpreted as a probability density.

```
> plot({psiHO(x,0),psiHO(x,1),psiHO(x,2),psiHO(x,3)},x=-4..4);
```

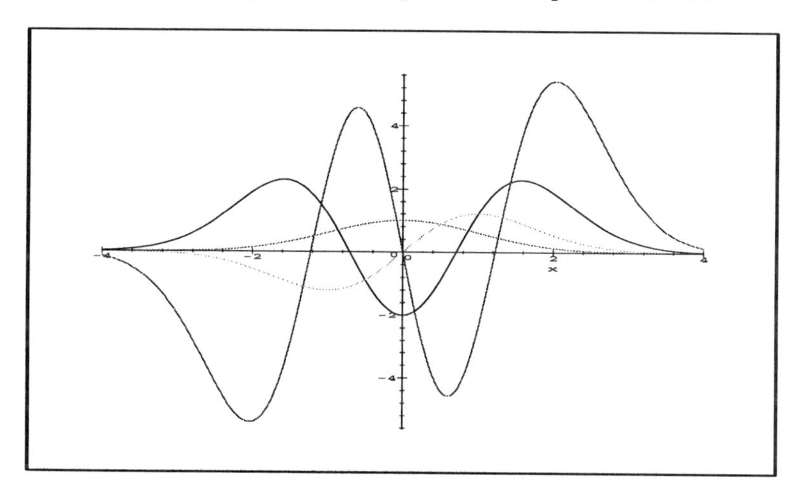

Finally we normalize and graph the states together with their energy eigenvalues inside the potential. With units specified by $m = \hbar = 1$, the choice $\lambda = 1$ implies $\omega = 1$ and the energy eigenvalues become

$$E_n = (n + 1/2)\hbar\omega = n + 1/2, \qquad \text{for} \qquad n = 0, 1, 2, \ldots .$$

We could program a formula that normalizes the states for arbitrary n. Maple's integration facilities cannot calculate the overlap between Hermite polynomials (with appropriate weight factor) automatically for arbitrary n:

```
> int(exp(-x^2)*H(n,x)*H(m,x),x=-infinity..infinity);
```

$$\int_{-\infty}^{\infty} \mathrm{e}^{(-x^2)}\, \mathrm{H}(n, x)\, \mathrm{H}(m, x)\, dx$$

However, for specific m, n a non-trivial (not symmetry-based) orthogonality relation works:

```
> int(exp(-x^2)*H(0,x)*H(2,x),x=-infinity..infinity);
```

$$0$$

The same applies to the normalization integral:

```
> assume(n,integer);
> int(exp(-x^2)*H(n,x)*H(n,x),x=-infinity..infinity);
```

$$\int_{-\infty}^{\infty} \mathrm{e}^{(-x^2)}\, \mathrm{H}(n\tilde{}, x)^2\, dx$$

```
> int(exp(-x^2)*H(2,x)*H(2,x),x=-infinity..infinity);
```

$$8\sqrt{\pi}$$

Thus, we define (the output is suppressed):

```
> psiHOn:=(x,n)->psiHO(x,n)/sqrt(int(psiHO(x,n)^2,
> x=-infinity..infinity)):
```

```
> psiHOn(x,0);          psiHOn(x,4);
```

$$\frac{e^{(-1/2\,x^2)}}{\pi^{1/4}} \qquad \frac{1}{48}\,\frac{e^{(-1/2\,x^2)}\,(16\,x^4-48\,x^2+12)\,\sqrt{6}}{\pi^{1/4}}$$

```
> psiHOn(x,3);
```

$$\frac{1}{12}\,\frac{e^{(-1/2\,x^2)}\,(8\,x^3-12\,x)\,\sqrt{3}}{\pi^{1/4}}$$

We are ready to assemble a sequence that contains $E_n + \psi_n(x)$ for $n = 0..3$:

```
> wfplusen:=seq(psiHOn(x,n)+(n+1/2),n=0..3);
```

$$\text{wfplusen} := \frac{e^{(-1/2\,x^2)}}{\pi^{1/4}} + \frac{1}{2},\ \frac{e^{(-1/2\,x^2)}\,x\,\sqrt{2}}{\pi^{1/4}} + \frac{3}{2},$$
$$\frac{1}{4}\,\frac{e^{(-1/2\,x^2)}\,(4\,x^2-2)\,\sqrt{2}}{\pi^{1/4}} + \frac{5}{2},\ \frac{1}{12}\,\frac{e^{(-1/2\,x^2)}\,(8\,x^3-12\,x)\,\sqrt{3}}{\pi^{1/4}} + \frac{7}{2}$$

```
> plot({x^2/2,wfplusen},x=-5..5,V=0..5);
```

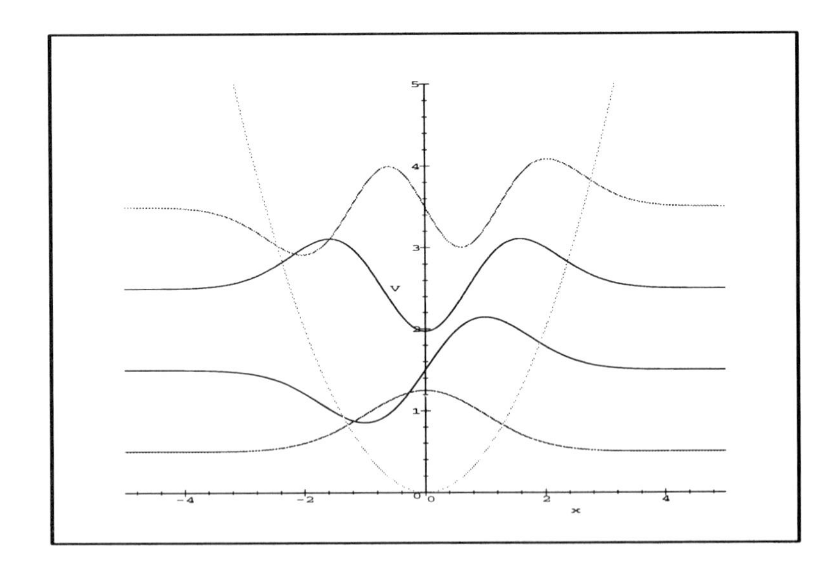

We note that the baselines for the eigenfunctions indicate the energy levels. Their intercept with the potential energy $V = (1/2)x^2$ specifies the classical turning point at which all the energy is in the form of potential energy, i.e., the kinetic energy vanishes.

Beyond the classically allowed region, i.e., outside the turning points, the classical kinetic energy would be negative. In this regime a wavefunction is attenuated, but there remains a finite probability of finding the particle in this range. This is more obvious from a graph of the probability density, which we indicate for a high-n state:

```
> plot({x^2/2,psiHOn(x,15)^2+(15+1/2)},x=-8..8,V=15.4..15.8);
```

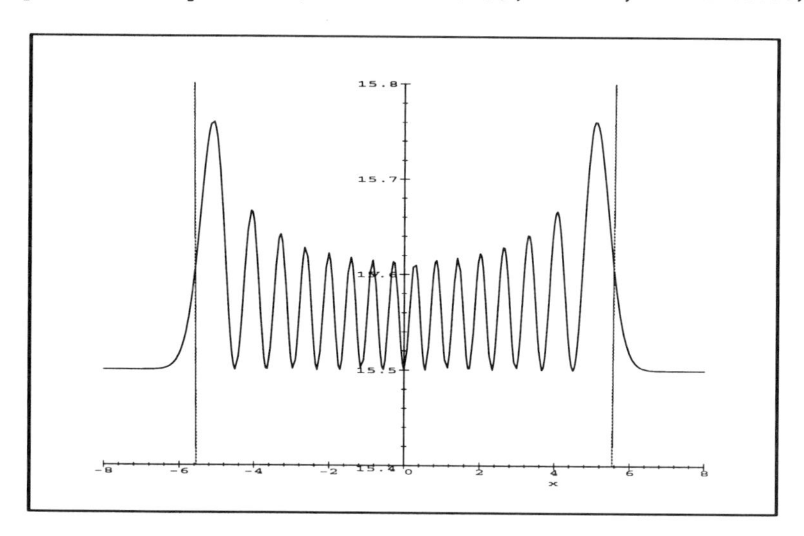

We note that the QM eigenstates for large n predict, in analogy with the classical answer, that the probability of finding the particle is largest near the turning point. Classically (consider the trajectory as a function of time) the probability of finding the particle as a function of time at a given place diverges at the turning point as the particle is at rest there.

Exercise 1.2.2: Compare the classical probability function that is based on the trajectory with the QM result for $n = 15$ shown above. The classical probability of finding the particle at location x is proportional to $(x_0^2 - x^2)^{(-1/2)}$, where x_0 is the maximum amplitude of the classical oscillator with a total energy equal to the QM eigenvalue.

1.3 A Time-Dependent Problem

So far we have considered the stationary Schrödinger equation

$$\hat{H}\psi(x) = -\frac{\hbar^2}{2m}\frac{\mathrm{d}^2\psi}{\mathrm{d}x^2} + V(x)\psi(x) = E\psi(x)\,.$$

This eigenvalue problem results from the separation of position and time variables in the time-dependent Schrödinger equation (TDSE)

$$\mathrm{i}\hbar\frac{\partial\Psi(x,t)}{\partial t} = \hat{H}\Psi(x,t)$$

with the separation ansatz $\Psi(x,t) = \psi(x)\exp(-\mathrm{i}Et/\hbar)$ where E plays the role of the separation constant. The argument works as long as the potential is time-independent and thus the energy remains conserved.

One can interpret $\exp(-\mathrm{i}\hat{H}t/\hbar)$ in this case as the time-evolution operator for a stationary state. When $\hat{H}$ acts upon an eigenstate $\Psi_i(x,t)$ with associated energy E_i, the phase factor becomes $\exp(-\mathrm{i}E_it/\hbar)$.

We wish to demonstrate that even this seemingly trivial time dependence (phase factors do not affect the probability density for a single state) can have real repercussions if the system is in a mixed state. Suppose we have a superposition of two harmonic-oscillator (HO) eigenstates with constant coefficients:

```
> with(orthopoly);
```

$$[G, H, L, P, T, U]$$

We define a procedure to evaluate the inner product. Note how one does not have to use explicit arguments in the procedure declaration, but can make use of the `args` variable. The `nargs` variable (not used in this example) can be used to query how many arguments are actually passed when the procedure is invoked. This allows one to make calls to the same procedure with a different number of arguments.

```
> InProd:=proc();

> int(evalc(conjugate(args[1])*args[2]),
> args[3]=-infinity..infinity);

> end;

  InProd := proc()
                int(evalc(conjugate(args[1])*args[2]),
                    args[3] = -infinity .. infinity)
            end
```

We define a stationary, properly normalized HO eigenstate labeled by the n quantum number:

```
> psiHO:=(n,x)->simplify(exp(-x^2/2)*H(n,x)/sqrt
> (InProd(exp(-x^2/2)*H(n,x),exp(-x^2/2)*H(n,x),x))):
```

```
> psiHO(2,y);
```

$$\frac{1}{2}\frac{(2y^2-1)\sqrt{2}\,e^{(-1/2y^2)}}{\pi^{1/4}}$$

Let us now consider a harmonic oscillator with $\omega = 1$ ($\hbar = m = 1$ is also assumed) and prepare a time-dependent state that is mixed from 50% $n = 0$ and 50% $n = 1$. The energies are given by $E_n = (n + 1/2)$, and thus:

```
> Psi01:=(x,t)->1/sqrt(2)*psiHO(0,x)*exp(-I*(1/2)*t)
> +1/sqrt(2)*psiHO(1,x)*exp(-I*(3/2)*t);
```

$$\Psi 01 := (x,t) \rightarrow \frac{\mathrm{psiHO}(0,x)\,e^{(-1/2It)}}{\mathrm{sqrt}(2)} + \frac{\mathrm{psiHO}(1,x)\,e^{(-3/2It)}}{\mathrm{sqrt}(2)}$$

```
> InProd(Psi01(x,1),Psi01(x,1),x);
```

$$\frac{1}{2}\cos\left(\frac{1}{2}\right)^2 + \frac{1}{2}\cos\left(\frac{3}{2}\right)^2 + \frac{1}{2}\sin\left(\frac{1}{2}\right)^2 + \frac{1}{2}\sin\left(\frac{3}{2}\right)^2$$

```
> combine(",trig);
```

$$1$$

The time evolution is unitary, i.e., the correct normalization is preserved (note the `1/sqrt(2)` factors in front of the normalized pure $n = 0$ and $n = 1$ states). We are interested in displaying the time evolution of the probability density. First we calculate some simple expectation values.

We follow the time evolution of the position expectation value. Note that we do not start at $t = 0$ with $\langle x \rangle = 0$ even though both states separately have $\langle x \rangle_i = 0$.

```
> Xoft:=t->InProd(Psi01(x,t),x*Psi01(x,t),x);
```

$$Xoft := t \rightarrow \mathrm{InProd}(\Psi 01(x,t), x\,\Psi 01(x,t), x)$$

```
> Xoft(0);
```

$$\frac{1}{2}\sqrt{2}$$

We begin the display of some results with a graph of the wavefunction at the initial time $t = 0$.

```
> plot(Psi01(x,0),x=-5..5);
```

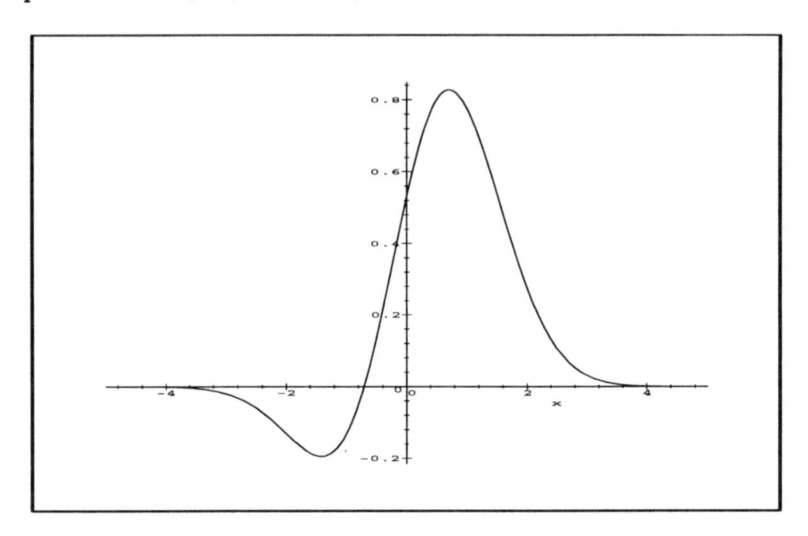

More relevant is the probability density:

```
> plot(evalc(conjugate(Psi01(x,0))*Psi01(x,0)),x=-5..5);
```

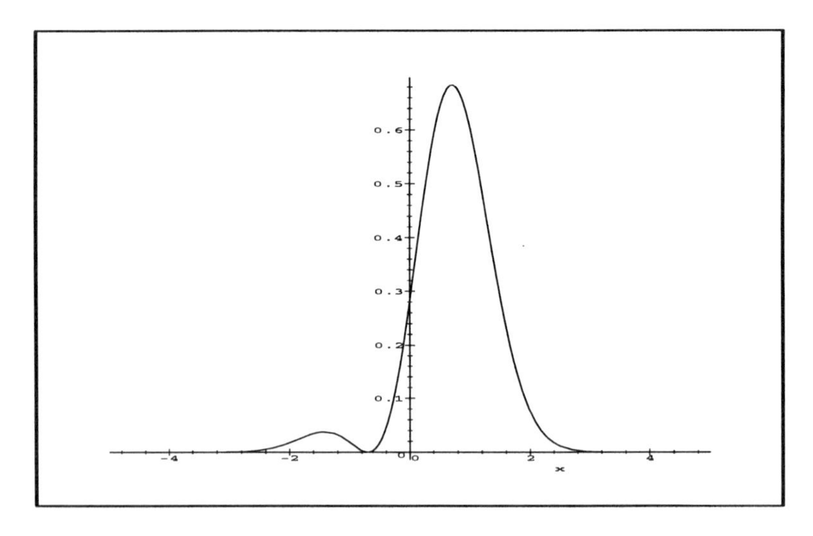


```
> Xoft(1);
```

$$\frac{1}{2}\sqrt{2}\cos\left(\frac{1}{2}\right)\cos\left(\frac{3}{2}\right)+\frac{1}{2}\sqrt{2}\sin\left(\frac{1}{2}\right)\sin\left(\frac{3}{2}\right)$$

```
> combine(",trig);
```

$$\frac{1}{2}\sqrt{2}\cos(1)$$

```
> plot(Xoft(t),t=0..10);
```

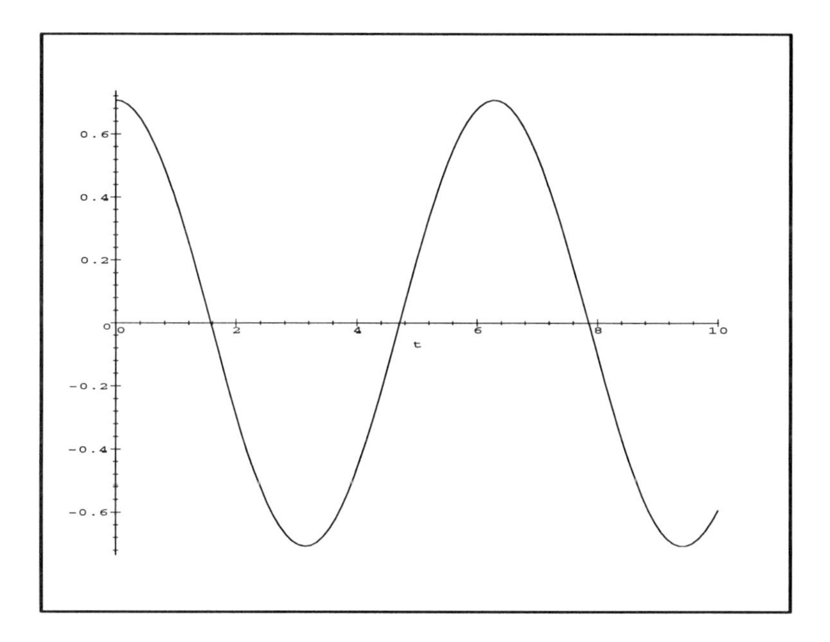

We observe a periodic oscillation in the position expectation value even though we have a static superposition of two eigenstates.

We can produce the detailed information about the probability density using the `animate` facility in Maple.

```
> with(plots):
```

```
> animate(evalc(conjugate(Psi01(x,t))*Psi01(x,t)),
> x=-5..5,t=0..10);
```

Exercise 1.3.1: Repeat the calculation for the position expectation value for different superpositions and different HO states. Figure out what determines the period in the oscillation of the density or $\langle x\rangle(t)$.

1.4 Variational Method

The Rayleigh-Ritz variational method represents a useful approximation technique for the simple determination of ground-state energies and eigenfunctions in a given symmetry sector of a Hamiltonian that does not admit an exact solution. For a formal proof of the property that the exact ground-state energies are approximated from above I refer the reader to the standard literature (e.g., [Sch68]).

We demonstrate the method for a 1D potential that has a long-range behaviour reminiscent of the Coulomb potential in 3D. For simplicity we make use of harmonic Gaussian trial wavefunctions which are not ideal for such a potential as they die out very quickly for large x. Nevertheless, they generate acceptable eigenenergies.

The method requires the calculation of expectation values for the kinetic and potential energies $\langle E\rangle = \langle T\rangle + \langle V\rangle$ for a given trial state $|\psi_T\rangle$ that depends on a family of parameters. In case $|\psi_T\rangle$ depends on a single parameter β, the estimate for the ground-state energy E_T becomes a function of this parameter (note that β is related to the circular frequency ω of the HO for which the trial state is an exact eigenstate, by $\beta^2 = \omega$)

$$E_T(\beta) = \langle \psi_T(\beta)|T + V|\psi_T(\beta)\rangle \,.$$

The best estimate for the ground-state energy within the given class of trial functions is obtained from the global minimum of $E_T(\beta)$. Usually this minimum is obtained at a stationary point, but it is also possible that end-point values of β (e.g., for $0 < \beta < \infty$ the points $\beta = 0$ or $\beta = \infty$) provide the global minimum.

We define the potential

```
> V:=x->-1/sqrt(1+x^2);
```

$$V := x \to -\frac{1}{\mathrm{sqrt}(1+x^2)}$$

and a Gaussian trial state:

```
> psi01:=(x,beta)->exp(-(beta*x)^2/2);
```

$$\psi 01 := (x, \beta) \to \mathrm{e}^{(-1/2\,\beta^2\,x^2)}$$

In order to calculate integrals over all x successfully we need an assumption on β that ensures the convergence of these (strictly the assumption that β is real should be sufficient).

```
> assume(beta>0);
```

Now we normalize the trial state:

```
> norm1:=int(psi01(x,beta)^2,x=-infinity..infinity);
```

$$\text{norm1} := \frac{\sqrt{\pi}}{\beta\tilde{}}$$

```
> norm1s:=sqrt(norm1):
```

While redefining the trial wavefunction we note that the mapping feature in Maple does not map the dummy variable to an expression that happens to depend on a variable with the same name. It would not be correct to use β as a dummy variable in the procedure declaration and to expect that the β dependence in **norm1s** would be mapped automatically. Instead one has to make use of a substitution.

```
> psi0:=(x,alpha)->exp(-(alpha*x)^2/2)/subs(beta=alpha,norm1s);
```

$$\psi 0 := (x, \alpha) \rightarrow \frac{e^{(-1/2\,\alpha^2\,x^2)}}{\text{subs}(\beta = \alpha, \text{norm1s})}$$

```
> psi0(x,beta);
```

$$\frac{e^{(-1/2\,\beta\tilde{}^{2}\,x^2)}}{\sqrt{\frac{\sqrt{\pi}}{\beta\tilde{}}}}$$

It is always worthwile to test the correctness of the mapping by using some other argument (e.g., γ instead of β) to test whether the substitution has worked properly. Maple is error-prone in this area, especially if assumptions about variables have been made and subsequently overwritten. There is potential for errors with the **assume** facility insofar as the user refers to the assumed variable by the same name (the thilde is only displayed, never entered), but the Maple system can reference these separately internally and sometimes loses the connection.

Now we define the energy expression:

```
> Energy:=beta->int(-psi0(x,beta)*diff(psi0(x,beta),x$2)/2
> +V(x)*psi0(x,beta)^2,x=-infinity..infinity);
```

$$\text{Energy} := \beta \rightarrow \int_{-\infty}^{\infty} -\frac{1}{2}\,\psi 0(x,\beta)\,\text{diff}(\psi 0(x,\beta), x\,\$\,2) + \text{V}(x)\,\psi 0(x,\beta)^2\,dx$$

We try to evaluate the energy expectation value for some choice of variational parameter, namely $\beta = 1$:

```
> Energy(1);
```

$$\int_{-\infty}^{\infty} -\frac{1}{2}\,\frac{e^{(-1/2\,x^2)}\left(-\frac{e^{(-1/2\,x^2)}}{\pi^{1/4}} + \frac{x^2\,e^{(-1/2\,x^2)}}{\pi^{1/4}}\right)}{\pi^{1/4}} - \frac{(e^{(-1/2\,x^2)})^2}{\sqrt{1+x^2}\,\sqrt{\pi}}\,dx$$

```
> simplify(");
```

$$-\frac{1}{2}\frac{\int_{-\infty}^{\infty}\frac{e^{(-x^2)}\left(-\sqrt{1+x^2}+\sqrt{1+x^2}\,x^2+2\right)}{\sqrt{1+x^2}}\,dx}{\sqrt{\pi}}$$

This is bad news: Maple tried harder and still couldn't do the integral. It forces us to resort to numerical techniques for the present potential function. While this is easily done in Maple, as shown below, it is much more time-consuming than in a compiled strictly numerical environment.

```
> evalf(");
```

$$-.6098866395$$

Given that we need numerical evaluation, we should redefine the integral as unevaluated (upper-case `Int` command) to avoid the attempt at a symbolic evaluation:

```
> Energy:=beta->Int(-psi0(x,beta)*diff(psi0(x,beta),x$2)/2
> +V(x)*psi0(x,beta)^2,x=-infinity..infinity);
```

$$\text{Energy} := \beta \to \int_{-\infty}^{\infty} -\frac{1}{2}\,\psi0(x,\beta)\,\text{diff}(\psi0(x,\beta),x\,\$\,2)+\text{V}(x)\,\psi0(x,\beta)^2\,dx$$

We are ready for the loop that evaluates $E(\beta_j)$ numerically. To save paper I have deleted values except for a few around the minimum.

```
> for j from 1 to 15 do

> bj:=0.1*j;

> Ej[j]:=evalf(Energy(bj));

> od;
```

$$\text{bj} := .6$$

$$\text{Ej}_6 := -.6612629018$$

$$\text{bj} := .7$$

$$\text{Ej}_7 := -.6648638221$$

$$\text{bj} := .8$$

$$\text{Ej}_8 := -.6564656009$$

```
> plot([[0.1*i,Ej[i]] $i=1..15],beta=0..1.5,style=point,
> symbol=circle,title=`Variational approximation to E0`);
```

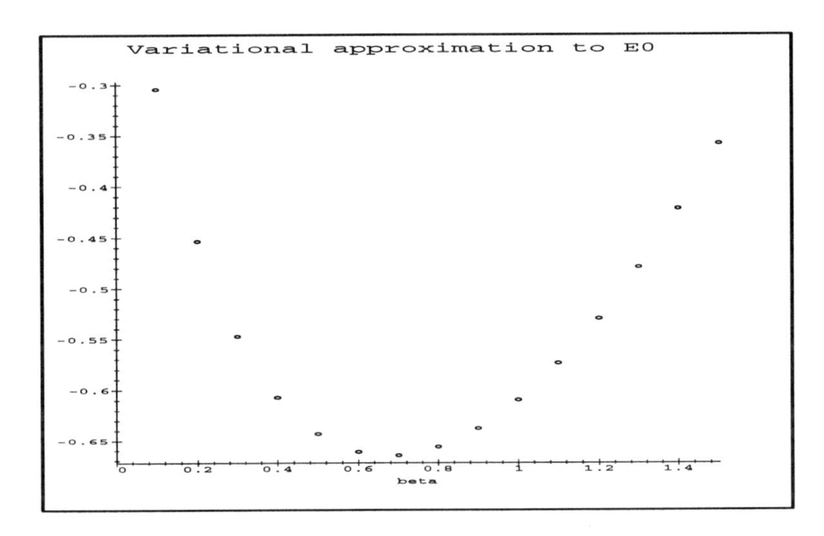

Thus, $\beta = 0.7$, $E_0 = -0.664864$ is approximately the best-possible value while using a Gaussian wavefunction. Let's graph it together with the potential. We shift the baseline for the wavefunction to reflect the estimated energy eigenvalue:

```
> plot({V(x),psi0(x,0.7)+Ej[7]},x=-10..10,y=-1..0,
> title='psi0 in V = -1/sqrt(1+x^2)');
```

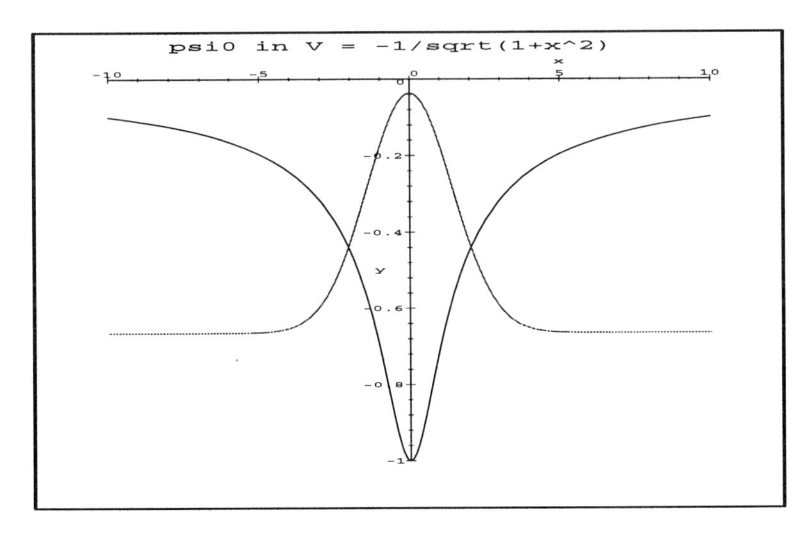

Let us discuss strengths and weaknesses of this calculation:

The Gaussian shape is not well-suited to describing the tail of the wavefunction in a long-range potential. We should expect the correct ground-state wavefunction to fall off more slowly.

We understand intuitively what makes the Gaussian choose its width: a strong localization at $x = 0$ would minimize the potential energy towards $V_{\min} = -1$. This can be achieved with the Gaussian in the limit of $\beta \to \infty$ when it becomes a delta-function (position eigenstate) centred at zero. The penalty for this choice (after all the energy shoots up as β increases) is a big increase in the kinetic energy. It is sensitive to the curvature in ψ and a strongly peaked function has a large second derivative.

At the other extreme ($\beta \to 0$), a totally delocalized state minimizes the kinetic energy (curvature). It pays a price for this in not utilizing the potential well, i.e., its potential energy is somewhere in between -1 and 0. The wider the wavefunction, the closer the number gets to 0. This illustrates why there is always an optimum.

For potentials for which the matrix elements of both the kinetic and the potential energies can be calculated in closed form, one can see this dependence analytically. We have just explored this intuitively.

For the kinetic energy we can prove that

$$T(\beta) \to 0 \qquad \text{for} \qquad \beta \to 0\,,$$

and

$$T(\beta) \to \infty \qquad \text{for} \qquad \beta \to \infty\,.$$

```
> Tkin:=beta->int(-psi0(x,beta)*diff(psi0(x,beta),x$2)/2,
> x=-infinity..infinity);
```

$$\mathrm{Tkin} := \beta \to \int_{-\infty}^{\infty} -\frac{1}{2}\,\psi 0(\,x,\beta\,)\,\mathrm{diff}(\,\psi 0(\,x,\beta\,),x\,\$\,2\,)\,dx$$

```
> Tkin(beta);
```

$$\frac{1}{4}\beta\tilde{}^{2}$$

Let us explore the potential energy expectation value for a simple square-well potential of width 2 and depth -1 as calculated with a Gaussian.

```
> Vpot:=beta->int(-1*psi0(x,beta)^2,x=-1..1);
```

$$\mathrm{Vpot} := \beta \to \int_{-1}^{1} -\psi 0(\,x,\beta\,)^2\,dx$$

```
> Vpot(beta);
```

$$-\mathrm{erf}(\,\beta\tilde{}\,)$$

```
> plot(",beta=0..5);
```

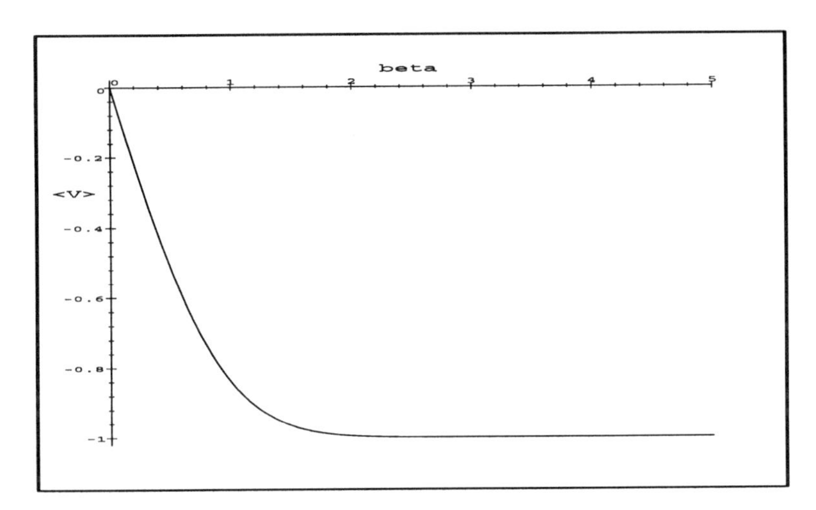

The graph demonstrates how the potential energy becomes minimal for a localized state (large β), while it approaches the asymptotic value of zero for a totally delocalized one (small β).

We now proceed to calculate the first excited state. It can still be obtained in the sense of an upper bound as it is the lowest state in the symmetry class of anti-symmetric solutions (negative-parity states).

```
> psi11:=(x,beta)->x*exp(-(beta*x)^2/2);
```

$$\psi 11 := (x, \beta) \to x\, \mathrm{e}^{\left(-1/2\, \beta^2 x^2\right)}$$

As before we normalize this state:

```
> no11:=int(psi11(x,beta)^2,x=-infinity..infinity);
```

$$\mathrm{no11} := \frac{1}{2} \frac{\sqrt{\pi}}{\beta^{\sim 3}}$$

```
> no1s:=sqrt(no11);
```

$$\mathrm{no1s} := \frac{1}{2} \frac{\sqrt{2}\sqrt{\dfrac{\sqrt{\pi}}{\beta^{\sim}}}}{\beta^{\sim}}$$

```
> psi1:=(x,alpha)->x*exp(-(alpha*x)^2/2)/subs(beta=alpha,no1s);
```

$$\psi 1 := (x, \alpha) \to \frac{x\, \mathrm{e}^{\left(-1/2\, \alpha^2 x^2\right)}}{\mathrm{subs}(\beta = \alpha, \mathrm{no1s})}$$

```
> psi1(x,beta);
```

$$\frac{x\,e^{\left(-1/2\,\beta^{\sim -2}\,x^2\right)}\,\sqrt{2}\,\beta^{\sim}}{\sqrt{\dfrac{\sqrt{\pi}}{\beta^{\sim}}}}$$

We define the energy expression for the first excited state

```
> Energy1:=beta->Int(-psi1(x,beta)*diff(psi1(x,beta),x$2)/2
> +V(x)*psi1(x,beta)^2,x=-infinity..infinity);
```

$$\mathrm{Energy1} := \beta \to \int_{-\infty}^{\infty} -\frac{1}{2}\,\psi 1(\,x,\beta\,)\,\mathrm{diff}(\,\psi 1(\,x,\beta\,),x\,\$\,2\,)+\mathrm{V}(\,x\,)\,\psi 1(\,x,\beta\,)^2\,dx$$

Again, we are ready for the loop, and I display only a few values to save space:

```
> for j from 1 to 15 do
> bj:=0.1*j;
> Ej[j]:=evalf(Energy1(bj));
> od;
```

$$\mathrm{bj} := .3$$

$$\mathrm{Ej}_3 := -.2339668317$$

$$\mathrm{bj} := .4$$

$$\mathrm{Ej}_4 := -.2604652448$$

$$\mathrm{bj} := .5$$

$$\mathrm{Ej}_5 := -.2618753211$$

$$\mathrm{bj} := .6$$

$$\mathrm{Ej}_6 := -.2393257784$$

$$\mathrm{bj} := .7$$

$$\mathrm{Ej}_7 := -.1939553160$$

$$\mathrm{bj} := .8$$

$$\mathrm{Ej}_8 := -.1268218546$$

```
> plot([[0.1*i,Ej[i]] $i=1..15],beta=0..1.5,style=point,
> symbol=diamond,title=`Variational approximation to E1`);
```

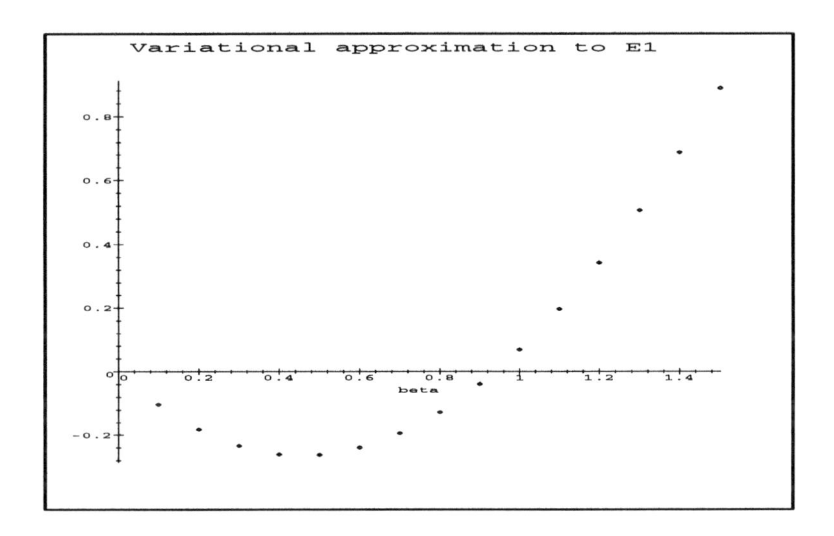


```
> plot({V(x),psi1(x,0.175)+Ej[5]},x=-10..10,y=-1..0.5,
> title='psi1 in V = -1/sqrt(1+x^2)');
```

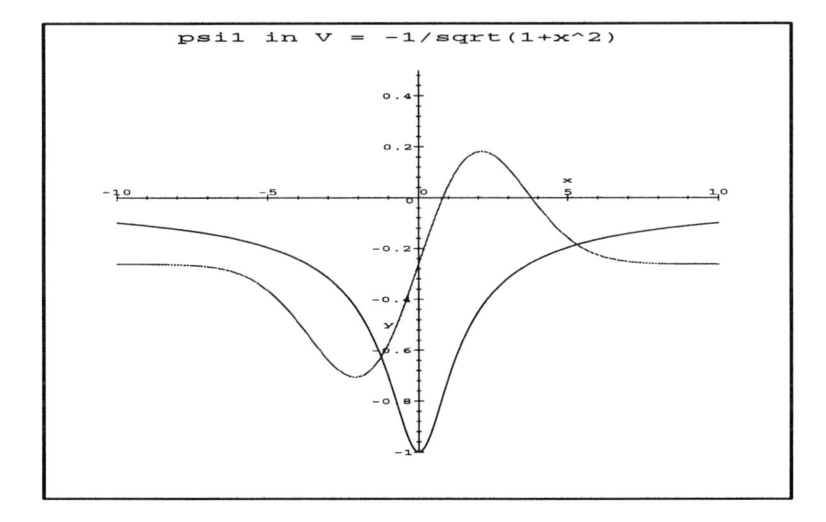

One can understand the smaller value of the optimum β value in this case: the higher energy value makes the particle want to live in a more extended region in space. Note that the crossing of the baseline for the wavefunction, i.e., the energy eigenvalue with the classical potential, again gives the classical turning points.

The excited state has a considerable desire to be beyond this turning point as can be seen from the square of the wavefunction. The latter fact plays an important role in the tunneling ionization mechanism for an atom exposed to a very intense laser field (cf. Sect. 3.3).

Exercise 1.4.1: Apply the variational method with a Gaussian trial state to calculate the lowest symmetric and anti-symmetric eigenfunctions and corresponding eigenenergies for the anharmonic oscillator (AHO) $V = \lambda x^4$ for various choices of the strength parameter λ. Note that the potential energy expectation value can be calculated analytically in this case.

Exercise 1.4.2: A symmetric, double-well anharmonic oscillator is defined by the potential

$$V(x) = -\mu^2 x^2 + \lambda x^4 ,$$

where μ and $\lambda > 0$ are two given coupling constants. Find the positions of the minima and add a constant to the potential such that the potential value at the minimum equals zero. Make your own parameter choices for μ and λ and observe under what conditions the two wells are more or less strongly separated by a barrier in between. For the case of a high barrier approximate one of the wells by a single HO and estimate the zero-point energy for a state confined to the single potential well.

Exercise 1.4.3: Solve the problem presented in exercise 1.4.2 by considering as trial wavefunctions the symmetric and anti-symmetric combinations of HO-eigenfunctions centered on the minima of the potential located at $x = \pm x_0$, *viz.*,

$$\psi_{\mathrm{T}\pm} = \mathrm{e}^{-\beta^2 (x-x_0)^2/2} \pm \mathrm{e}^{-\beta^2 (x+x_0)^2/2} .$$

Note that this trial wavefunction is not normalized. For which potential parameter choices do the energies obtained for the states $\psi_{\mathrm{T}+}$ and $\psi_{\mathrm{T}-}$ approach each other? Note that $\psi_{\mathrm{T}+}$ is associated with the ground state as it obeys the symmetry of the Hamiltonian.

Exercise 1.4.4: Use the variational method with a simple HO trial wavefunction to calculate the ground state of a potential that is not bound from below, such as

$$V(x) = -\frac{1}{|x|^{1/3}} .$$

Does the result constitute a proof that the ground-state energy is bound from below? Describe the motion of a classical particle in this potential. What makes it possible for the quantum particle to possess stable motion in the unbound potential? A numerical solution to the problem is suggested as exercise 2.6.4.

1.5 Commutation Relations

In this section I demonstrate how to implement differential operators and their commutation relations in Maple. First we define a 1D momentum operator and try it out by acting on a plane wave.

```
> wavef:=x->exp(-I/hbar*p0*x);
```

$$\text{wavef} := x \to e^{\left(-\frac{I\, p0\, x}{\text{hbar}}\right)}$$

```
> popx:=proc();
> simplify(-I*hbar*(diff(args[1],x)));
> end;
```

```
popx := proc() simplify(-I*hbar*diff(args[1],x)) end
```

```
> popx(wavef(x));
```

$$-p0\, e^{\left(-\frac{I\, p0\, x}{\text{hbar}}\right)}$$

```
> popx(wavef(x))/wavef(x);
```

$$-p0$$

This proves that the above wavefunction has an eigenvalue of $-p0$ for the momentum operator, i.e., that it represents a left-travelling wave. A procedure for the free Hamiltonian is given by

```
> FreeH:=proc();
> -hbar^2*diff(args[1],x$2)/(2*m);
> end;
```

```
FreeH := proc() -1/2*hbar^2*diff(args[1],x $ 2)/m end
```

```
> FreeH(wavef(x))/wavef(x);
```

$$\frac{1}{2}\frac{p0^2}{m}$$

Here we have shown that `wavef(x)` is an eigenfunction of the free Hamiltonian with eigenvalue $E_0 = p_0^2/(2m)$. The harmonic oscillator (HO) potential applied to a wavefunction is given by

```
> VHO:=f->f*m*(omega*x)^2/2;
```

$$\mathrm{VHO} := f \to \frac{1}{2}\, f\, m\, \omega^2\, x^2$$

```
> VHO(wavef(x));
```

$$\frac{1}{2}\, \mathrm{e}^{\left(-\frac{I\,\mathrm{p0}\,x}{\mathrm{hbar}}\right)}\, m\, \omega^2\, x^2$$

We can verify whether ψ is an eigenstate of $\hat{H}$ from the evaluation of the expression $(\hat{\psi})/\psi$:

```
> (FreeH+VHO)(wavef(x))/wavef(x);
```

$$\frac{\frac{1}{2}\,\frac{p0^2\,\mathrm{e}^{\left(-\frac{I\,\mathrm{p0}\,x}{\mathrm{hbar}}\right)}}{m} + \frac{1}{2}\,\mathrm{e}^{\left(-\frac{I\,\mathrm{p0}\,x}{\mathrm{hbar}}\right)}\, m\, \omega^2\, x^2}{\mathrm{e}^{\left(-\frac{I\,\mathrm{p0}\,x}{\mathrm{hbar}}\right)}}$$

```
> simplify(");
```

$$\frac{1}{2}\,\frac{p0^2 + m^2\,\omega^2\,x^2}{m}$$

The plane wave is not an eigenfunction of the harmonic oscillator Hamiltonian as the result of the last operation is a function of x and not a number, i.e., an eigenvalue.

Let's try the ground-state HO eigenfunction with $\omega = \sqrt{k/m} = 1$

```
> wfHO:=x->exp(-(omega*m/hbar)*x^2/2);
```

$$\mathrm{wfHO} := x \to \mathrm{e}^{\left(-1/2\,\frac{\omega\, m\, x^2}{\mathrm{hbar}}\right)}$$

```
> FreeH(wfHO(x))/wfHO(x);
```

$$-\frac{1}{2}\,\frac{hbar^2\left(-\frac{\omega\, m\, \mathrm{e}^{\left(-1/2\,\frac{\omega\, m\, x^2}{\mathrm{hbar}}\right)}}{\mathrm{hbar}} + \frac{\omega^2\, m^2\, x^2\, \mathrm{e}^{\left(-1/2\,\frac{\omega\, m\, x^2}{\mathrm{hbar}}\right)}}{\mathrm{hbar}^2}\right)}{m\,\mathrm{e}^{\left(-1/2\,\frac{\omega\, m\, x^2}{\mathrm{hbar}}\right)}}$$

```
> simplify(");
```

$$-\frac{1}{2}\,\omega\,(-\mathrm{hbar} + m\,\omega\,x^2)$$

The HO ground state is not an eigenfunction of the free Hamiltonian, but of the HO Hamiltonian:

```
> (FreeH+VHO)(wfHO(x))/wfHO(x);
```

$$\frac{-\frac{1}{2}\frac{\text{hbar}^2\left(-\frac{\omega\, m\,\%1}{\text{hbar}}+\frac{\omega^2\, m^2\, x^2\,\%1}{\text{hbar}^2}\right)}{m}+\frac{1}{2}\%1\, m\,\omega^2\, x^2}{\%1}$$

$$\%1 := \mathrm{e}^{\left(-1/2\,\frac{\omega\, m\, x^2}{\text{hbar}}\right)}$$

```
> simplify(");
```

$$\frac{1}{2}\,\text{hbar}\,\omega$$

The above result is the ground-state energy of the QM harmonic oscillator. It shows that in the ground state the particle is not at rest (i.e., at the bottom of the well, $E = 0$), but fluctuates.

Note that Planck's constant $\hbar$ controls the amount of ground-state fluctuations. Also the latter depend on the value of ω: for a stronger spring constant k or a lighter particle (smaller mass, $\omega^2 = k/m$) the zero-point energy E_0 is larger.

We now proceed with the commutation relations. Note that we don't use abstract operators in Maple, but always let them act on functions. Remember that the HO eigenfunction is not an eigenfunction of the momentum operator. For the commutation relations it is important to remember that the operators act on functions. It is irrelevant whether the functions are eigenfunctions or not.

```
> popx(wfHO(x))/wfHO(x);
```

$$I\,\omega\, m\, x$$

The position operator is easily defined:

```
> xop:=f->x*f;
```

$$xop := f \to x\, f$$

Neither the plane wave nor the HO ground state are eigenfunctions of the position operator:

```
> xop(wavef(x))/wavef(x);
```

$$x$$

```
> xop(wfHO(x))/wfHO(x);
```

$$x$$

Now we try the Dirac delta-function:

```
> xop(Dirac(x-x0))/Dirac(x-x0);
```

$$x$$

Maple isn't clever enough to realize that this should give $x0$ as an eigenvalue. Let us see what we obtain for the average position from the position-eigenstate and from a Gaussian, respectively.

```
> int(xop(Dirac(x-x0)),x=-infinity..infinity)/int(Dirac(x-x0),
> x=-infinity..infinity);
```

$$x0$$

```
> int(xop(wfHO(x-x0)),x=-infinity..infinity)/int(wfHO(x-x0),
> x=-infinity..infinity);
```

$$x0$$

Now try the plane wave:

```
> int(xop(wavef(x-x0)),x=-infinity..infinity)/int(wavef(x-x0),
> x=-infinity..infinity);
```

$$1$$

The above results show that:

- for the average positions the Dirac delta and the Gaussian return x_0;
- the plane wave has no average position, the particle is equally likely to be anywhere.

We check the commutation relation $[x, \hat{p}_x] := (x\hat{p}_x - \hat{p}_x x)$ as applied to the Dirac-delta and Gaussian functions:

```
> (xop(popx(wavef(x)))-popx(xop(wavef(x))))/wavef(x);
```

$$I\,\mathrm{hbar}\tilde{}$$

```
> (xop(popx(wfHO(x)))-popx(xop(wfHO(x))))/wfHO(x);
```

$$\frac{I\,x^2\,\omega\tilde{}\;m\tilde{}\;\%1 - I\,(-\mathrm{hbar}\tilde{}\;\%1 + x^2\,m\tilde{}\;\omega\tilde{}\;\%1)}{\%1}$$

$$\%1 := \mathrm{e}^{\left(-1/2\,\frac{\omega\tilde{}\;m\tilde{}\;x^2}{\mathrm{hbar}\tilde{}}\right)}$$

```
> simplify(");
```

$$I\,\mathrm{hbar}\tilde{}$$

We have shown that the commutation relation $[x, \hat{p}_x] = i\hbar$ holds, while acting on two specific functions. The following question arises: *can we show commutation relations (CR) without applying the operators to functions?* We define a CR procedure:

```
> CR:=proc();
> (args[1])@(args[2])-(args[2])@(args[1]);
> end;
```

CR := proc() args[1] @ args[2]-args[2] @ args[1] end

```
> CR(xop,popx)(wavef(x));
```

$$I\,\mathrm{e}^{\left(-\frac{I\,\mathrm{p0}\,x}{\mathrm{hbar\tilde{}}}\right)}\,\mathrm{hbar\tilde{}}$$

Thus, we have found a short-hand way to obtain the CR. We are, however, far from our goal: we still have to act on something in order to evaluate the commutator. Observe how Maple understands the abstract statement

```
> CR(xop,popx);
```

$$xop@popx - popx@xop$$

```
> CR(xop,popx)(wfHO(x))/wfHO(x);
```

$$\frac{I\,x^2\,\omega\tilde{}\,m\tilde{}\,\%1 - I\,(-\mathrm{hbar\tilde{}}\,\%1 + m\tilde{}\,\omega\tilde{}\,x^2\,\%1)}{\%1}$$

$$\%1 := \mathrm{e}^{\left(-1/2\,\frac{\omega\tilde{}\,m\tilde{}\,x^2}{\mathrm{hbar\tilde{}}}\right)}$$

```
> simplify(");
```

$$I\,\mathrm{hbar\tilde{}}$$

Nevertheless, we can use many objects to satisfy ourselves that the CR holds independent of what we act upon:

```
> CR(xop,popx)(x^2)/x^2;
```

$$I\,\mathrm{hbar\tilde{}}$$

Now we generalize by letting the operators act on some arbitrary function $f(x)$:

```
> CR(xop,popx)(f(x));
```

$$-I\,x\,\mathrm{hbar}\left(\frac{\partial}{\partial x}f(x)\right)+I\,\mathrm{hbar}\left(f(x)+x\left(\frac{\partial}{\partial x}f(x)\right)\right)$$

```
> simplify(");
```

$$I\,\mathrm{hbar}\,f(x)$$

This completes the proof that

$$[x,\hat{p}_x]f(x)=\mathrm{i}\hbar f(x)\,.$$

Note that, a priori, $[x,\hat{p}_x]$ was not just a simple multiplicative operator.

Maple has a built-in package to deal with non-commuting algebras. We have not figured out how to use it for our purposes in a straightforward way. Our definition of CR is more direct and allows the simplification made above to calculate commutators. I encourage some exploration of the built-in package. Help is obtained from the command

```
> ?commutat
```

which shows that once one reads in the package

```
> readlib(commutat):
```

a structure `c(A,B)` is defined in Maple:

```
> c(xop,popx);
```

$$[\,xop,popx\,]$$

Alternatively it is possible to use:

```
> commutat(");
```

$$[\,xop,popx\,]$$

```
> c(xop,popx)(x^2);
```

$$[\,xop,popx\,](x^2)$$

```
> simplify(");
```

$$[\,xop,popx\,](x^2)$$

Maybe this doesn't simplify since Maple doesn't know what $\hbar$ is?

```
> subs(hbar=1,");
```

$$[\,xop,popx\,](x^2)$$

```
> simplify(");
```

$$[\mathit{xop}, \mathit{popx}](x^2)$$

While we can't get it to evaluate, `c(A,B)` satisfies the right properties:

```
> c(x,x);
```

$$0$$

```
> c(a+b,d);
```

$$[a, d] + [b, d]$$

Let's proceed with testing a few known CRs. We begin with the commutativity of the free-particle Hamiltonian `FreeH` and the momentum operator

```
> CR(FreeH,popx)(f(x));
```

$$0$$

This property enables one to find common eigenfunctions to the kinetic energy and the momentum operator. It is obvious since `FreeH` $= T = \hat{p}^2/(2m)$ is a simple function of $\hat{p}$.

```
> CR(FreeH,VHO)(f(x));
```

$$-\frac{1}{2}\frac{\mathrm{hbar}^2\left(\frac{1}{2}\left(\frac{\partial^2}{\partial x^2}f(x)\right)m\,\omega^2 x^2 + 2\left(\frac{\partial}{\partial x}f(x)\right)m\,\omega^2 x + f(x)\,m\,\omega^2\right)}{m}$$
$$+\frac{1}{4}\mathrm{hbar}^2\left(\frac{\partial^2}{\partial x^2}\mathrm{f}(x)\right)\omega^2 x^2$$

```
> simplify(");
```

$$-\frac{1}{2}\mathrm{hbar}^2\omega^2\left(2\left(\frac{\partial}{\partial x}f(x)\right)x + f(x)\right)$$

This is obviously non-vanishing and the operators for $\hat{T}$ and the HO potential cannot be diagonalized separately by a common set of eigenfunctions. Also, position eigenstates do not diagonalize the free Hamiltonian:

```
> CR(xop,FreeH)(f(x));
```

$$-\frac{1}{2}\frac{x\,\mathrm{hbar}^2\left(\frac{\partial^2}{\partial x^2}f(x)\right)}{m} + \frac{1}{2}\frac{\mathrm{hbar}^2\left(2\left(\frac{\partial}{\partial x}f(x)\right) + \left(\frac{\partial^2}{\partial x^2}f(x)\right)x\right)}{m}$$

```
> simplify(");
```

$$\frac{\text{hbar}^2 \left(\frac{\partial}{\partial x} f(x)\right)}{m}$$

This latter result shows, however, that the answer is related to the momentum operator. We check this using:

```
> "-popx(f(x))*hbar*I/m;
```

$$0$$

Exercise 1.5.1: Verify your own examples of commuting and non-commuting operators. Investigate the space displacement operator $\exp(-\mathrm{i}\Delta x\hat{p}/\hbar)$ for small displacements Δx using a low-order Taylor approximation.

Exercise 1.5.2: Use Maple to investigate sequences of functions that have the Dirac-delta function as a limit, such as the simple rectangular box-shaped function

$$\delta_n(x) = \frac{1}{n} \quad \text{for} \quad -\frac{1}{2n} \leq x \leq \frac{1}{2n}.$$

Verify how in the limit $n \to \infty$ properties of the Dirac-delta function emerge, such as

$$\int_{-\infty}^{\infty} f(x)\delta(x-a)\mathrm{d}x = f(a).$$

Consider different sequences of functions with this property (cf. [Ar85]), such as, e.g., the Gaussian bell-shaped functions.

Exercise 1.5.3: Consider the integral representation of the Dirac-delta function(cf. [Ar85])

$$\delta(x-x_0) = \frac{1}{2\pi}\int_{-\infty}^{\infty} \mathrm{e}^{ip(x-x_0)}\mathrm{d}p\,,$$

and determine the momentum space representation of position eigenstates.

Exercise 1.5.4: Use Maple's Taylor expansion capabilities to investigate the problem of how to evaluate

$$\mathrm{e}^{\lambda\hat{A}\hat{B}}\,,$$

where $\hat{A}$ and $\hat{B}$ are non-commuting observables and their commutator is defined as $C = [\hat{A}, \hat{B}]$ for low order in λ.

1.6 Anharmonic Oscillator

We discuss a quartic anharmonic oscillator as an example for some approximation techniques for bound states in QM. First we apply stationary perturbation theory for non-degenerate levels, and subsequently we introduce matrix diagonalization in a finite, truncated basis set.

We split the Hamiltonian into a zero-order part and additional potential $H = H_0 + V$, where the zero-order problem is the well-known harmonic oscillator

$$H_0 = \hat{T} + \frac{1}{2}m\omega^2 x^2 .$$

For the additional interaction that can't be diagonalized exactly we take a quartic potential

$$V = \lambda x^4 .$$

We are interested in the determination of the corrected energy levels (and wavefunctions) as compared to the H_0 results for various values of λ. Of course, we can think of corrections to the exactly known H_0 results only as long as λ is small.

Note that this is not a particularly good example for perturbation theory, and that we are interested in showing both the weakness and strength of this, in general, very useful technique.

We begin with the oscillator states and choose $\omega = 1$ for simplicity.

```
> psi0:=exp(-x^2/2);
```

$$\psi 0 := e^{(-1/2\,x^2)}$$

```
> psi1:=x*exp(-x^2/2);
```

$$\psi 1 := x\, e^{(-1/2\,x^2)}$$

We could build them up from scratch, normalize, etc, but we prefer to make use of the built-in polynomials.

```
> with(orthopoly);
```

$$[G, H, L, P, T, U]$$

```
> psi:=(x,n)->sqrt(sqrt(1/Pi)/(2^n*n!))*exp(-x^2/2)*H(n,x);
```

$$\psi := (x, n) \to \mathrm{sqrt}\left(\frac{\mathrm{sqrt}\left(\frac{1}{\pi}\right)}{2^n\, n!}\right) e^{(-1/2\,x^2)}\, \mathrm{H}(n, x)$$

```
> int(psi(x,3)^2,x=-infinity..infinity);
```

$$1$$

The normalization appears to be in order, we check a non-trivial orthogonality:

```
> int(psi(x,2)*psi(x,4),x=-infinity..infinity);
```

$$0$$

The spectrum has no degeneracies, $E_n = (n + 1/2)\hbar\omega = (n + 1/2)$ (in atomic units) with $\omega = 1$ for $n = 0, 1, \ldots$.

In stationary first-order perturbation theory (PT) the energy shift due to the perturbation is simply given as the expectation (average) value between the zeroth order states, i.e., the eigenstates of the H_0 problem:

$$\mathrm{DE}(n) = \langle n|V|n\rangle \, .$$

More explicitly the expectation value is calculated in coordinate representation as

$$\langle n|V|n\rangle = \int_{-\infty}^{\infty} \psi_n^*(x)V(x)\psi_n(x)\mathrm{d}x \, .$$

Thus, we calculate the matrix element $\lambda\langle n|x^4|n\rangle$ taking into account that the HO eigenfunctions are real valued:

```
> DE:=n->int(x^4*psi(x,n)^2,x=-infinity..infinity);
```

$$\mathrm{DE} := n \to \int_{-\infty}^{\infty} x^4 \, \psi(\, x, n \,)^2 \, dx$$

```
> DE(0);      DE(1);
```

$$\frac{3}{4} \qquad \frac{15}{4}$$

We now define a function that provides the energy for any n and λ accurate to first order in λ:

```
> EPT1:=(n,lambda)->(n+1/2)+lambda*DE(n);
```

$$\mathrm{EPT1} := (\, n, \lambda \,) \to n + \frac{1}{2} + \lambda \, \mathrm{DE}(\, n \,)$$

```
> plot({EPT1(0,lambda),EPT1(1,lambda),EPT1(2,lambda)},
> lambda=0..2,En=0..10);
```

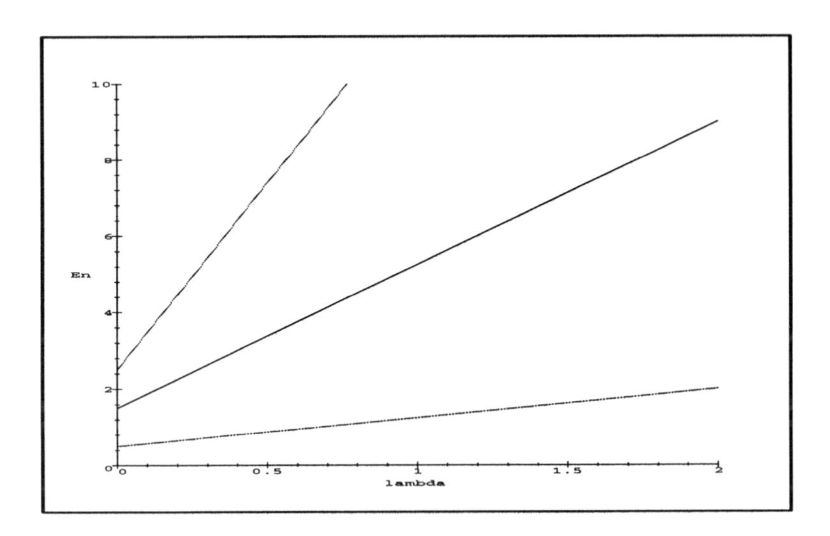

First-order PT tells us that the higher levels are more strongly affected by the perturbing potential. Even for weak λ the steep rise of the x^4 potential puts the squeeze on the excited states and they move up in energy rapidly.

We now investigate the behaviour of these levels more accurately by choosing a matrix representation and diagonalizing the Hamiltonian matrix. We should make use of the symmetry and calculated the even-parity and odd-parity states separately, but let's try to be lazy and just assemble the matrix. We do, however, make use of the fact that the functions are eigenfunctions of H_0. We truncate the problem at $N = 5$ (convergence as a function of N should be demonstrated). The linear algebra package has to be loaded explicitly in Maple:

```
> with(linalg):
> N:=5; Np:=N+1;
```

$$N := 5 \qquad Np := 6$$

We make the assumption that λ is positive to ensure that the spectrum of the Hamiltonian is bounded from below.

```
> assume(lambda>0);
> Hmat:=matrix(Np,Np,0):
> for i1 from 1 to Np do
> n1:=i1-1;
> for i2 from 1 to Np do
```

```
> n2:=i2-1;
> Vmat:=int(psi(x,n1)*psi(x,n2)*x^4,x=-infinity..infinity);
> if n1=n2 then Hmat[i1,i2]:=(n1+1/2)+lambda*Vmat;
> else  Hmat[i1,i2]:=lambda*Vmat;
> fi;
> od:
> od:
> evalm(Hmat);
```

$$\left[\begin{array}{c}
\frac{1}{2}+\frac{3}{4}\lambda^{\sim}\,,\,0\,,\,\frac{3}{2}\sqrt{2}\,\lambda^{\sim}\,,\,0\,,\,\frac{1}{2}\sqrt{6}\,\lambda^{\sim}\,,\,0 \\
0\,,\frac{3}{2}+\frac{15}{4}\lambda^{\sim}\,,0\,,\frac{5}{2}\sqrt{2}\,\sqrt{3}\,\lambda^{\sim}\,,0\,,\frac{1}{2}\sqrt{2}\,\sqrt{15}\,\lambda^{\sim} \\
\frac{3}{2}\sqrt{2}\,\lambda^{\sim}\,,\,0\,,\,\frac{5}{2}+\frac{39}{4}\lambda^{\sim}\,,\,0\,,\,7\,\lambda^{\sim}\sqrt{3}\,,\,0 \\
0\,,\,\frac{5}{2}\sqrt{2}\,\sqrt{3}\,\lambda^{\sim}\,,\,0\,,\,\frac{7}{2}+\frac{75}{4}\lambda^{\sim}\,,\,0\,,\,9\,\lambda^{\sim}\sqrt{5} \\
\frac{1}{2}\sqrt{6}\,\lambda^{\sim}\,,\,0\,,\,7\,\lambda^{\sim}\sqrt{3}\,,\,0\,,\,\frac{9}{2}+\frac{123}{4}\lambda^{\sim}\,,\,0 \\
0\,,\,\frac{1}{2}\sqrt{2}\,\sqrt{15}\,\lambda^{\sim}\,,\,0\,,\,9\,\lambda^{\sim}\sqrt{5}\,,\,0\,,\,\frac{11}{2}+\frac{183}{4}\lambda^{\sim}
\end{array}\right]$$

```
> Ediag6:=eigenvals(Hmat);
  Error, (in evala)
  reducible RootOf detected.  Substitutions are, {RootOf(_Z^2-
  15) = -RootOf(_Z^2-5)*RootOf(_Z^2-3), RootOf(_Z^2-15) =
  RootOf(_Z^2-5)*RootOf(_Z^2-3)}
```

The eigenvalues were not obtained by Maple. We can try to evaluate the characteristic polynomial and attempt a factorization. To save space we suppress the next output.

```
> charpoly(Hmat,epsilon)
> factor(");
  Error, (in evala)
  reducible RootOf detected.  Substitutions are, {RootOf(_Z^2-
  15) = -RootOf(_Z^2-5)*RootOf(_Z^2-3), RootOf(_Z^2-15) =
  RootOf(_Z^2-5)*RootOf(_Z^2-3)}
```

We learn that the factorization has run into difficulty. Our laziness didn't pay off. We should define two submatrices:

```
> Hmateven:=matrix(2,2,0);
```

$$\text{Hmateven} := \begin{bmatrix} 0 & 0 \\ 0 & 0 \end{bmatrix}$$

```
> Hmatodd:=matrix(2,2,0);
```

$$\text{Hmatodd} := \begin{bmatrix} 0 & 0 \\ 0 & 0 \end{bmatrix}$$

We reorganize the non-zero entries into separate matrices as they decouple completely. The even label `i1ev=2*i1` labels the odd-parity states and vice versa.

```
> for i1 from 1 to 2 do
> i1ev:=2*i1;
> i1od:=2*i1-1;
> for i2 from 1 to 2 do
> i2ev:=2*i2;
> i2od:=2*i2-1;
> Hmatodd[i1,i2]:=Hmat[i1ev,i2ev];
> Hmateven[i1,i2]:=Hmat[i1od,i2od];
> od:
> od:
> evalm(Hmateven);
```

$$\begin{bmatrix} \frac{1}{2} + \frac{3}{4}\lambda^{\sim} & \frac{3}{2}\sqrt{2}\,\lambda^{\sim} \\ \frac{3}{2}\sqrt{2}\,\lambda^{\sim} & \frac{5}{2} + \frac{39}{4}\lambda^{\sim} \end{bmatrix}$$

```
> evalm(Hmatodd);
```

$$\begin{bmatrix} \frac{3}{2} + \frac{15}{4}\lambda^{\sim} & \frac{5}{2}\sqrt{2}\,\sqrt{3}\,\lambda^{\sim} \\ \frac{5}{2}\sqrt{2}\,\sqrt{3}\,\lambda^{\sim} & \frac{7}{2} + \frac{75}{4}\lambda^{\sim} \end{bmatrix}$$

```
> soleven:=eigenvals(Hmateven);
```

$$\text{soleven} := \frac{3}{2} + \frac{21}{4}\lambda^{\sim} + \frac{1}{2}\sqrt{4 + 36\,\lambda^{\sim} + 99\,\lambda^{\sim 2}},$$

$$\frac{3}{2}+\frac{21}{4}\lambda^{\sim}-\frac{1}{2}\sqrt{4+36\lambda^{\sim}+99\lambda^{\sim 2}}$$

```
> solodd:=eigenvals(Hmatodd);
```

$$\text{solodd} := \frac{5}{2}+\frac{45}{4}\lambda^{\sim}+\frac{1}{2}\sqrt{4+60\lambda^{\sim}+375\lambda^{\sim 2}},$$
$$\frac{5}{2}+\frac{45}{4}\lambda^{\sim}-\frac{1}{2}\sqrt{4+60\lambda^{\sim}+375\lambda^{\sim 2}}$$

```
> taylor(soleven[2],lambda=0);
```

$$\frac{1}{2}+\frac{3}{4}\lambda^{\sim}-\frac{9}{4}\lambda^{\sim 2}+\frac{81}{8}\lambda^{\sim 3}-\frac{1377}{32}\lambda^{\sim 4}+\frac{10935}{64}\lambda^{\sim 5}+\mathrm{O}\left(\lambda^{\sim 6}\right)$$

This is the ground state. Note that PT obtained the $(3/4)\lambda$ correction in first order.

```
> taylor(solodd[2],lambda=0);
```

$$\frac{3}{2}+\frac{15}{4}\lambda^{\sim}-\frac{75}{4}\lambda^{\sim 2}+\frac{1125}{8}\lambda^{\sim 3}-\frac{28125}{32}\lambda^{\sim 4}+\frac{253125}{64}\lambda^{\sim 5}+\mathrm{O}\left(\lambda^{\sim 6}\right)$$

This is the first excited state. Again the linear coefficient (in λ) agrees with the perturbative result.

```
> plot({soleven[2],solodd[2],soleven[1]},lambda=0..2,En=0..10);
```

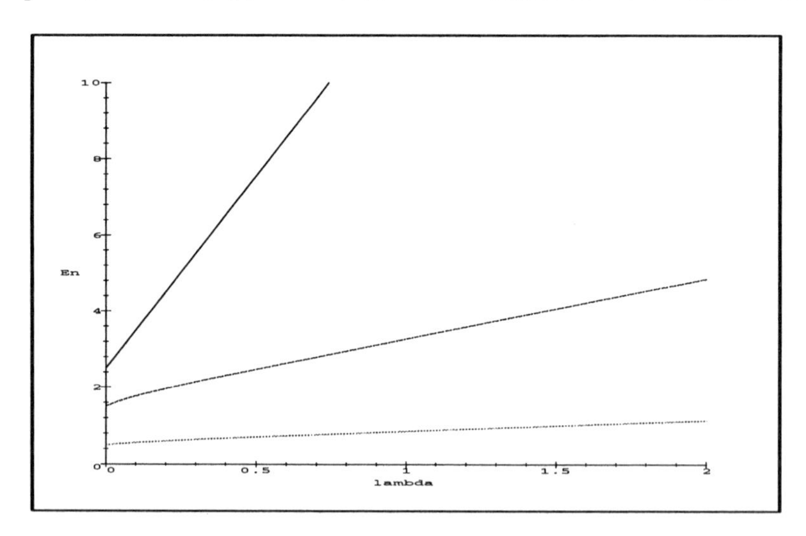

For large λ the PT result to first order is not useful as can be seen from a comparison with the previous figure. The first two levels grow more slowly with λ than indicated by first-order PT. The third level looks similar, but is likely to be still affected by the finite size of the diagonalized matrix. We find that the perturbative expansion does not work well for this problem. There are, however, numerous important problems in QM for which it is suited perfectly. We illustrate the differences between PT and matrix diagonalization results with some numbers. The ground state for coupling $\lambda = 2$ is approximated by

```
> evalf(subs(lambda=2,soleven[2]));
                                   1.13721951
```

while first-order PT gives

```
> evalf(EPT1(0,2));
                                       2.
```

For $\lambda = 1$ the discrepancy becomes smaller

```
> evalf(subs(lambda=1,soleven[2]));    evalf(EPT1(0,1));
                         .855086940          1.250000000
```

For $\lambda = 1/2$ we might have naively expected better agreement than is found:

```
> evalf(subs(lambda=1/2,soleven[2]));    evalf(EPT1(0,1/2));
                         .706301417          .8750000000
```

Only for anharmonic coupling strength, $\lambda < 0.1$, can the first-order results for the ground-state energy be deemed acceptable:

```
> evalf(subs(lambda=1/8,soleven[2]));    evalf(EPT1(0,1/8));
                         .571409708          .5937500000
```

```
> evalf(subs(lambda=1/64,soleven[2]));   evalf(EPT1(0,1/64));
                         .511205643          .5117187500
```

In fact, the 2-by-2 submatrix diagonalizations aren't perfect either. We can check this out by going to 3-by-3 (and larger) matrices. This can be done by minor modifications of the original code a few pages back. The numerical evaluation of eigenvalues doesn't mind the fact that reducible roots appear. For 3-by-3 submatrices or 6-by-6 complete Hamiltonian matrices we obtain:

```
> evalf(Eigenvals(subs(lambda=2,op(Hmat))));
        [.9821546 3.761116 12.06707 24.19495 76.95077 119.0439]
```

```
> evalf(Eigenvals(subs(lambda=1,op(Hmat))));
```

$$[.8082285\ 2.843872\ 7.382825\ 13.86749\ 40.55895\ 62.03864]$$

```
> evalf(Eigenvals(subs(lambda=1/2,op(Hmat))));
```

$$[.6975358\ 2.338140\ 5.051965\ 8.738981\ 22.37550\ 33.54788]$$

```
> evalf(Eigenvals(subs(lambda=1/8,op(Hmat))));
```

$$[.5713187\ 1.820729\ 3.282768\ 4.962996\ 8.802164\ 12.24752]$$

```
> evalf(Eigenvals(subs(lambda=1/64,op(Hmat))));
```

We observe that the convergence of matrix diagonalization also depends on the value of λ.

Exercise 1.6.1: Demonstrate the convergence of the matrix diagonalization method by increasing the rank for the matrix representation of the Hamiltonian for fixed choices of λ.

Exercise 1.6.2: Solve the double-well AHO problem of Exercises 1.4.2 and 1.4.3 using the HO basis representation and matrix diagonalization.

Exercise 1.6.3: Solve the problem of a linear potential

$$V(x) = \alpha|x| \qquad (\alpha > 0),$$

using a HO basis representation and matrix diagonalization. Draw conclusions about the behaviour of energy level as a function of the quantum number for low n by comparing the results to the HO.

Exercise 1.6.4: Solve the problem of an asymmetric potential, such as

$$V(x) = \alpha x + \mu^2 x^2,$$

using a HO basis representation and matrix diagonalization. Graph the eigenfunctions for the two lowest states and draw your conclusions for various parameter choices α and μ.

2. Bound States in 1D

2.1 Local Energy

This chapter deals with methods for obtaining eigenfunctions of QM Hamiltonians. Various approximation techniques are presented for simple one-dimensional (1D) problems even though their usefulness becomes more evident in multi-dimensional many-body problems. To provide some insight into what it takes for a wavefunction to be an eigenfunction of a differential operator I discuss the idea of a local energy density in some detail.

When we find the eigenfunctions of a Hamiltonian, the energy is returned as a constant number. Nevertheless, we can ask the question of whether locally kinetic and potential energies are split up in analogy with classical mechanics. In classical mechanics we consider that the particle's energy along a trajectory in a conservative force field is given as

$$E = T_{\text{kin}}(t) + V_{\text{pot}}(t) = \text{const}\,,$$

where the connection between position and time is given by means of the trajectory $x(t)$.

For a stationary bound state in QM we begin with the expectation values of kinetic and potential energy. They are defined as

$$\langle T \rangle = \int_{-\infty}^{\infty} \psi^*(x)\hat{T}\psi(x)\mathrm{d}x\,,$$

and

$$\langle V \rangle = \int_{-\infty}^{\infty} \psi^*(x)V(x)\psi(x)\mathrm{d}x\,.$$

For the total energy we have $\langle H \rangle = \langle T \rangle + \langle V \rangle$, which for an eigenstate ψ_0 is equal to the eigenvalue E_0.

We can ask the following: is it meaningful to define a local energy density by means of the relevant integrands in the above expectation values? Note that for an eigenstate ψ_ν one can factor the single-particle density ρ_ν in the integrand

$$\psi_\nu^*(x)\hat{H}\psi_\nu(x) = E_\nu\psi_\nu^*(x)\psi_\nu(x) = E_\nu\rho_\nu(x)\,.$$

Alternatively, we could define for the total local energy the quantity

$$\frac{(\hat{H}\psi_\nu(x))}{\psi_\nu(x)},$$

which becomes a constant for an eigenstate. This quantity represents a good check of whether ψ is an eigenstate or not: for all x one should obtain a constant if ψ is indeed a true eigenstate.

Let us experiment with an exact harmonic oscillator eigenstate first. As usual, we take units with $\hbar = m = 1$ and consider an oscillator with $\omega = 1$. We define our first attempt at a local (x dependent) kinetic energy simply as the integrand of the kinetic energy expectation value $\langle T \rangle$, and a corresponding expression for the potential energy.

```
> Tkin:=(arg,x)->evalc(-conjugate(arg)*diff(arg,x$2)/2);
```

$$\text{Tkin} := (\text{arg}, x) \to \text{evalc}\left(-\frac{1}{2}\,\text{conjugate}(\text{arg})\,\text{diff}(\text{arg}, x\,\$\,2)\right)$$

```
> VHO:=x->x^2/2:
```

```
> Vpot:=(arg,x)->evalc(conjugate(arg)*VHO(x)*arg);
```

$$\text{Vpot} := (\text{arg}, x) \to \text{evalc}(\text{conjugate}(\text{arg})\,\text{VHO}(x)\,\text{arg})$$

Note that the wavefunction is passed as a Maple expression (not a function) into the routines.

```
> plot({VHO(x),Tkin(exp(-x^2/2),x)},x=-2.5..2.5,y=-0.2..1);
```

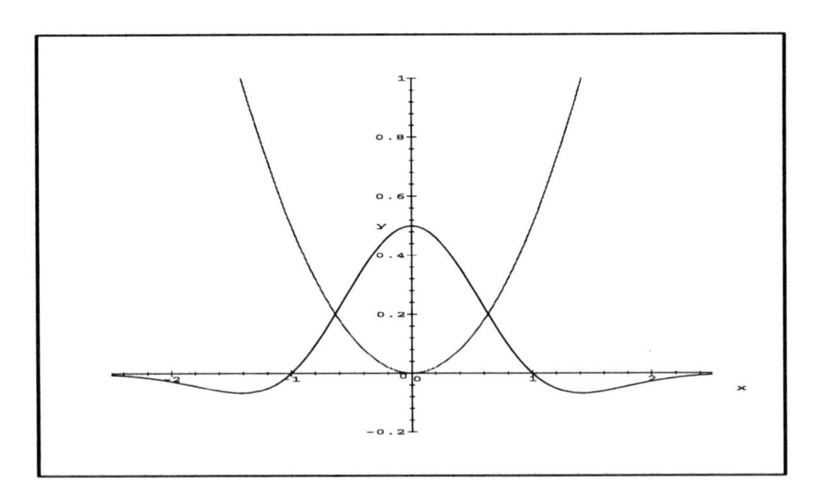

Indeed, the local kinetic energy (energy density), which has no direct physical interpretation, becomes zero at the classical turning points and negative in the classically forbidden region. This means that the wavefunction's curvature has an inflection point at the turning point.

Exercise 2.1.1: Repeat this analysis for excited states of the HO and find the turning points using $E_n = (n + 1/2) = \mathrm{VHO}(x)$.

The defined expression for a local potential energy indicates how the eigenfunction samples the classical potential $\mathrm{VHO}(x)$:

```
> plot({VHO(x),Vpot(exp(-x^2/2),x)},x=-2.5..2.5,y=-0.2..1);
```

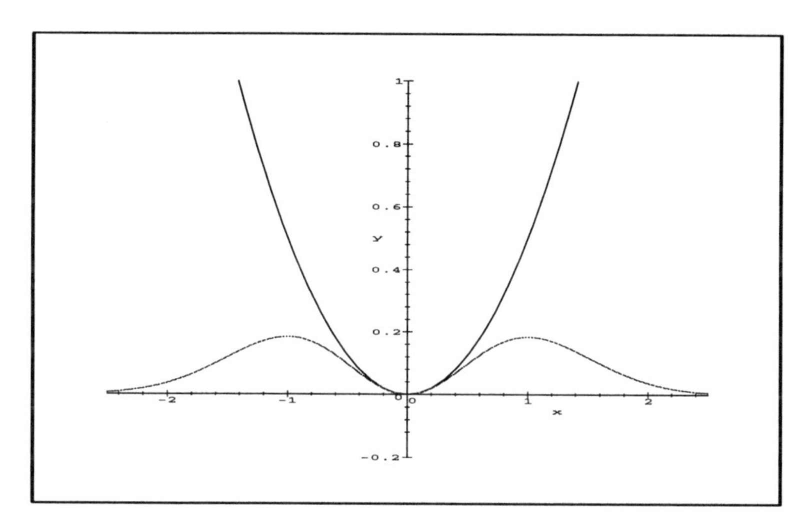

Now we combine both contributions to the integrand of the energy expectation value

```
> plot({VHO(x),Tkin(exp(-x^2/2),x)+Vpot(exp(-x^2/2),x)},
> x=-2.5..2.5,y=-0.2..1);
```

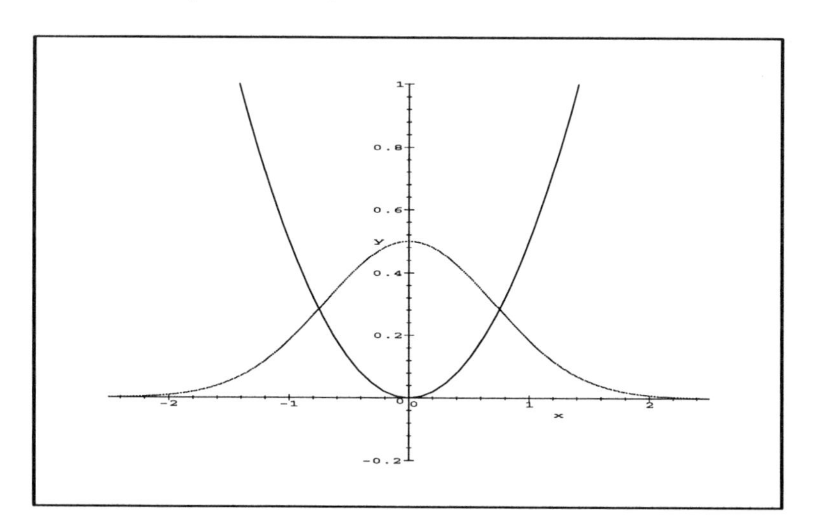

This diagram shows that we should divide out the density in the graph of

the kinetic and potential energy densities. The total energy density should be constant everywhere and equal to the eigenvalue. We define them through:

$$\langle T \rangle = \int_{-\infty}^{\infty} \tau(x)\rho(x)\mathrm{d}x \,,$$

and

$$\langle H \rangle = \int_{-\infty}^{\infty} \epsilon(x)\rho(x)\mathrm{d}x \,.$$

We revise our functions accordingly. The local potential energy is simply the potential function, i.e., $\epsilon(x) = \tau(x) + V(x)$.

```
> Tkin:=(arg,x)->evalc(-conjugate(arg)*diff(arg,x$2)/2/
> (conjugate(arg)*arg));
```

$$\mathrm{Tkin} := (\,\mathrm{arg}, x\,) \to \mathrm{evalc}\left(-\frac{1}{2}\,\frac{\mathrm{diff}(\,\mathrm{arg}, x\,\$\,2\,)}{\mathrm{arg}}\right)$$

```
> plot({VHO(x),Tkin(exp(-x^2/2),x)},x=-2.5..2.5,y=-1..1);
```

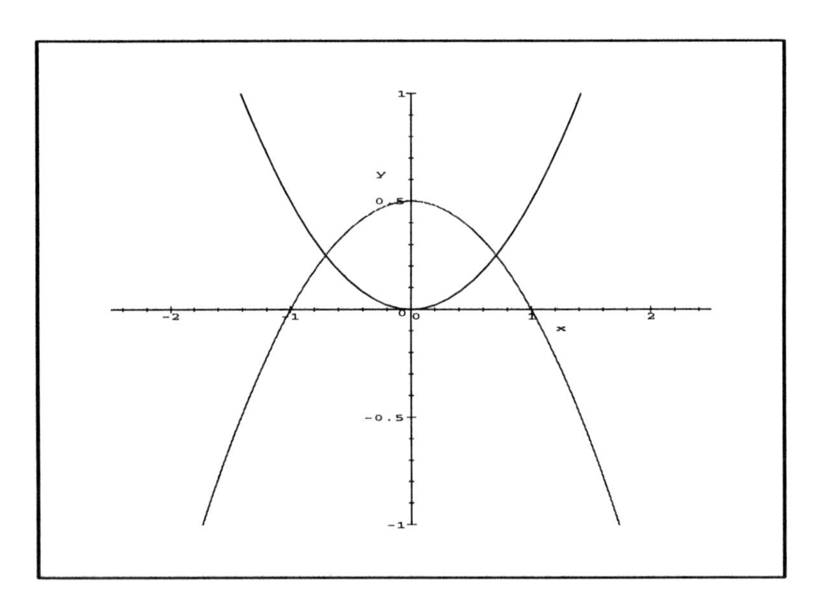

The local kinetic energy density is just a flipped parabola, which together with the classical potential results in a constant energy density that is equal to the energy eigenvalue:

```
> plot({VHO(x),Tkin(exp(-x^2/2),x)+VHO(x)},
> x=-2.5..2.5,y=-0.2..1);
```

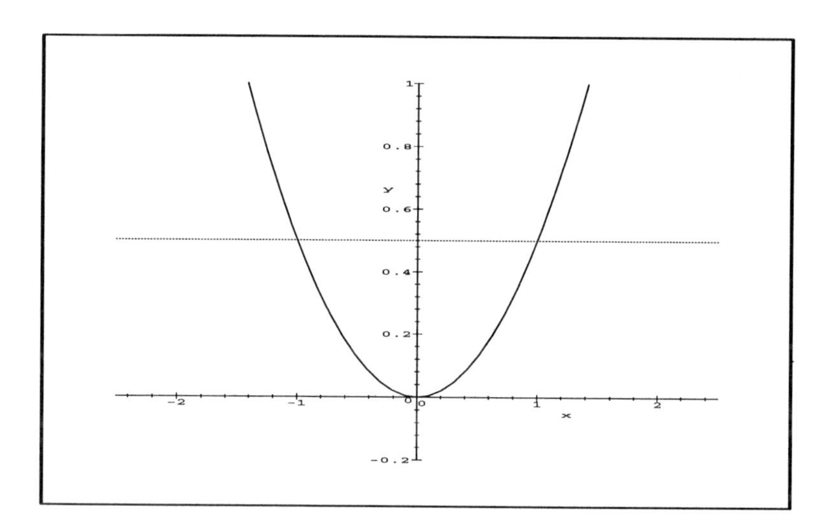

We can now demonstrate what happens in a variational approximation. Consider a quartic anharmonic oscillator to be solved approximately for the symmetric ground state by a Gaussian trial function:

```
> VAHO:=x->x^4/4;
```

$$\mathrm{VAHO} := x \to \frac{1}{4}\,x^4$$

We define an energy expression as in Sect. 1.4. We use an unnormalized Gaussian trial wavefunction and divide the expectation value of the Hamiltonian by the normalization integral.

```
> En:=alpha->int(exp(-alpha^2*x^2/2)*(-1/2*diff(exp(-alpha^2*
> x^2/2),x$2)+VAHO(x)*exp(-alpha^2*x^2/2)),x=-infinity..
> infinity)/int(exp(-alpha^2*x^2),x=-infinity..infinity):
```

```
> assume(beta>0);
```

```
> Enb:=En(beta);
```

$$\mathrm{Enb} := \frac{\left(\frac{1}{4}\,\beta\tilde{}\,\sqrt{\pi} + \frac{3}{16}\,\frac{\sqrt{\pi}}{\beta\tilde{}^{5}}\right)\beta\tilde{}}{\sqrt{\pi}}$$

Remember from Sect. 1.4 that there are two competing terms in the Hamiltonian: the kinetic energy is minimized for small β (flat wavefunction, i.e., no curvature), whereas the potential energy decreases with growing β (localized state at the bottom of the potential well). The optimal variational solution finds a compromise between the two opposing sides.

```
> plot(Enb,beta=0..3,y=0..5);
```

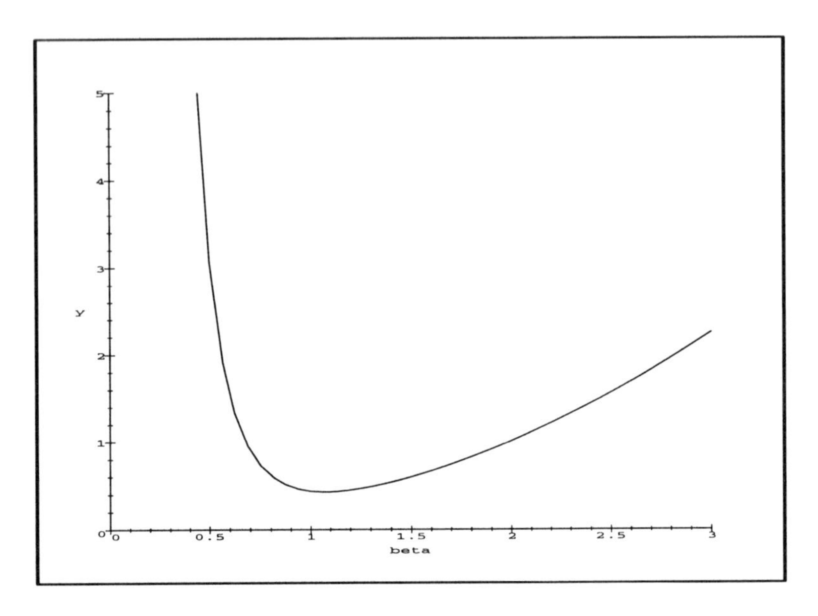

We find the optimum value for β. The first command could be completed with a call to **allvalues**, but we use a numerical determination instead.

```
> solve(diff(Enb,beta)=0,beta);
```

$$\mathrm{RootOf}(2\_Z^6 - 3)$$

```
> beta0:=fsolve(diff(Enb,beta)=0,beta=0..3);
```

$$\beta 0 := 1.069913194$$

```
> E0:=simplify(subs(beta=beta0,Enb));
```

$$E0 := .4292678410$$

A graph of the energy density reveals by how much it deviates locally in x from the variationally obtained approximate energy value. Ideally, we would want a constant line at a height indicating the approximate eigenvalue as obtained for the exact solution of the harmonic oscillator. In practice, however, we have to settle for much less.

```
> plot({VAHO(x),Tkin(exp(-(beta0*x)^2/2),x)+VAHO(x)},
> x=-2.5..2.5,y=-0.2..1);
```

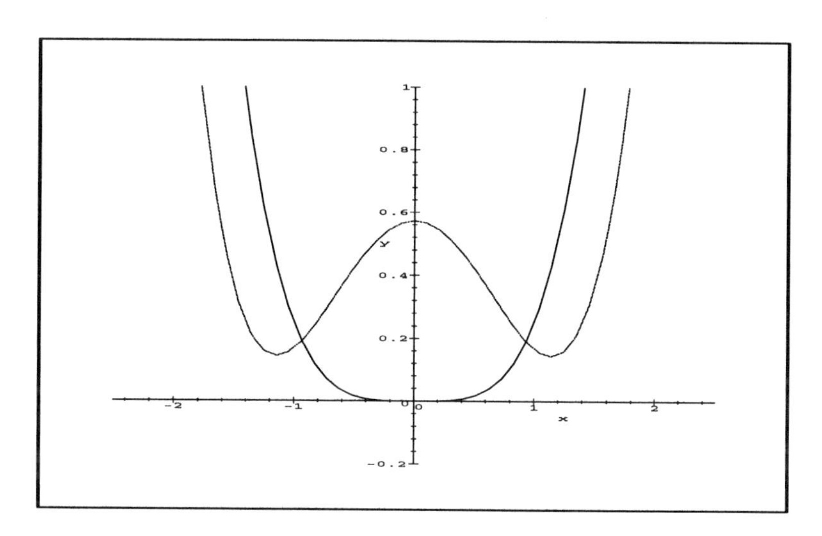

It is clear that a Gaussian trial state approximates the true ground state only poorly. It has problems at $x = 0$ (where the local energy overshoots the eigenvalue) and near the classical turning points. It fails miserably in the asymptotic region where it falls off too slowly, as it is appropriate for a more slowly growing classical potential. We graph the potential together with the (unnormalized) approximate eigenstate:

```
> plot({VAHO(x),exp(-(beta0*x)^2/2)},x=-2.5..2.5,y=0..1.5);
```

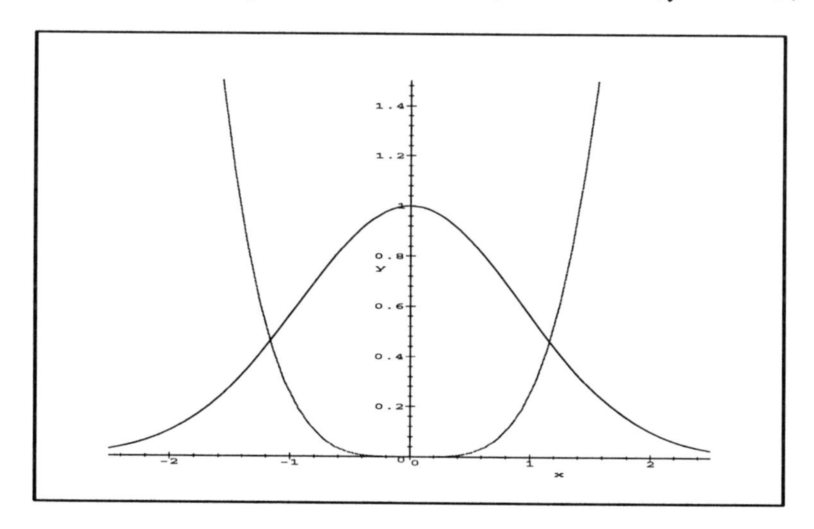

An accurate determination of the eigenstates of the AHO requires numerical techniques such as:

- (1) The variational method with more elaborate trial states (more parameters);
- (2) Expansion in a complete set of states, such as, e.g., a Gaussian multiplied by a Hermite polynomial of increasing order n. The solution of the resulting matrix eigenvalue problem provides approximations to many eigenstates;
- (3) Discretization of the differential equation by finite differences (or finite elements) and solution of the resulting matrix eigenvalue problem;
- (4) Numerical calculation of individual eigenfunctions using a shooting method.

We experiment a little with method 1, and use a non-Gaussian trial state

$$(1 + c_1 x^4)\exp(-\frac{\beta^2 x^2}{2}).$$

First, we define an energy expression that takes an arbitrary wavefunction as its argument:

```
> En1:=arg->int(arg*(-1/2*diff(arg,x$2)+VAHO(x)*arg),
> x=-infinity..infinity)/int(arg^2,x=-infinity..infinity);
```

$$\mathrm{En1} := \mathrm{arg} \to \frac{\int_{-\infty}^{\infty} \mathrm{arg}\left(-\frac{1}{2}\,\mathrm{diff}(\,\mathrm{arg}, x\,\$\,2\,) + \mathrm{VAHO}(\,x\,)\,\mathrm{arg}\right) dx}{\int_{-\infty}^{\infty} \mathrm{arg}^2\, dx}$$

```
> En1c:=En1((1+c1*x^4)*exp(-(beta*x)^2/2));
```

$$\mathrm{En1c} := \left(\frac{1}{2}\beta^{\sim}\sqrt{\pi} + \frac{3}{16}\,\frac{(20\,c1\,\beta^{\sim 2}+1)\sqrt{\pi}}{\beta^{\sim 5}} - \frac{1}{4}\,\frac{(12\,c1 + \beta^{\sim 4})\sqrt{\pi}}{\beta^{\sim 3}} - \frac{15}{8}\,\frac{c1\,(\beta^{\sim 4} + 6\,c1)\sqrt{\pi}}{\beta^{\sim 7}} + \frac{105}{32}\,\frac{c1\,(1 + 9\,c1\,\beta^{\sim 2})\sqrt{\pi}}{\beta^{\sim 9}} + \frac{10395}{256}\,\frac{c1^2\sqrt{\pi}}{\beta^{\sim 13}} - \frac{945}{64}\,\frac{c1^2\sqrt{\pi}}{\beta^{\sim 7}}\right) \Big/ \left(\frac{\sqrt{\pi}}{\beta^{\sim}} + \frac{3}{2}\,\frac{c1\sqrt{\pi}}{\beta^{\sim 5}} + \frac{105}{16}\,\frac{c1^2\sqrt{\pi}}{\beta^{\sim 9}}\right)$$

Let us freeze for a moment the Gaussian width parameter β to the previously obtained optimum value β_0.

```
> En1c0:=subs(beta=beta0,En1c);
```

$$\mathrm{En1c0} := \big(.5349565970\sqrt{\pi} + .1337391492\,(\,22.89428486\,c1 + 1\,)\sqrt{\pi} - .2041241452\,(\,12\,c1 + 1.310370697\,)\sqrt{\pi} - 1.168319081\,c1\,(\,1.310370697 + 6\,c1\,)\sqrt{\pi} + 1.786086269\,c1\,(\,1 + 10.30242819\,c1\,)\sqrt{\pi} + 7.667093960\,c1^2\sqrt{\pi}\big) \big/ \big(.9346552651\sqrt{\pi} + 1.069913193\,c1\sqrt{\pi} + 3.572172539\,c1^2\sqrt{\pi}\big)$$

The graph shows that by choosing a non-zero coefficient c_1 we can lower somewhat the ground-state energy:

```
> plot(En1c0,c1=-0.02..0.02);
```

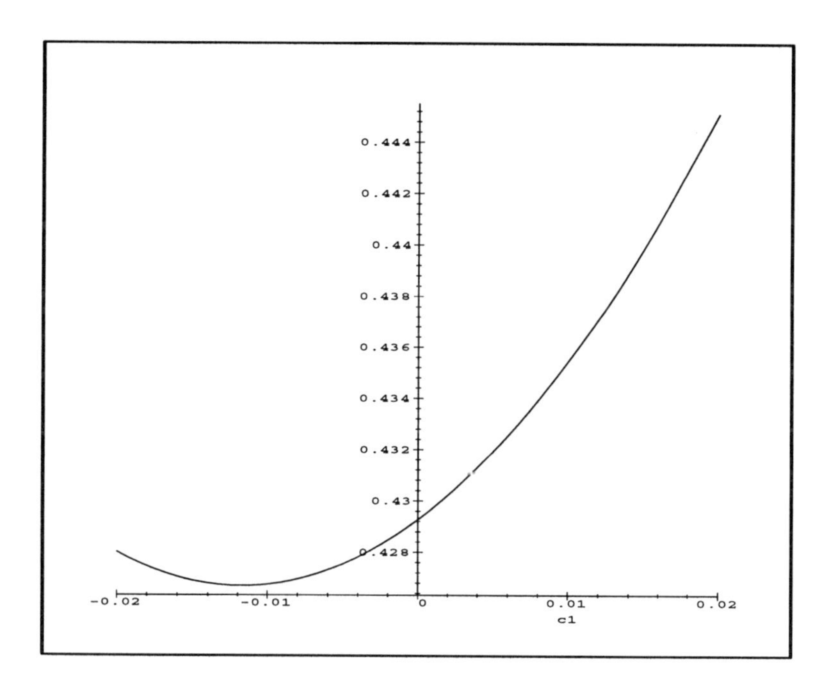

We determine the optimum parameter value

```
> c10:=fsolve(diff(En1c0,c1)=0,c1=-0.1..0);
```

$$c10 := -.01172024593$$

The new energy estimate amounts to $E_0 \approx 0.4267$ as compared to the previous value of 0.4293. In the next diagram we consider the improvement in the local energy due to the more elaborate trial wavefunction:

```
> plot({VAHO(x),Tkin((1+c10*x^4)*exp(-(beta0*x)^2/2),x)+VAHO(x)},
> x=-2.5..2.5,y=-0.2..1);
```

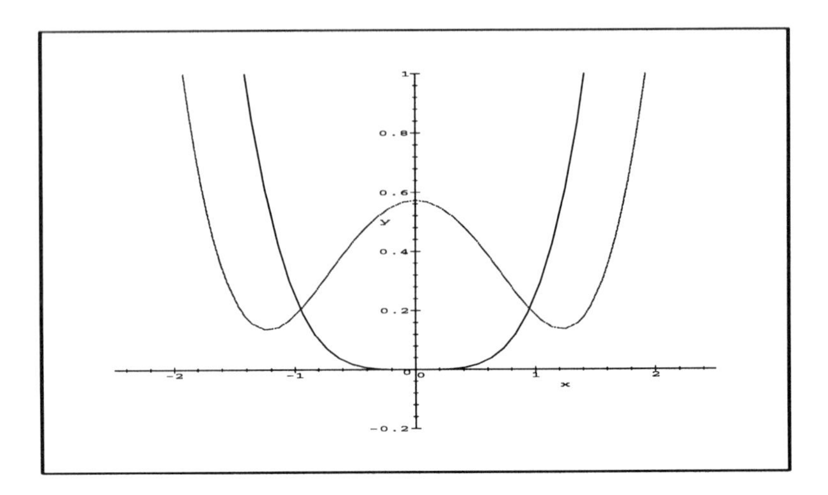

The improvement is rather marginal and in accord with the small decrease in the ground-state energy. We now lift the restriction $\beta = \beta_0$ and consider a minimization of the energy with respect to both parameters β and c_1. We simply draw the contour diagram of the energy as a height over the two parameters displayed in the $x - y$ plane and look for the minimum.

```
> with(plots):
> contourplot(En1c,c1=-0.03..0.01,beta=0.95..1.1,axes=boxed);
```

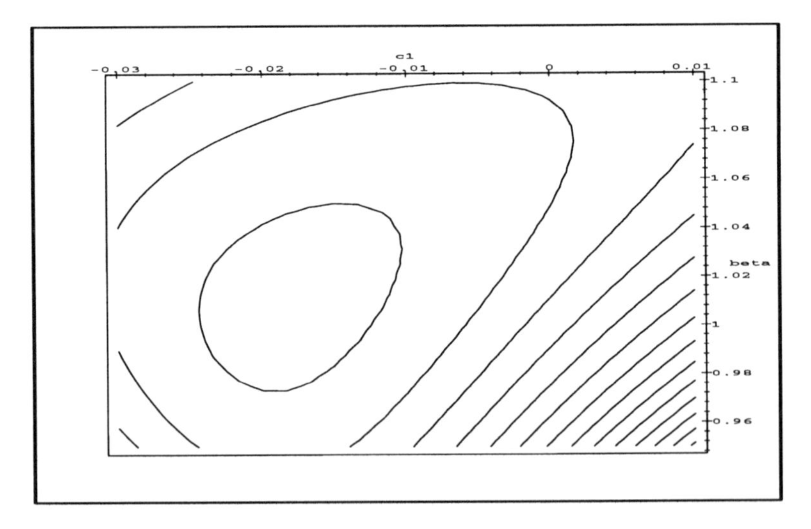

The values $\beta_0 = 1.01$, $c_1 = -0.02$ are close to the minimum. Since the contours are unlabeled we have to verify that they indeed show a minimum and not a maximum. This can be done with a surface plot:

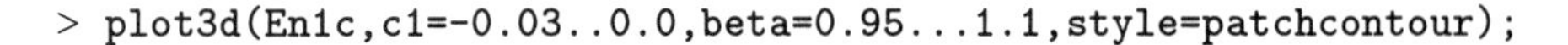

```
> plot3d(En1c,c1=-0.03..0.0,beta=0.95...1.1,style=patchcontour);
```

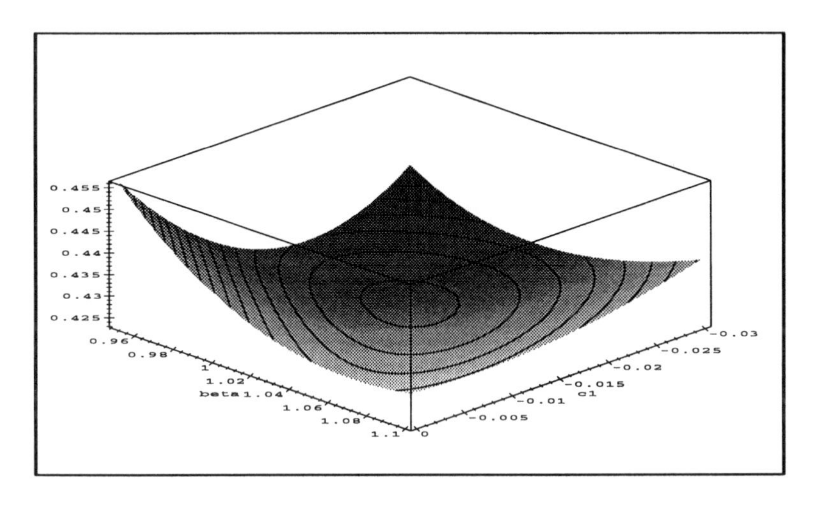

The best energy value within this approximate trial space is calculated to be

```
> subs(beta=1.01,c1=-0.02,En1c);
```

$$.4235856492$$

Now observe the improvement in the local energy:

```
> plot({VAHO(x),Tkin((1-0.02*x^4)*exp(-(1.01*x)^2/2),x)
> +VAHO(x)},x=-2.5..2.5,y=-0.2..1);
```

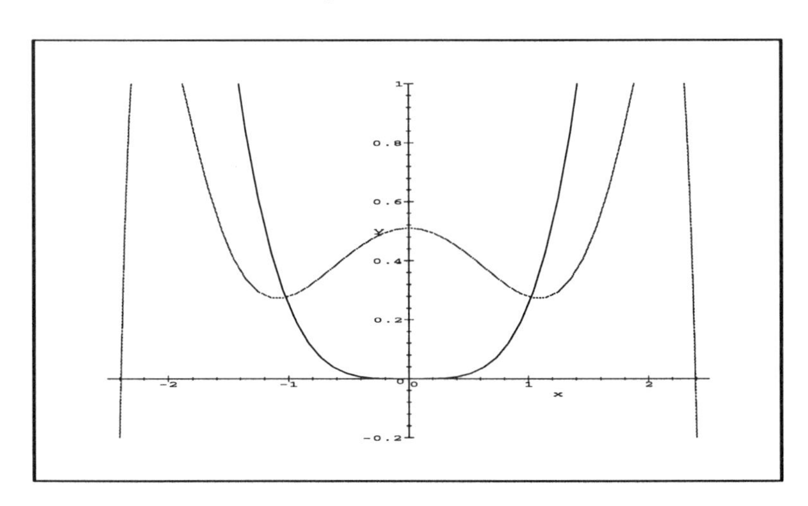

Even though the improvement in the ground-state energy is small (between 1 % and 2 %), we see that the local energy density oscillates less around the eigenvalue. The asymptotic region is still poorly represented, and, as a matter of fact, no improvement can be expected without a change in the Gaussian. The asymptotic region is, however, not too important for many purposes, as there is a negligible probability of finding the particle there. As an aside I mention that this region is extremely sensitive to the eigenvalue and that an analysis of the exact solution of differential eigenvalue problems shows that the eigenvalue appears directly in the part of the wavefunction that defines its asymptotic behaviour.

Another check is given by the virial theorem. We can investigate whether known relationships between the expectation values of kinetic, potential, and total energies are satisfied. For simple power-law potentials of type λx^n the virial theorem relates the expectation values for kinetic and potential energies for exact eigenstates as follows:

$$2\langle T\rangle = \langle x\frac{\mathrm{d}V}{\mathrm{d}x}\rangle ,$$

and, thus, $2\langle T\rangle = n\langle V\rangle$. Furthermore, a relationship exists to the total energy, i.e., $\langle E\rangle = \langle T\rangle + \langle V\rangle$, which permits us to determine what fraction of the total energy resides in the kinetic and the potential energies, respectively. It is a virtue of the variational method that the optimum, yet approximate, variational solutions satisfy the exact virial conditions. For numerical approximation techniques the check of the virial theorem represents a real test.

The kinetic and potential energy expectation values for a simple Gaussian variational trial state are given by:

```
> Texp:=alpha->int(exp(-alpha^2*x^2/2)*(-1/2
> *diff(exp(-alpha^2*x^2/2),x$2)),x=-infinity..infinity)
> /int(exp(-alpha^2*x^2),x=-infinity..infinity);
```

$$\mathrm{Texp} := \alpha \to \frac{\int_{-\infty}^{\infty} -\frac{1}{2}\mathrm{e}^{(-1/2\,\alpha^2\,x^2)}\,\mathrm{diff}\left(\mathrm{e}^{(-1/2\,\alpha^2\,x^2)}, x\,\$\,2\right)\,dx}{\int_{-\infty}^{\infty}\mathrm{e}^{(-\alpha^2\,x^2)}\,dx}$$

```
> Vexp:=alpha->int(exp(-alpha^2*x^2/2)*(VAHO(x)*exp(-alpha^2
> *x^2/2)),x=-infinity..infinity)/int(exp(-alpha^2*x^2),
> x=-infinity..infinity);
```

$$\mathrm{Vexp} := \alpha \to \frac{\int_{-\infty}^{\infty}\left(\mathrm{e}^{(-1/2\,\alpha^2\,x^2)}\right)^2\,\mathrm{VAHO}(\,x\,)\,dx}{\int_{-\infty}^{\infty}\mathrm{e}^{(-\alpha^2\,x^2)}\,dx}$$

```
> Texpb:=Texp(beta);
```

$$\mathrm{Texpb} := \frac{1}{4}\,\beta^{\sim 2}$$

```
> Vexpb:=Vexp(beta);
```

$$\text{Vexpb} := \frac{3}{16}\,\frac{1}{\beta^{\sim 4}}$$

```
> evalf(subs(beta=beta0,Texpb)/E0);
> evalf(subs(beta=beta0,Vexpb)/E0);
```

$$.6666666670 \qquad .3333333333$$

We verified that the variational method imposes the relationship of the kinetic and potential energy expectation values to the same ratios as satisfied by the exact solution.

The above example showed only a small improvement in the energy by a generalization of the variational ansatz. Often the convergence towards the exact answer is slow. This is a reflection of the fact that the chosen ansatz is inappropriate for the problem at hand, such as demonstrated in the asymptotic region. Numerical techniques are more reliable in such a case, as they are not biased. Variational approximations tend to provide good wavefunctions in those regions of space that contribute significantly to the energy. They can be completely wrong in the irrelevant parts, such as the asymptotic domain. Yet sometimes, such as in scattering from atoms, calculated properties depend strongly on precisely this domain. Thus, one has to be careful with variationally obtained wavefunctions. Mathematically this becomes obvious from an error analysis. The error in the eigenvalue is of higher order than in the wavefunction, i.e., usually the eigenvalue is estimated more accurately than the eigenfunction.

Exercise 2.1.2: Derive the virial theorem for bound-state problems from the time evolution of the expectation value $\langle x p_{\mathrm{op}}\rangle$ determined by the evolution equation

$$\frac{\mathrm{d}}{\mathrm{d}t}\langle x p_{\mathrm{op}}\rangle = \frac{1}{\mathrm{i}\hbar}\langle [x p_{\mathrm{op}}, \frac{p_{\mathrm{op}}^2}{2m} + V(x)]\rangle = 0\,.$$

Exercise 2.1.3: Perform the analysis of local energy in a variational approximation to other power-law anharmonic potentials, such as μx^6.

Exercise 2.1.4: Consider some potential function $V(x)$ of your own choice that has a minimum at $x = x_0$. Such a function can describe the motion of a physical system (e.g., vibration of a diatomic molecule) around the equilibrium position x_0. Calculate a Taylor expansion of $V(x)$ around $x = x_0$. Observe how a harmonic oscillator problem arises in the lowest non-trivial order and what role will be played by anharmonic corrections.

2.2 External Source

It is of interest to consider a quantum system that is exposed to an external potential source term of type Jx. We calculate the response of the ground state as a function of the strength of the source J using the variational method discussed in section 1.4. Note that the system's symmetry that is present in the kinetic energy and usually also in the potential $V(x) = V(-x)$ is broken by the source term.

We consider a Gaussian trial state that is centered around the mean position $\langle x \rangle = \xi$ and do not worry about normalization (we normalize in the expectation values). Note that the trial function is real and we make use of a simplified calculation of expectation values without complex conjugation.

```
> psi:=(xi,alpha)->exp(-(alpha*(x-xi))^2/2);
```

$$\psi := (\xi, \alpha) \to e^{(-1/2\,\alpha^2\,(x-\xi)^2)}$$

```
> Tkin:=(xi,alpha)->-(1/2)*int(psi(xi,alpha)*diff
> (psi(xi,alpha),x$2),x=-infinity..infinity)
> /int(psi(xi,alpha)^2,x=-infinity..infinity);
```

$$\text{Tkin} := (\xi, \alpha) \to -\frac{1}{2}\frac{\int_{-\infty}^{\infty} \psi(\xi,\alpha)\,\text{diff}(\psi(\xi,\alpha), x\,\$\,2)\,dx}{\int_{-\infty}^{\infty} \psi(\xi,\alpha)^2\,dx}$$

A quartic anharmonic oscillator with strength $\lambda = 1$ is chosen for the unperturbed Hamiltonian as this represents a non-trivial example.

```
> Vpot:=(xi,alpha)->int(psi(xi,alpha)^2*x^4,x=-infinity..
> infinity)/int(psi(xi,alpha)^2,x=-infinity..infinity);
```

$$\text{Vpot} := (\xi, \alpha) \to \frac{\int_{-\infty}^{\infty} \psi(\xi,\alpha)^2\,x^4\,dx}{\int_{-\infty}^{\infty} \psi(\xi,\alpha)^2\,dx}$$

The source term yields the expectation value

```
> Vsrc:=(J,xi,alpha)->J*int(psi(xi,alpha)^2*x,x=-infinity..
> infinity)/int(psi(xi,alpha)^2,x=-infinity..infinity);
```

$$\text{Vsrc} := (J, \xi, \alpha) \to \frac{J\int_{-\infty}^{\infty} \psi(\xi,\alpha)^2\,x\,dx}{\int_{-\infty}^{\infty} \psi(\xi,\alpha)^2\,dx}$$

```
> Etot:=(J,xi,alpha)->Tkin(xi,alpha)+Vpot(xi,alpha)
> +Vsrc(J,xi,alpha);
```

$$\text{Etot} := (J, \xi, \alpha) \rightarrow \text{Tkin}(\xi, \alpha) + \text{Vpot}(\xi, \alpha) + \text{Vsrc}(J, \xi, \alpha)$$

```
> assume(beta>0);
```

First we analyze the quartic oscillator variationally without the source term. We minimize the energy with respect to the width parameter β as well as the displacement parameter ξ.

```
> with(plots):
> contourplot(Etot(0,xi,beta),xi=-1..1,beta=1..2,axes=boxed);
```

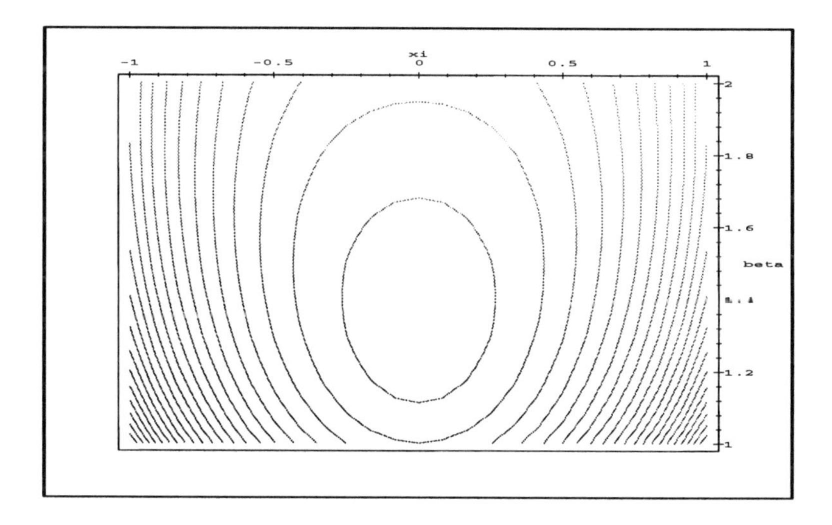

The minimum in the approximate ground-state energy is achieved for zero average displacement $\langle x \rangle = \xi = 0$. We define a simple function for the estimated ground-state energy as a function of the variational width parameter.

```
> E0:=alpha->Etot(0,0,alpha);    E0(beta);
```

$$E0 := \alpha \rightarrow \text{Etot}(0, 0, \alpha) \qquad \frac{1}{4}\beta^{\sim 2} + \frac{3}{4}\frac{1}{\beta^{\sim 4}}$$

Now we calculate the approximate ground state with a non-zero source term $J = 1$ and observe how the optimum wavefunction shifts its center position to $\xi \neq 0$. To obtain an accurate location for the minimum of the energy in the plane of trial parameters β and ξ we increase the number of contours by the optional `contours` parameter. Hopefully in future releases of Maple more optional control will be given to the user in the `contourplot` command, such as, e.g., a specification of contour heights. These can be read off currently only approximately from a surface plot with requested contours.

```
> contourplot(Etot(1,xi,beta),xi=-1..1,beta=1..2,
> axes=boxed,contours=15);
```

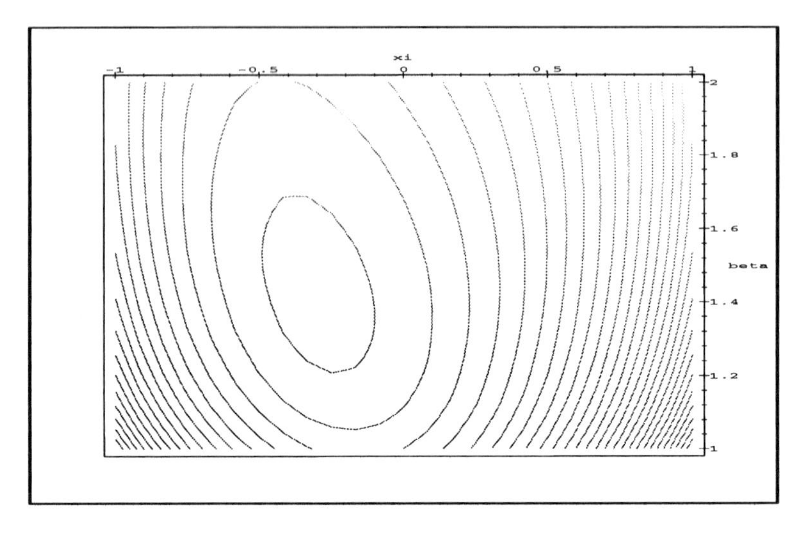

Next we consider the position expectation value

```
> PosX:=(xi,alpha)->int(psi(xi,alpha)^2*x,x=-infinity..
> infinity)/int(psi(xi,alpha)^2,x=-infinity..infinity);
```

$$\mathrm{PosX} := (\,\xi, \alpha\,) \to \frac{\displaystyle\int_{-\infty}^{\infty} \psi(\,\xi, \alpha\,)^2\, x\, dx}{\displaystyle\int_{-\infty}^{\infty} \psi(\,\xi, \alpha\,)^2\, dx}$$

```
> PosX(xi,beta);
```

$$\xi$$

We have established the expected result. The parameter ξ plays the role of position expectation value, since the Gaussian is symmetric around it. We investigate the relationship between the ξ value that minimizes the energy and the strength parameter of the source term J. We fix β to the value that optimizes the energy for $J = 0$.

```
> beta0:=fsolve(diff(E0(beta),beta)=0,beta=0.5..3);
```

$$\beta 0 := 1.348006155$$

```
> solve(diff(E0(beta),beta)=0,beta);
```

$$6^{1/6},\ \left(\frac{1}{2} + \frac{1}{2}\,I\,\sqrt{3}\right)\,6^{1/6},\ \left(-\frac{1}{2} + \frac{1}{2}\,I\,\sqrt{3}\right)\,6^{1/6},\ -6^{1/6},$$

$$\left(-\frac{1}{2}-\frac{1}{2}I\sqrt{3}\right)6^{1/6},\left(\frac{1}{2}-\frac{1}{2}I\sqrt{3}\right)6^{1/6}$$

```
> beta0a:="[1];
```

$$\text{beta0a} := 6^{1/6}$$

```
> EJ0:=Etot(J,xi,beta0a);
```

$$EJ0 := \frac{1}{4}6^{1/3}+\frac{\left(\frac{1}{6}\xi^4\,6^{5/6}\sqrt{\pi}+\frac{1}{2}\xi^2\sqrt{6}\sqrt{\pi}+\frac{1}{8}6^{1/6}\sqrt{\pi}\right)6^{1/6}}{\sqrt{\pi}}+J\,\xi$$

To understand why the optimum variational ground-state wavefunction chooses a non-zero average position $\langle x\rangle = \xi$ we plot the combined potential and source term for, e.g., $J = 2$:

```
> plot(x^4+2*x,x=-3..3,V=-2..10);
```

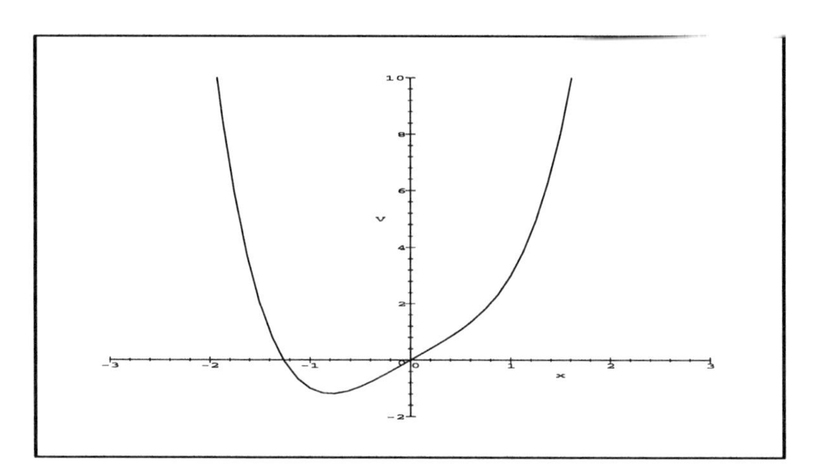

Clearly, we expect the ground state of the Schrödinger equation for such a potential to have a non-zero position expectation value. Now we calculate the relationship between the strength of the source term J and the expected position $\langle x\rangle = \xi$. The answer for ξ_0 can be found numerically for given J:

```
> fsolve(diff(subs(J=1,EJ0),xi)=0,xi=-3..3);
```

$$-.2770829799$$

However, we try for the full answer and ignore the complex roots:

```
> solve(diff(EJ0,xi)=0,xi);
```

$$\%1^{1/3} - \frac{1}{12}\frac{6^{2/3}}{\%1^{1/3}}, -\frac{1}{2}\%1^{1/3} + \frac{1}{24}\frac{6^{2/3}}{\%1^{1/3}} + \frac{1}{2} I\sqrt{3}\left(\%1^{1/3} + \frac{1}{12}\frac{6^{2/3}}{\%1^{1/3}}\right),$$
$$-\frac{1}{2}\%1^{1/3} + \frac{1}{24}\frac{6^{2/3}}{\%1^{1/3}} - \frac{1}{2} I\sqrt{3}\left(\%1^{1/3} + \frac{1}{12}\frac{6^{2/3}}{\%1^{1/3}}\right)$$
$$\%1 := -\frac{1}{8}J + \frac{1}{24}\sqrt{12+9J^2}$$

```
> xi0:="[1];
```

$$\xi 0 := \left(-\frac{1}{8}J + \frac{1}{24}\sqrt{12+9J^2}\right)^{1/3} - \frac{1}{12}\frac{6^{2/3}}{\left(-\frac{1}{8}J + \frac{1}{24}\sqrt{12+9J^2}\right)^{1/3}}$$

```
> plot(xi0,J=-10..10,xi=-1.5..1.5);
```

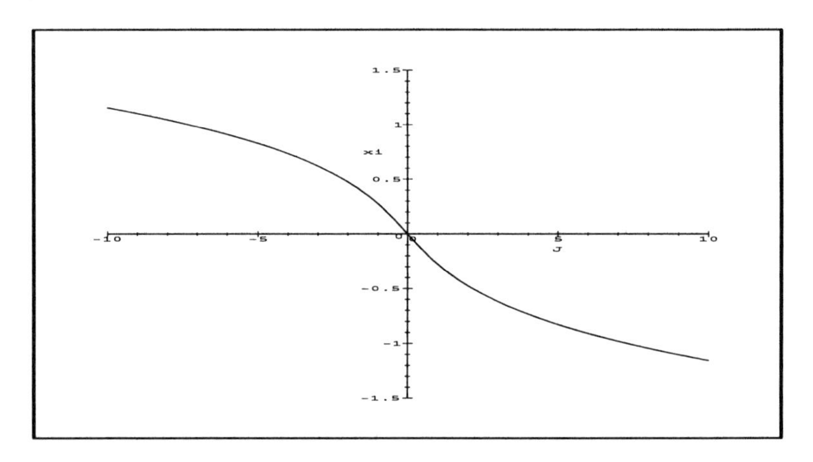

We see that for small J the wavefunction shifts readily away from $\xi = 0$. For large J it has to work itself up against the steeply rising potential. The slope near $J = 0$ is related to the excitation energy $E_1 - E_0$.

For many-particle quantum systems one uses this idea to extract information about the so-called mass gap without constructing explicitly the first excited state. This idea leads to the Gaussian effective potential [St85, BG80] in quantum field theory. In solid-state physics an analogous problem is given by the polarization of a system in an external field with a linear regime as well as saturation effects (nonlinear response).

Exercise 2.2.1: Obtain the energies E_0 and E_1 for a quartic oscillator λx^4 for some λ values from a variational calculation without source ($\psi_1 = x\psi_0$ with $\xi = 0$). Find the precise relationship between the slope at the origin in the function $\langle x\rangle(J)$ and the gap $\Delta = E_1 - E_0$.

2.3 Periodic Lattice

The periodic lattice problem ([Fl71, problem 15]) serves to demonstrate the remarkable feature of what happens to the discrete eigenvalues of bound-state problems, if the potential well is replicated many times to form an infinite lattice. This problem is relevant in solid-state physics, particularly in 3D and 2D in order to explain electron band structure and conductivity.

From a very simple mathematical toy problem in 1D we can learn how electronic delocalization comes about in a metal or even a macromolecule. The periodic arrangement of square-well potentials can be imagined to represent the potential provided by a lattice of ions to the outermost electrons. The idealization of a realistic, smoothly varying attractive potential by a discontinuous piecewise constant potential is implemented in order to allow for a simple solution by matching, as in the single square-well problem.

We consider a periodic potential in 1D that represents a sequence of square wells of width a separated by segments of length b. We set a basic potential region originating at 0 to be of the following type:

```
> Vpot:=(x,a,b,V0)->Heaviside(x+b)*Heaviside(-x);
```

$$\text{Vpot} := (x, a, b, V0) \to \text{Heaviside}(x+b)\,\text{Heaviside}(-x)$$

```
> plot(Vpot(x,2,1,1),x=-2..4);
```

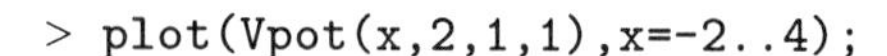

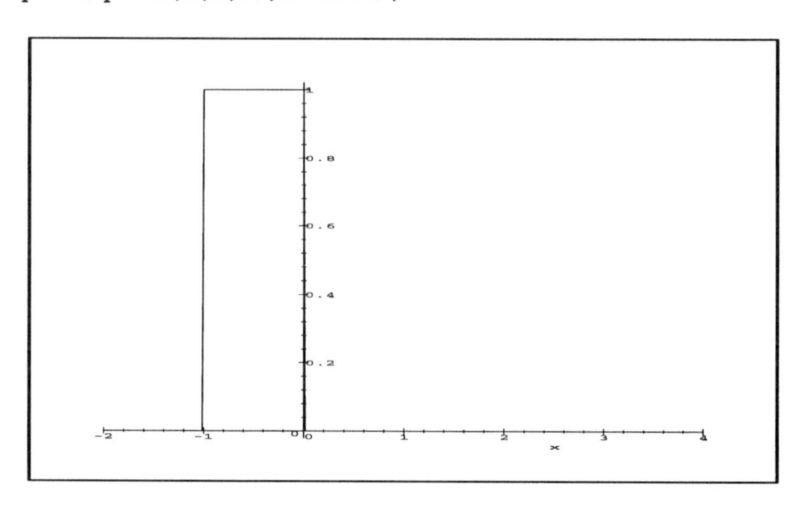

The interval of periodicity equals $l = a + b$. We plot three adjacent segments of the potential.

```
> plot(Vpot(x,2,1,1)+Vpot(x+3,2,1,1)+Vpot(x-3,2,1,1),x=-5..5);
```

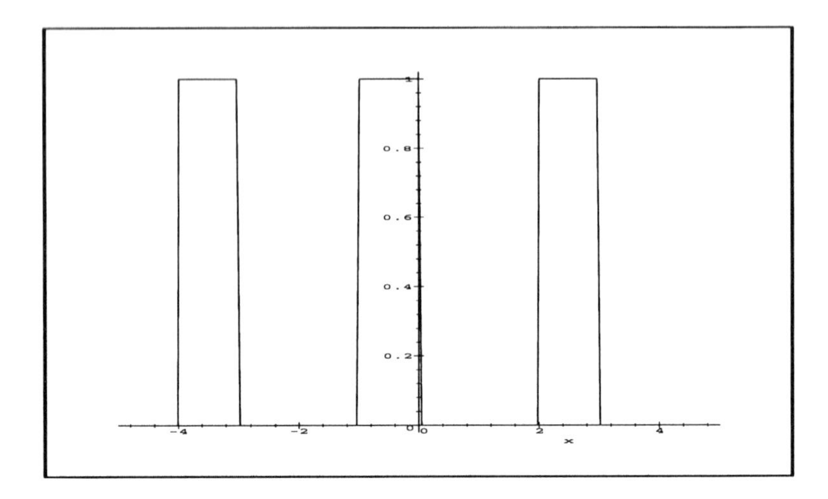

Naively one might think that this is a replica of the standard square-well problem. But there are some surprises. First, one uses an argument to relate the neighbouring problems, by requesting that the probability distribution be the same for the electron within each cell. Thus, their wavefunctions can differ by at most a phase factor

$$\psi(x+l) = \beta\psi(x) \rightarrow \psi(x+nl) = \beta^n \psi(x)\,.$$

The phase factor can be expressed in terms of the lattice period l and a constant to cancel the dimension: $\beta = \exp(\mathrm{i}Kl)$, with K dimensionally a wave number. The periodicity in the function $\exp(\mathrm{i}\phi)$ allows us to restrict the range of the propagation number (which is a vector if considered in 3D) to the base interval

$$-\pi < Kl < \pi\,,$$

thereby introducing a reduced propagation number.

With the assumption that $E < V0$, $k^2 = 2mE/\hbar^2$, $q^2 = 2m(V0-E)/\hbar^2$ we write down the wavefunctions for the first and second cells (in units in which $m = \hbar = 1$):

```
> V0:=1;    a:=2;    b:=1;
```

$$V0 := 1 \quad a := 2 \quad b := 1$$

```
> ks:=sqrt(2*En);     qs:=sqrt(2*(V0-En));
```

$$ks := \sqrt{2}\sqrt{En} \quad qs := \sqrt{2-2\,En}$$

```
> assume(En>0);
```

ψ_{11} is the inside-the-barrier part, ψ_{21} is the free solution – both are for cell 1.

```
> psi11:=x->A1*exp(qs*x)+B1*exp(-qs*x);
```

$$\psi 11 := x \to A1\, e^{(qs\, x)} + B1\, e^{(-qs\, x)}$$

```
> psi21:=x->A2*exp(I*ks*x)+B2*exp(-I*ks*x);
```

$$\psi 21 := x \to A2\, e^{(I\, ks\, x)} + B2\, e^{(-I\, ks\, x)}$$

Now we define the same states in the neighbouring cell: lp is the periodicity length, Kp the propagation number. Note how the argument in parts of ψ_{12} and ψ_{22} is shifted!

```
> lp:=a+b;
```

$$lp := 3$$

```
> psi12:=x->exp(I*Kp*lp)*(A1*exp(qs*(x-lp))+B1*exp(-qs*(x-lp)));
```

$$\psi 12 := x \to e^{(I\, Kp\, lp)} \left(A1\, e^{(qs\,(x-lp))} + B1\, e^{(-qs\,(x-lp))} \right)$$

```
> psi22:=x->exp(I*Kp*lp)*(A2*exp(I*ks*(x-lp))
> +B2*exp(-I*ks*(x-lp)));
```

$$\psi 22 := x \to e^{(I\, Kp\, lp)} \left(A2\, e^{(I\, ks\,(x-lp))} + B2\, e^{(-I\, ks\,(x-lp))} \right)$$

Now we set up the matching conditions. The four constants Ai, Bi are determined from 4 matching conditions, which, however, represent a homogeneous system of equations.

```
> eq1:=psi11(0)=psi21(0);
```

$$\mathrm{eq1} := A1 + B1 = A2 + B2$$

```
> eq2:=D(psi11)(0)=D(psi21)(0);
```

$$\mathrm{eq2} := A1\, \sqrt{2 - 2\, En^{\sim}} - B1\, \sqrt{2 - 2\, En^{\sim}} = I\, A2\, \sqrt{2}\, \sqrt{En^{\sim}} - I\, B2\, \sqrt{2}\, \sqrt{En^{\sim}}$$

Now we match onto the next cell at $x = a$:

```
> eq3:=psi21(a)=psi12(a);
```

$$\mathrm{eq3} := A2\, e^{\left(2\, I\, \sqrt{2}\, \sqrt{En^{\sim}}\right)} + B2\, e^{\left(-2\, I\, \sqrt{2}\, \sqrt{En^{\sim}}\right)} = e^{(3\, I\, Kp)} \left(A1\, e^{\left(-\sqrt{2-2\, En^{\sim}}\right)} + B1\, e^{\left(\sqrt{2-2\, En^{\sim}}\right)} \right)$$

```
> eq4:=D(psi21)(a)=D(psi12)(a);
```

$$eq4 := I\,A2\,\sqrt{2}\,\sqrt{En^{\sim}}\,e^{(2\,I\,\sqrt{2}\,\sqrt{En^{\sim}})} - I\,B2\,\sqrt{2}\,\sqrt{En^{\sim}}\,e^{(-2\,I\,\sqrt{2}\,\sqrt{En^{\sim}})} = e^{(3\,I\,Kp)} \left(A1\,\sqrt{2-2\,En^{\sim}}\,e^{(-\sqrt{2-2\,En^{\sim}})} - B1\,\sqrt{2-2\,En^{\sim}}\,e^{(\sqrt{2-2\,En^{\sim}})}\right)$$

Now we try something naive by attempting a direct solution:

```
> solve({eq1,eq2,eq3,eq4},{A1,A2,B1,B2});
```

$$\{B1 = 0, B2 = 0, A2 = 0, A1 = 0\}$$

Is this just an uninteresting trivial solution? No, we have to use the energy En to make the determinant of the coefficient matrix vanish. Whether we can find energies will depend on the choice of the propagation number Kp. In fact, the continuous variable Kp allows us to find a continuum of solutions En, i.e., the problem is qualitatively different from a single well with the potential high at the boundaries that has a discrete spectrum. It is also different from a potential that goes to zero asymptotically, which admits continuous scattering solutions at all energies $En > 0$.

How do we extract the determinant for the system of equations? We order the unknowns into $(A1, B1, A2, B2)$

```
> collect(eq2,A1);
```

$$A1\,\sqrt{2-2\,En^{\sim}} - B1\,\sqrt{2-2\,En^{\sim}} = I\,A2\,\sqrt{2}\,\sqrt{En^{\sim}} - I\,B2\,\sqrt{2}\,\sqrt{En^{\sim}}$$

```
> coeff(",A1);
Error, unable to compute coeff
```

It appears that `collect/coeff` doesn't work on equations, one can use it only on expressions. So let us redefine the equations by arranging all terms on the left: row 1 in the coefficient matrix is trivial: $1, 1, -1, -1$.

```
> eq2:=D(psi11)(0)-D(psi21)(0);
```

$$eq2 := A1\,\sqrt{2-2\,En^{\sim}} - B1\,\sqrt{2-2\,En^{\sim}} - I\,A2\,\sqrt{2}\,\sqrt{En^{\sim}} + I\,B2\,\sqrt{2}\,\sqrt{En^{\sim}}$$

```
> eq3:=expand(psi21(a)-psi12(a));
```

$$eq3 := A2\,\left(e^{(I\,\sqrt{2}\,\sqrt{En^{\sim}})}\right)^2 + \frac{B2}{\left(e^{(I\,\sqrt{2}\,\sqrt{En^{\sim}})}\right)^2} - \frac{(e^{(I\,Kp)})^3\,A1}{e^{(\sqrt{2-2\,En^{\sim}})}} - (e^{(I\,Kp)})^3\,B1\,e^{(\sqrt{2-2\,En^{\sim}})}$$

```
> eq4:=expand(D(psi21)(a)-D(psi12)(a));
```

$$eq4 := I\,A2\,\sqrt{2}\,\sqrt{En\tilde{}}\,\left(e^{(I\sqrt{2}\sqrt{En\tilde{}})}\right)^2 - \frac{I\,B2\,\sqrt{2}\,\sqrt{En\tilde{}}}{\left(e^{(I\sqrt{2}\sqrt{En\tilde{}})}\right)^2} - \frac{(e^{(I\,Kp)})^3\,A1\,\sqrt{2-2\,En\tilde{}}}{e^{(\sqrt{2-2\,En\tilde{}})}} + (e^{(I\,Kp)})^3\,B1\,\sqrt{2-2\,En\tilde{}}\,e^{(\sqrt{2-2\,En\tilde{}})}$$

```
> coeff(eq2,A2);
```

$$-I\,\sqrt{2}\,\sqrt{En\tilde{}}$$

This seems to work.

```
> with(linalg):
```

```
> row1:=vector([1,1,-1,-1]);
```

$$row1 := [1\;1\;-1\;-1]$$

```
> row2:=vector([coeff(eq2,A1),coeff(eq2,B1),coeff(eq2,A2),
> coeff(eq2,B2)]);
```

$$row2 := \left[\sqrt{2-2\,En\tilde{}}\quad -\sqrt{2-2\,En\tilde{}}\quad -I\,\sqrt{2}\,\sqrt{En\tilde{}}\quad I\,\sqrt{2}\,\sqrt{En\tilde{}}\right]$$

```
> row3:=vector([coeff(eq3,A1),coeff(eq3,B1),coeff(eq3,A2),
> coeff(eq3,B2)]);
```

$$row3 := \left[-\frac{(e^{(I\,Kp)})^3}{e^{(\sqrt{2-2\,En\tilde{}})}}\quad -(e^{(I\,Kp)})^3\,e^{(\sqrt{2-2\,En\tilde{}})}\quad \left(e^{(I\sqrt{2}\sqrt{En\tilde{}})}\right)^2\quad \frac{1}{\left(e^{(I\sqrt{2}\sqrt{En\tilde{}})}\right)^2}\right]$$

```
> row4:=vector([coeff(eq4,A1),coeff(eq4,B1),coeff(eq4,A2),
> coeff(eq4,B2)]);
```

$$row4 := \left[-\frac{(e^{(I\,Kp)})^3\,\sqrt{2-2\,En\tilde{}}}{e^{(\sqrt{2-2\,En\tilde{}})}}\quad (e^{(I\,Kp)})^3\,\sqrt{2-2\,En\tilde{}}\,e^{(\sqrt{2-2\,En\tilde{}})}\right.$$

$$I\sqrt{2}\sqrt{En\tilde{}}\left(e^{(I\sqrt{2}\sqrt{En\tilde{}})}\right)^2 - \frac{I\sqrt{2}\sqrt{En\tilde{}}}{\left(e^{(I\sqrt{2}\sqrt{En\tilde{}})}\right)^2}\Bigg]$$

```
> Coeff:=matrix(4,4,0):
> for i from 1 to 4 do
> Coeff[1,i]:=row1[i];   Coeff[2,i]:=row2[i];
> Coeff[3,i]:=row3[i];   Coeff[4,i]:=row4[i];
> od:
> evalm(Coeff);
```

$$[1, 1, -1, -1]$$
$$[\sqrt{2-2\,En\tilde{}}, -\sqrt{2-2\,En\tilde{}}, -I\sqrt{2}\sqrt{En\tilde{}}, I\sqrt{2}\sqrt{En\tilde{}}]$$
$$\left[-\frac{(e^{(I\,Kp)})^3}{\%1}, -(e^{(I\,Kp)})^3\,\%1, \%2^2, \frac{1}{\%2^2}\right]$$
$$\Bigg[-\frac{(e^{(I\,Kp)})^3\sqrt{2-2\,En\tilde{}}}{\%1}, (e^{(I\,Kp)})^3\sqrt{2-2\,En\tilde{}}\,\%1,$$
$$I\sqrt{2}\sqrt{En\tilde{}}\,\%2^2, -\frac{I\sqrt{2}\sqrt{En\tilde{}}}{\%2^2}\Bigg]$$
$$\%1 := e^{(\sqrt{2-2\,En\tilde{}})}$$
$$\%2 := e^{(I\sqrt{2}\sqrt{En\tilde{}})}$$

```
> chareq:=det(Coeff);
```

$$\mathrm{chareq} := 2\Big(2\,I\,\%2\sqrt{2-2\,En\tilde{}}\sqrt{2}\sqrt{En\tilde{}}\,\%1^2$$
$$-2\,(e^{(I\,Kp)})^3\,\%2^2\,En\tilde{} + 2\,(e^{(I\,Kp)})^3\,\%2^2\,En\tilde{}\,\%1^4$$
$$-I\,(e^{(I\,Kp)})^3\sqrt{2-2\,En\tilde{}}\,\%2^2\sqrt{2}\sqrt{En\tilde{}}$$
$$-I\,(e^{(I\,Kp)})^3\sqrt{2-2\,En\tilde{}}\,\%2^2\sqrt{2}\sqrt{En\tilde{}}\,\%1^4$$
$$+(e^{(I\,Kp)})^3\,\%2^2 - \%2^2\,(e^{(I\,Kp)})^3\,\%1^4$$
$$+2\,(e^{(I\,Kp)})^3\,En\tilde{} - 2\,(e^{(I\,Kp)})^3\,En\tilde{}\,\%1^4$$
$$-I\,(e^{(I\,Kp)})^3\sqrt{2-2\,En\tilde{}}\sqrt{2}\sqrt{En\tilde{}}$$
$$-I\,(e^{(I\,Kp)})^3\sqrt{2-2\,En\tilde{}}\sqrt{2}\sqrt{En\tilde{}}\,\%1^4$$
$$+2\,I\,(e^{(I\,Kp)})^6\sqrt{2-2\,En\tilde{}}\,\%2\sqrt{2}\sqrt{En\tilde{}}\,\%1^2$$
$$-(e^{(I\,Kp)})^3 + (e^{(I\,Kp)})^3\,\%1^4\Big)\,/(\%2\,\%1^2)$$
$$\%1 := e^{(I\sqrt{2}\sqrt{En\tilde{}})}$$

$$\%2 := e^{\left(\sqrt{2-2\,En^{\sim}}\right)}$$

```
> sol:=solve(chareq,En);
```

sol :=

That didn't work. Suspected problem: `Kp` isn't specified (the next fruitless attempt takes quite a while to complete).

```
> sol:=solve(subs(Kp=0,chareq),En);
```

sol :=

Now we should morally be allowed to use a numeric solver:

```
> sol:=fsolve(subs(Kp=0,chareq),En=0..V0);
```

sol := .5000000000

Apart from the occasionally occurring undesired numerical noise of an imaginary part, we have a numerical answer. However, we aren't too happy about the large value of `En`, as we are mostly interested in `En` < `V0`.

```
> sol:=Re(fsolve(subs(Kp=0,chareq),En=0.51..V0));
```

sol := .7550000000

Now we are ready to explore and set up a loop to obtain a list of values for plotting purposes. The numerical output of pairs (Kp_i, E_i) is removed below for $i > 5$. We explain below why we cannot solve for the roots of the magnitude of the complex-valued characteristic equation.

```
> for i from 1 to 20 do

> Kpi[i]:=evalf(-Pi+6.28/21*i)/lp;

> Ensol[i]:=fsolve(subs(Kp=Kpi[i],Re(chareq)+Im(chareq)),
> En=0.01..V0-0.01);

> print(Kpi[i],Ensol[i]);      od:
```

−.9475150116, .5849117411

−.8478324719, .5456879542

−.7481499323, .4959266352

−.6484673926, .4447491723

−.5487848526, .3970435050

```
> plot(abs(subs(Kp=-0.9475,chareq)),En=0..V0);
```

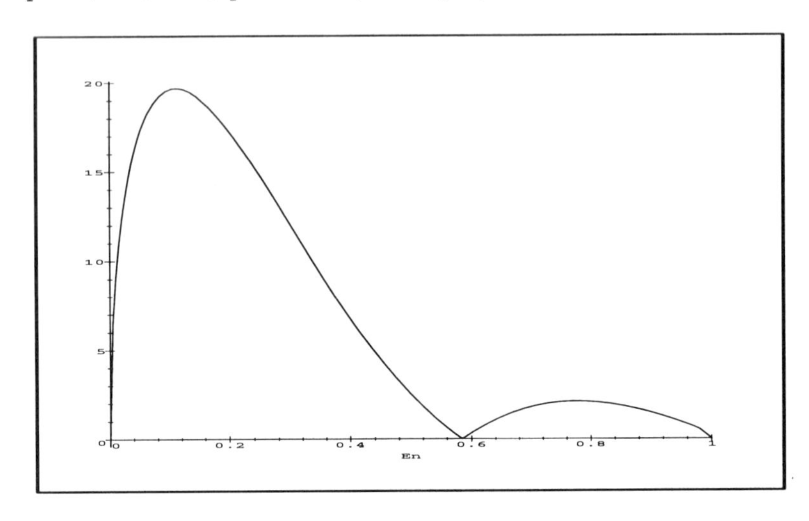

Bisection does not work on this function, as there is no sign change in $\sqrt{f^*f}$ at the root! That forces us to use $\mathrm{Re}(f) + \mathrm{Im}(f)$, which is real-valued and changes sign. In addition, we can't use the range $En = 0..V0$, since the function vanishes at the endpoints. We plot the numerical solution obtained in the above loop.

```
> plot([[Kpi[j]*lp,Ensol[j]] $j=1..20],phi=-Pi..Pi,En=0..V0,
> style=line,title=`1st Brioullin zone`);
```

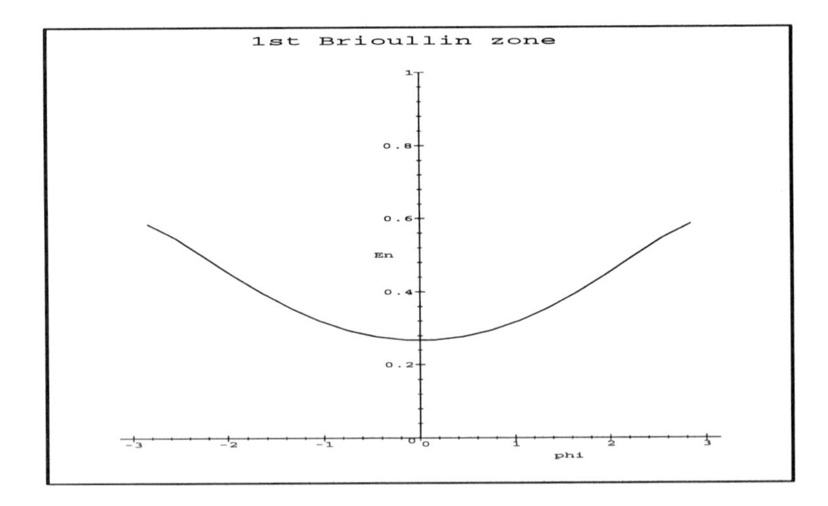

Now we explore the existence of higher bands, i.e, we seek additional energy levels, possibly for $E > V_0$. Again, the printout is removed for $i > 3$.

```
> for i from 1 to 20 do
> Kpi[i]:=evalf(-Pi+6.28/21*i)/lp;
> Emax:=2:
> if abs(Kpi[i])<0.25 then Emax:=2.35 ; fi:
> Ensol1[i]:=fsolve(subs(Kp=Kpi[i],Re(chareq)+Im(chareq)),
> En=1.01..Emax);
> print(Kpi[i],Ensol1[i]);
> od:
```

$$-.9475150116, 1.139654583$$

$$-.8478324719, 1.208099797$$

$$-.7481499323, 1.306509246$$

```
> plot({[[Kpi[j]*lp,Ensol[j]] $j=1..20],[[Kpi[j]*lp,Ensol1[j]]
> $j=1..20]},phi=-Pi..Pi,En=0..2.5,title=`1st Brioullin zone`);
```

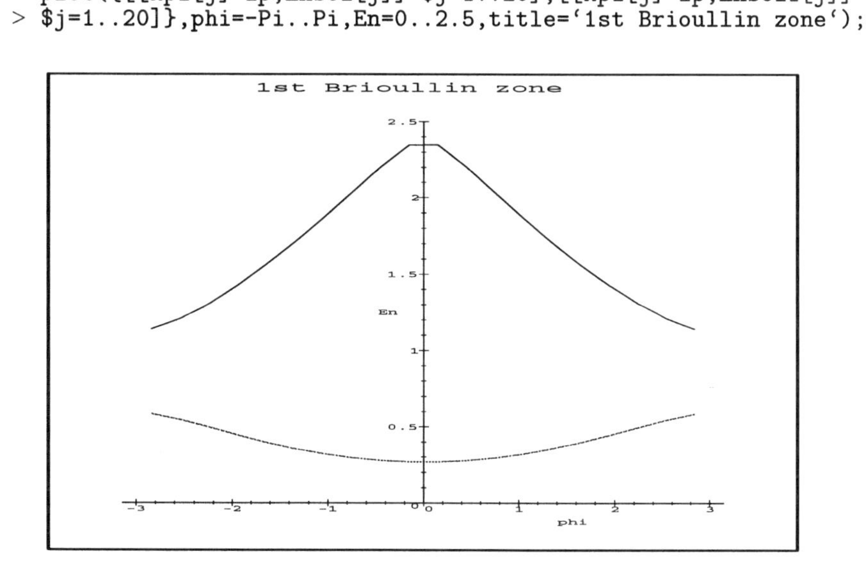

Notice the gap between the first two allowed zones (energy bands), and also that the upper zone corresponds to states with $E > V0$.

Exercise 2.3.1: Repeat the calculation of the allowed energy bands for potentials with a changed barrier height. Vary the cell size.

2.4 Lanczos Tridiagonalization

The Lanczos technique represents a powerful method for finding extreme eigenvalues of multi-dimensional Sturm-Liouville problems. It is adapted from linear algebra, where the idea originated from extensions of the power method.

In the power method one makes use of the fact that repeated application of the Hamiltonian onto an arbitrary start vector (that has an admixture of the symmetry in which one is interested) projects out the ground state of a given symmetry sector (e.g., even and odd parity in 1D problems).

The idea of a Krylov space goes further: the states $|v_n\rangle = \hat{H}|v_{n-1}\rangle$ are linearly independent and the sequence $\{|v_0\rangle, |v_1\rangle, ..., |v_n\rangle\}$ spans a subspace of the infinite-dimensional Hilbert space. A diagonalization of the Hamiltonian in this subspace certainly quickly acquires knowledge about the ground state even for small n, particularly if the start vector is chosen properly.

In the Lanczos construction the Krylov states are orthogonalized in such a way that the matrix representation of the Hamiltonian becomes tridiagonal in this basis. This permits a straightforward calculation of eigenvalues and eigenvectors (the expansion coefficients of the eigenfunctions in the Lanczos basis). While I present no formal derivation of the method, I emphasize that it is straightforwardly obtained. Beginning with an arbitrary, but normalized start vector $|\phi_1\rangle$ one calculates $|\psi_2\rangle = \hat{H}|\phi_1\rangle$. From Gram-Schmidt orthonormalization one obtains $|\phi_2\rangle$. The procedure is repeated and yields a tridiagonal representation of $\hat{H}$ in a subspace that contains the directions of extreme (low-lying and high-lying) eigenvalues. The details of the implementation are explained in the program below.

The power of the Lanczos method becomes apparent in the application to multi-dimensional Hamiltonians. The resulting tridiagonal matrix is always simple, even though the wavefunctions depend on many degrees of freedom.

We use a Gaussian state of arbitrary width to seed the Lanczos iteration. Alternatively an $\exp(-x^4)$-type basis could be used, but the expressions become more cumbersome and the results are not necessarily better for a given interaction potential. We can vary the scale parameter (width) in the start vector and observe how the iteration towards the exact low-lying eigenenergies can be optimized by a judicious choice.

In this worksheet, for clarity the matrix elements (inner products) are calculated every time as needed. It is possible to calculate the basic integrals beforehand and to assemble the inner products from a data table [Ka94].

```
> assume(n,integer);     N:=3;
```

$$N := 3$$

We limit the number of Lanczos iterations to three in the test run. The linear algebra package is required for matrix functions.

```
> with(linalg):
```

Now we initialize empty vectors to hold the diagonal and its neighbours of the Hamiltonian matrix in Lanczos representation.

```
> Hdiag:=vector(N,0): Hodg:=vector(N,0):
```

For the potential we choose a combination of a harmonic and a quartic oscillator.

```
> Vop:=arg->1/2*arg^2+1*arg^4;
```

$$\mathrm{Vop} := \arg \to \frac{1}{2}\arg^2 + \arg^4$$

We work in units in which $\hbar = m = 1$, and thus the Hamiltonian is given by:

```
> Hop:=(arg,x)->-1/2*diff(arg,x$2)+Vop(x)*arg;
```

$$\mathrm{Hop} := (\arg, x) \to -\frac{1}{2}\,\mathrm{diff}(\arg, x\,\$\,2) + \mathrm{Vop}(x)\arg$$

The scale (width) parameter is set to a constant to simplify matters. It can be changed in a re-run.

```
> alpha:=1;
```

$$\alpha := 1$$

The unnormalized start vector $|\psi_1\rangle$ is declared as a subscripted function

```
> psi[1]:=x->exp(-alpha*x^2);
```

$$\psi_1 := x \to e^{(-\alpha\, x^2)}$$

and its normalization is given by:

```
> N1:=sqrt(int(psi[1](x)^2,x=-infinity..infinity));
```

$$N1 := \frac{1}{2}\sqrt{2}\sqrt{\sqrt{2}\,\sqrt{\pi}}$$

The normalized start vector $|\psi_1\rangle$ is defined again as an indexed function assignment:

```
> phi[1]:=unapply(psi[1](x)/(N1),x);
```

$$\phi_1 := x \to \frac{e^{(-x^2)}\sqrt{2}}{\sqrt{\sqrt{2}\,\sqrt{\pi}}}$$

The unnormalized first iterate $|\psi_2\rangle = \hat{H}|\phi_1\rangle$ follows as:

```
> psi[2]:=unapply(simplify(Hop(phi[1](x),x)),x);
```

$$\psi_2 := x \to \frac{1}{2}\,\frac{e^{(-x^2)}\,2^{1/4}\,(2 - 3x^2 + 2x^4)}{\pi^{1/4}}$$

The first diagonal matrix element is given by $\langle\phi_1|\psi_2\rangle = \langle\phi_1|\hat{H}|\phi_1\rangle$:

```
> Hd:=int(collect(phi[1](x)*psi[2](x),x),x=-infinity..infinity);
```

$$Hd := \frac{13}{16}\,\frac{2^{1/4}\,\pi^{1/4}}{\sqrt{\sqrt{2}\,\sqrt{\pi}}}$$

```
> Hdiag[1]:=Hd:
```

Next we need, for the Gram-Schmidt process, the vector

$$|\xi_2\rangle = |\psi_2\rangle - \langle\phi_1|\psi_2\rangle|\psi_1\rangle \quad :$$

```
> xi[2]:=unapply(simplify(psi[2](x)-Hd*phi[1](x)),x);
```

$$\xi_2 := x \to \frac{1}{16}\,\frac{2^{1/4}\,e^{(-x^2)}\,(3 - 24x^2 + 16x^4)}{\pi^{1/4}}$$

In order to calculate $\langle\xi_2|\xi_2\rangle$ without difficulty it helps to collect the integrand in x:

```
> N2:=sqrt(int(collect(xi[2](x)^2,x),x=-infinity..infinity));
```

$$N2 := \frac{1}{8}\sqrt{6}$$

The first off-diagonal element is given by the normalizing factor $\sqrt{\langle\xi_2|\xi_2\rangle}$:

```
> Hodg[2]:=N2:
```

The next Lanczos vector $|\phi_2\rangle$ is obtained from $|\xi_2\rangle$ upon normalization

```
> phi[2]:=unapply(simplify(xi[2](x)/N2),x);
```

$$\phi_2 := x \to \frac{1}{12}\,\frac{2^{3/4}\,e^{(-x^2)}\,(3 - 24x^2 + 16x^4)\sqrt{3}}{\pi^{1/4}}$$

The above procedure is repeated for as many iterations as one wishes. However, to ensure a tridiagonal Hamiltonian matrix one has to calculate the new Lanczos vectors according to

$$|\xi_{n+1}\rangle = |\psi_{n+1}\rangle - \langle\phi_n|\hat{H}|\phi_n\rangle|\phi_n\rangle - \langle\phi_n|\hat{H}|\phi_{n+1}\rangle|\phi_{n-1}\rangle\,.$$

```
> for n from 2 to N-1 do
> psi[n+1]:=unapply(simplify(Hop(phi[n](x),x)),x);
> Hd:=int(collect(phi[n](x)*psi[n+1](x),x),
> x=-infinity..infinity);
> Hdiag[n]:=Hd;
> xi[n+1]:=unapply(simplify(psi[n+1](x)-Hd*phi[n](x)-Hodg[n]
> *phi[n-1](x)),x);
> N2:=sqrt(int(collect(xi[n+1](x)^2,x),x=-infinity..infinity));
> Hodg[n+1]:=N2;
> phi[n+1]:=unapply(simplify(xi[n+1](x)/N2),x);
> od:
```

Now complete the matrix by calculation of the last diagonal element:

```
> psi[N+1]:=unapply(simplify(Hop(phi[N](x),x)),x):
> Hd:=int(collect(phi[N](x)*psi[N+1](x),x),
> x=-infinity..infinity):
> Hdiag[N]:=Hd:
> evalm(Hdiag);
```

$$\left[\frac{13}{16} \quad \frac{2^{1/4}\,\pi^{1/4}}{\sqrt{\sqrt{2}\,\sqrt{\pi}}} \quad \frac{213}{16} \quad \frac{112999}{3376}\right]$$

```
> evalm(Hodg);
```

$$\left[0 \quad \frac{1}{8}\sqrt{6} \quad \frac{1}{4}\sqrt{633}\right]$$

```
> Hmat:=matrix(N,N,0):
```

Now assemble the diagonal and neighbouring rows into the tridiagonal Hamiltonian matrix:

```
> for i from 1 to N do
> for j from 1 to N do
> if i=j then Hmat[i,j]:=Hdiag[i] fi;
> if i=j-1 then Hmat[i,j]:=Hodg[j] fi;
> if i=j+1 then Hmat[i,j]:=Hodg[i] fi;
> od:
```

```
> od:
> evalm(map(evalf,Hmat));
```

$$\begin{bmatrix} .8125000000 & .3061862179 & 0 \\ .3061862179 & 13.31250000 & 6.289872813 \\ 0 & 6.289872813 & 33.47126777 \end{bmatrix}$$

For best efficiency we should make use of a special algorithm to find the eigenvalues/eigenvectors of a tridiagonal matrix. Since we consider small matrices, we simply use Maple's built-in numerical diagonalization routine.

```
> evalf(Eigenvals(Hmat));
```

$$[.8042014956 \;\; 11.51904690 \;\; 35.27301941]$$

Having performed only three iterations we can hope only for the ground state to be accurate. We illustrate the Lanczos iteration by a graph of the start vector and two iterates.

```
> plot({Vop(x),psi[1](x)},x=-2..2,y=0..5,title=
> `Quartic potential and Lanczos start vector`);
```

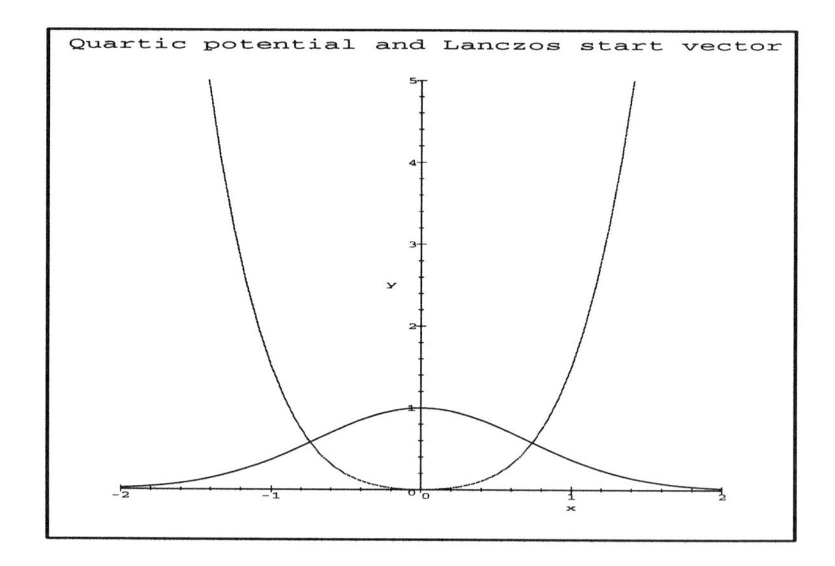


```
> plot({Vop(x),phi[1](x),phi[2](x),phi[3](x)},x=-4..4,y=-1..2,
> title=`Quartic potential and 3 Lanczos vectors`,color=black);
```

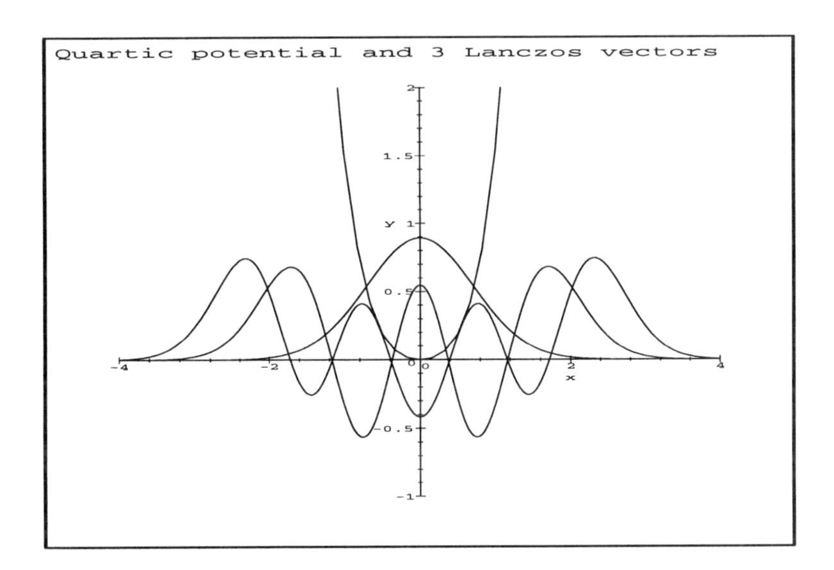

It can be seen that the Lanczos basis picks up nodes very quickly. Even-symmetry start vectors build up states of the same symmetry (as this is the symmetry of the Hamiltonian). To obtain odd-symmetry solutions one has to use a start vector that is not symmetric. It is, however, more economical to calculate the different symmetry sectors separately.

Exercise 2.4.1: Perform a convergence analysis for the ground-state eigenvalue as a function of the number of Lanczos iterations.

Exercise 2.4.2: Find the optimum value of the width parameter in the start vector to obtain a lowest-possible ground state after 5 Lanczos iterations.

Exercise 2.4.3: Obtain an estimate for the energy of the first excited state. Investigate double-well potentials of type $V(x) = -ax^2 + bx^4$ with $a > 0$, and $b > 0$. Observe how the energy gap $\Delta = E_1 - E_0$ decreases for parameter choices $\{a, b\}$ such that the barrier between the two wells increases.

Exercise 2.4.4: Consider a double-well potential such as $V(x) = -2x^2 + (1/4)x^4$. Use a Gaussian centered over one of the minima. Observe how the Lanczos algorithm copes with the problem of obtaining the lowest symmetric and anti-symmetric eigenfunctions respectively.

Exercise 2.4.5: Extract the expansion coefficients for the ground state in the Lanczos basis from the corresponding eigenvector of the matrix diagonalization and graph the result.

2.5 Shooting Method

I show a simple version of the shooting method to solve numerically for individual eigenfunction/eigenvalue pairs of differential operators. The aim of the exercise is mostly pedagogical. It is shown how the request for normalizable solutions to the Schrödinger equation is tied in with the condition for the energy eigenvalue.

The idea works in its presented form only for symmetric potentials $V(x) = V(-x)$. There are two separate symmetry sectors, i.e., the eigenfunctions are either symmetric (even parity) or anti-symmetric (odd parity). They can be distinguished by their behaviour at the origin:

Symmetric solutions satisfy

$$\psi(0) = \text{const} \neq 0 \quad , \qquad \frac{d\psi}{dx}\Big|_{x=0} = 0 \quad ,$$

while anti-symmetric ones satisfy

$$\psi(0) = 0 \quad , \qquad \frac{d\psi}{dx}\Big|_{x=0} = \text{const} \neq 0 \quad .$$

The constants can be adjusted such that the eigenfunctions are properly normalized and admit a probability density interpretation. We can choose them arbitrarily at first and normalize the final results. Thus, our task is to solve the differential equation starting from $x = 0$ with two initial conditions. The eigenvalue E can be obtained iteratively from the requirement that $\psi(x_f)$ has to vanish for a finite, but sufficiently large value x_f. One can determine E by trial and error, or more systematically by an adaptation of the bisection algorithm.

There are two ways to proceed technically. One can work by imposing a finite-difference discretization of the second derivative, e.g., the one exact to second order in Δx. For an equidistant discretization $x_i := i\Delta x$, $\psi_i := \psi(x_i)$ we have

$$\psi''(x_i) = \psi_i'' \approx \frac{\psi_{i+1} + \psi_{i-1} - 2\psi_i}{(\Delta x)^2} \ .$$

Insertion of this approximation turns the differential equation into a recursion formula that allows one to calculate $\psi(x)$ by a stepping procedure based on $\psi_{i+1} = f(\psi_i, \psi_{i-1})$. One has to choose a step-size Δx small enough to minimize the error in the extracted eigenvalue, or calculate for a number of Δx values of reasonable size and perform an extrapolation to $\Delta x = 0$.

Instead of such a fixed step-size procedure that displays the essence of the method, one can choose a canned ordinary differential equation solver. I choose the latter to familiarize the reader with Maple's `numeric` option in the `dsolve` facility.

We work in units with $\hbar = m = 1$ and solve an anharmonic oscillator with potential $V(x) = x^4/4$. The Schrödinger equation becomes (with the energy value denoted as a trial energy):

$$-\frac{1}{2}\psi''(x) + (V(x) - E_T)\psi(x) = 0\,.$$

We define the anharmonic potential and an expression for the Schrödinger equation:

```
> VAHO:=x->x^4/4;
```

$$\mathrm{VAHO} := x \to \frac{1}{4}\,x^4$$

```
> Schr:=-diff(psi(x),x$2)/2+(VAHO(x)-ET)*psi(x)=0;
```

$$\mathrm{Schr} := -\frac{1}{2}\left(\frac{\partial^2}{\partial x^2}\,\psi(\,x\,)\right) + \left(\frac{1}{4}\,x^4 - ET\right)\,\psi(\,x\,) = 0$$

The initial conditions for a symmetric eigenstate are:

```
> ICsym:=psi(0)=1,D(psi)(0)=0;
```

$$\mathrm{ICsym} := \psi(\,0\,) = 1, \mathrm{D}(\,\psi\,)(\,0\,) = 0$$

while for the anti-symmetric case they are:

```
> ICasym:=psi(0)=0,D(psi)(0)=1;
```

$$\mathrm{ICasym} := \psi(\,0\,) = 0, \mathrm{D}(\,\psi\,)(\,0\,) = 1$$

The latter set of initial conditions would be used to generate the wavefunction for the first excited state, i.e., a function with a node at $x = 0$. Maple's Runge-Kutta differential equation solver is invoked by using the `numeric` option in `dsolve`:

```
> Solsym:=dsolve({subs(ET=0.4,Schr),ICsym},psi(x),numeric);

  Solsym := proc(rkf45_x) ... end
```

```
> Solsym(1.);
```

$$\left[x = 1., \psi(\,x\,) = .6390539165149802, \frac{\partial}{\partial x}\,\psi(\,x\,) = -.6256053932623299\right]$$

For graphing of solutions to ODEs it is convenient to make use of the special `odeplot` procedure from the `plots` package.

```
> with(plots):
> odeplot(Solsym,[x,psi(x)],0..3,-2..2,title=`ET=0.4`);
```

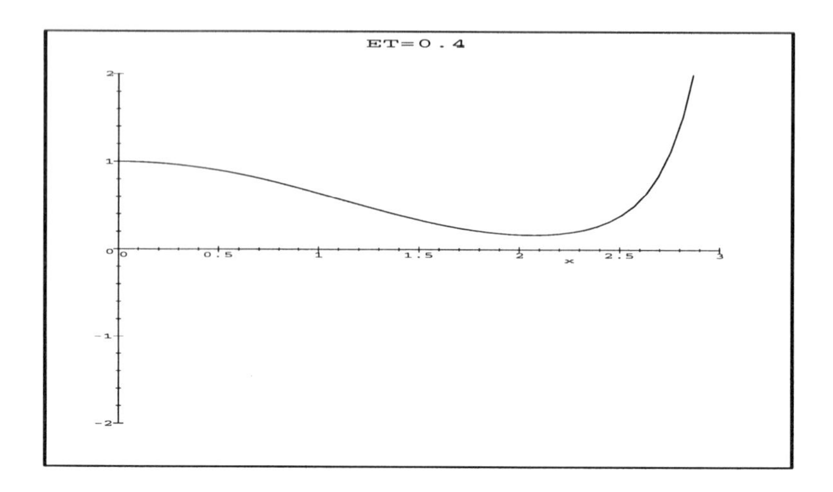

The energy trial value is clearly too small. The wavefunction diverges for large x before a touchdown to the x axis. We redefine the ODE procedure with a higher trial value for the energy.

```
> Solsym:=dsolve({subs(ET=0.45,Schr),ICsym},psi(x),numeric);

  Solsym := proc(rkf45_x) ... end
```

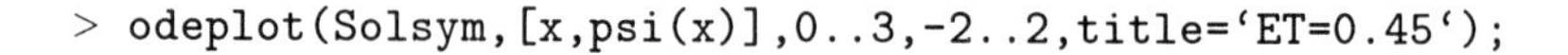

```
> odeplot(Solsym,[x,psi(x)],0..3,-2..2,title=`ET=0.45`);
```

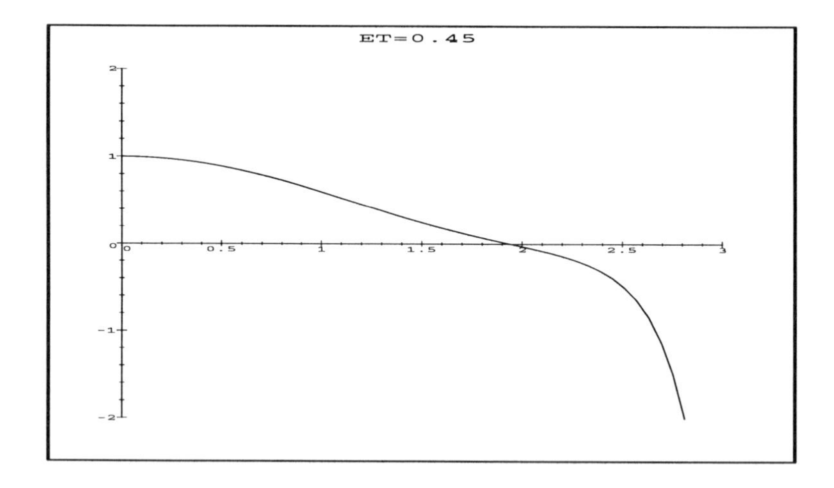

This energy trial value is too large. The wavefunction crosses the x axis and then diverges to negative infinity. Nevertheless we have succeeded in bracketing an eigenvalue. In fact, this will be the ground-state energy value as we are aiming for a nodeless wavefunction.

```
> Solsym:=dsolve({subs(ET=0.425,Schr),ICsym},psi(x),numeric):
> odeplot(Solsym,[x,psi(x)],0..3,-2..2,title='ET=0.425');
```

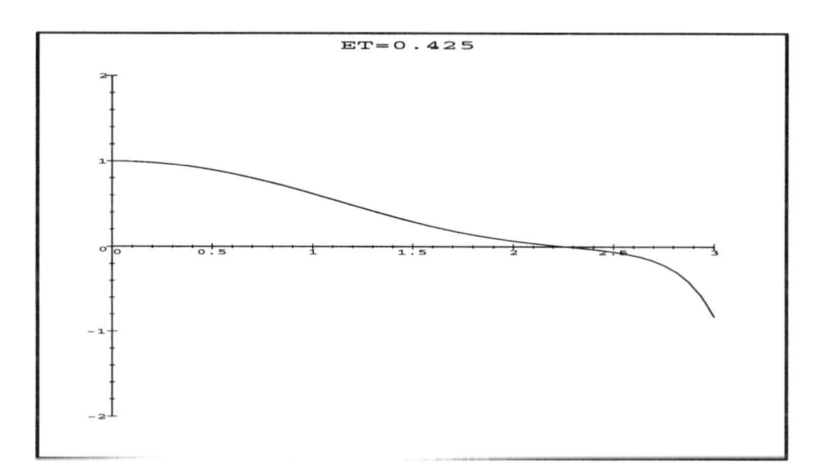

Clearly the correct eigenvalue lies between 0.4 and 0.425 as $E_T = 0.425$ gives a node at finite x, thus we try $E_T = 0.4125$.

```
> Solsym:=dsolve({subs(ET=.4125,Schr),ICsym},psi(x),numeric):
> odeplot(Solsym,[x,psi(x)],0..3,-2..2,title='ET=0.4125');
```

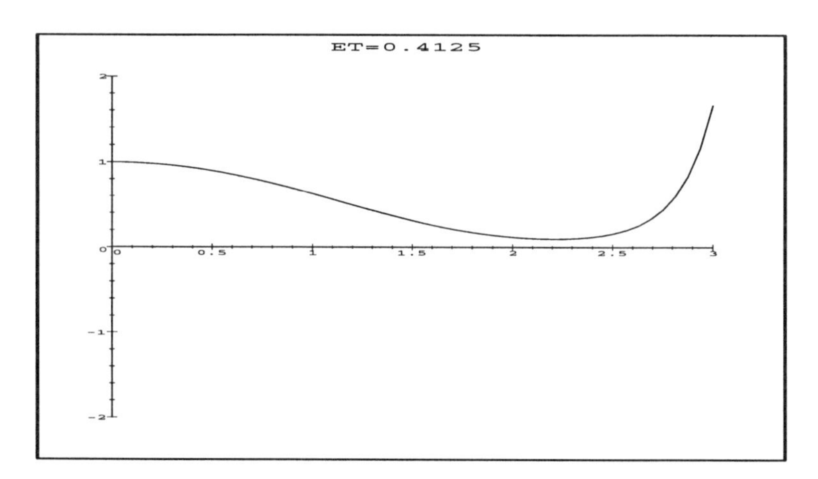

Now the eigenvalue is bracketed by $0.4125 < E0 < 0.425$. For the method to be practical an automated root search is required. An alternative to straightforward bisection would be to make use of the value of the wavefunction at

$x = x_f$ from two iterations in order to make an educated guess at the next trial value E_T. Instead of engaging in this Maple programming problem we finish the section with a comparison of the variational results from Sect. 2.1 for this potential. We try the improved variationally obtained energy value from Sect. 2.1:

```
> Solsym:=dsolve({subs(ET=.423586,Schr),ICsym},psi(x),numeric):
> odeplot(Solsym,[x,psi(x)],0..3,-2..2,title=`ET=0.423586`);
```

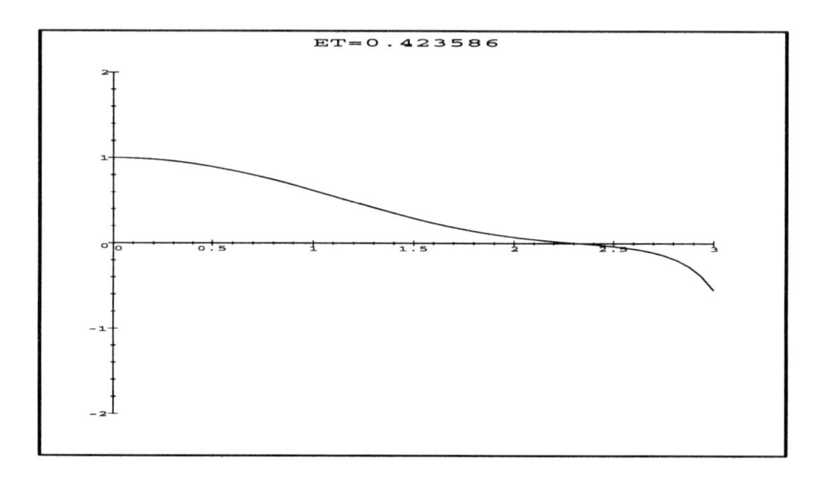

and confirm that it is above the true ground-state energy. A few more bisection steps allowed us to find an accurate solution for $E \approx 0.4208$. We skip the commands and just display the graph:

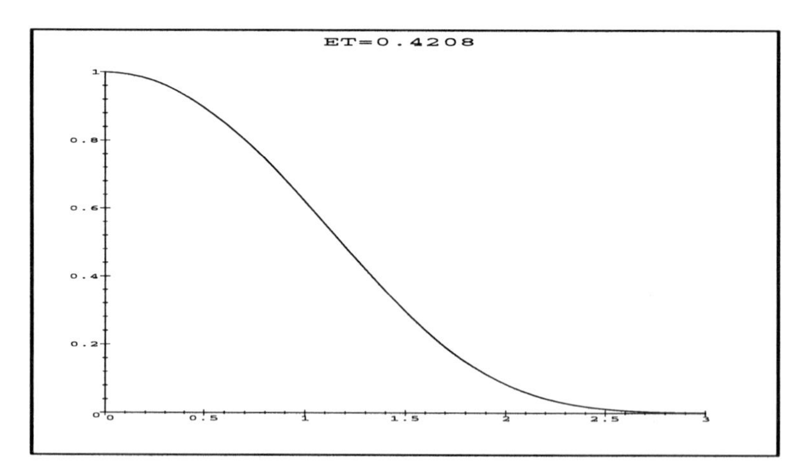

Note that the solution now follows smoothly along the x axis for a short interval and diverges beyond a finite value of x_f. It makes little sense to spend much time on pushing harder as this affects only the less significant digits in the eigenvalue. The numerical solution in the asymptotic region is highly unstable. In order to draw the solution together with the potential we can make use of the display function that is part of the plots package:

```
> P1:=plot(VAHO(x),x=0..4,y=0..4):
```

```
> P2:=odeplot(Solsym,[x,psi(x)],0..3,0..1,title='ET=0.4208',
> color=red):        display({P1,P2});
```

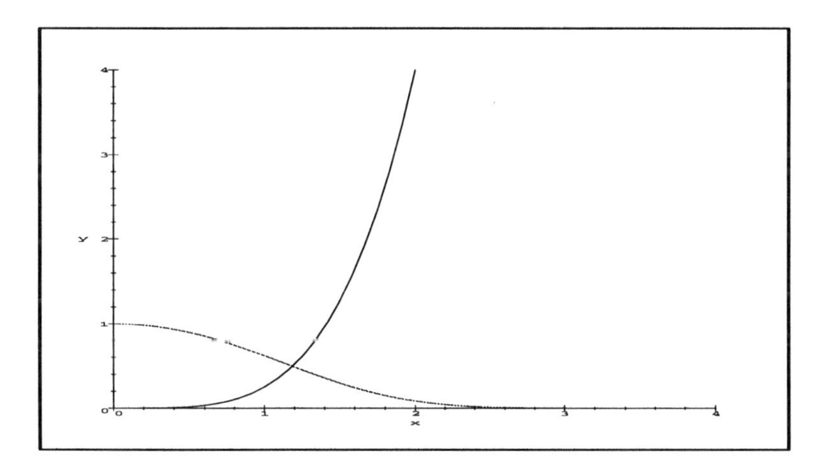

We can also display a comparison with the simple single-Gaussian variational result from Sect. 2.1:

```
> P3:=plot(exp(-(1.069913*x)^2/2),x=0..4):
```

```
> display({P2,P3});
```

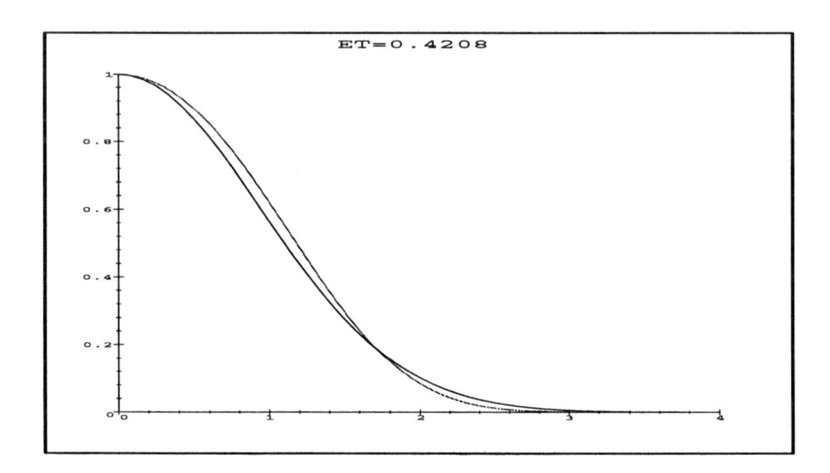

We note that the two solutions, in principle, are normalized differently. They are both unnormalized and share $\psi(0) = 1$. It can be seen that compared to a harmonic potential the quartic AHO pushes the wavefunction towards smaller x values as the potential is flat there. The larger-x region is avoided due to the steeper rise of the x^4 potential function.

Note that the approximate variational energy for the simple Gaussian wavefunction was found in Sect. 2.1 to be $E_0 = 0.42927$, i.e., about 2 % above the energy found here numerically. The graph shows that around $x = 1$ the relative error in the wavefunction reaches 8 %, and in the tail region it becomes very large.

We can make one more comparison to understand the nature of the variational approximation. The variational solution represents the exact eigensolution for a harmonic oscillator with a particular oscillator constant related to the optimum variational parameter. Thus, we plot this comparison potential together with the actual anharmonic potential:

```
> plot({x^4/4,1.069913^2*x^2/2},x=0..3,V=0..4);
```

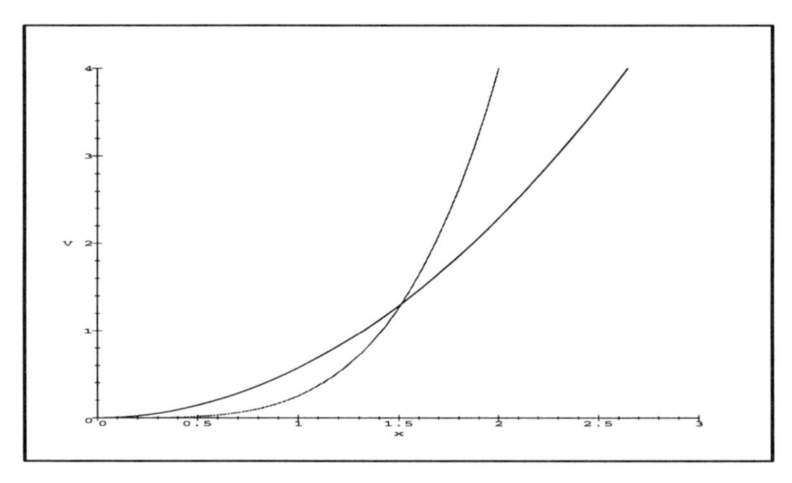

We conclude that the comparison potential is quite different from the quartic AHO.

Exercise 2.5.1: Obtain the first two symmetric and first two anti-symmetric eigenfunctions and corresponding eigenenergies for the quartic AHO. Observe how the spacing of the energy eigenvalues E_n increases with the label n compared to the harmonic case.

Exercise 2.5.2: Study the eigenfunctions of double-well potentials of type $V(x) = -ax^2 + bx^4$ for positive a and b. Investigate for which levels the eigenfunctions are composed of well-separated humps versus broadly distributed humps over the region where the potential is not too large.

2.6 Discrete Space

In this section we obtain a numerical approximation to the Schrödinger equation by restricting continuous space to a discrete lattice $\{x_j = j\Delta x, j = -J, -J+1, ..., -1, 0, 1, ..., J\}$. This represents a continuation of the previous section in which the finite-difference approximation to the second derivative was discussed in the context of the shooting method for individual eigenstates. Here we take the point of view that the discretization of space reduces the infinite-dimensional Hilbert space problem for the Sturm-Liouville operator to a finite-dimensional matrix eigenvalue problem.

In fact, we find that in one spatial dimension the matrix is tridiagonal. For more dimensions it remains sparse as the discretized Laplacian couples only neighbouring x sites. One can also arrive at the Hamiltonian matrix beginning with the basis state expansion method discussed briefly in Sect. 1.6, but using a Kronecker-delta basis (which isn't a basis in the usual sense and one looses the upper-bound property of the method stated in the Hylleraas-Undheim theorem)

$$\phi_n(x_j) = \delta_{nj} \, .$$

The matrix diagonalization provides $2J+1$ eigenvalues and eigenvectors. The eigenvectors can be interpolated to provide continuous eigenfunctions for all values of x. To make proper contact with the continuum limit one has to make sure that both the spacing is reduced to zero $\Delta x \to 0$, and a sufficiently large section of the x axis is covered by the mesh. For potentials that increase as $x \to \pm\infty$ this latter condition does not pose a problem and one can truncate space at a finite value $|x| = x_\mathrm{f}$.

We investigate in this section a double-well potential with deep well-separated wells to show that a near-degeneracy can occur in quantum mechanics. This has important consequences in atomic and molecular physics, such as the oscillation of the ammonia molecule (the nitrogen atom oscillating through the plane formed by three hydrogen atoms between the two equivalent configurations), and particularly in solid-state physics.

We set up the discrete spatial lattice:

```
> ipmax:=15; npmax:=2*ipmax+1;
```

$$ipmax := 15 \qquad npmax := 31$$

```
> with(linalg):
> Hmat:=matrix(npmax,npmax,0):
> Hmat[ipmax+1,ipmax+1]:=0:
```

We choose a maximum x value beyond which space is truncated:

```
> xmax:=8; dx:=xmax/ipmax;
```

$$\text{xmax} := 8 \qquad dx := \frac{8}{15}$$

We define a vector VCl that contains the x value and the potential value in adjacent elements. We store the potential in the diagonal of the Hamiltonian matrix.

```
> for ip from 1 to ipmax do:
> x:=ip*dx:
> VCl[2*ip-1]:=x:
> VCl[2*ip]:=-0.8*x^2+0.025*x^4;
> np1:=ipmax+1+ip: np2:=ipmax+1-ip:
> Hmat[np2,np2]:=VCl[2*ip];
> Hmat[np1,np1]:=VCl[2*ip];
> od:
> plot([[VCl[2*i-1],VCl[2*i]]$i=1..ipmax],0..xmax,
> style=point,title='Vclass');
```

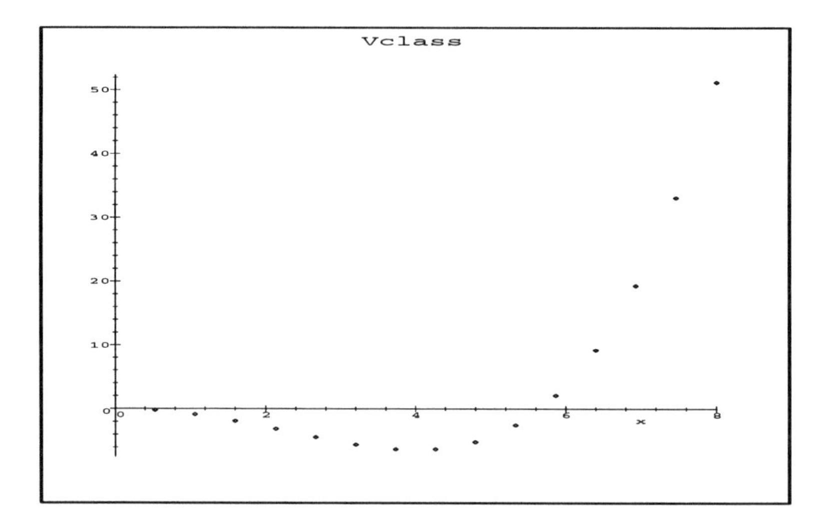

Now we set up the Hamiltonian matrix. The kinetic energy couples the diagonal that contains the potential to the nearest neighbours to reflect the discretization of the second derivative

$$\psi_i'' \approx \frac{\psi_{i+1} + \psi_{i-1} - 2\psi_i}{(\Delta x)^2} .$$

```
> dd:=1/(2*dx^2);
```

$$dd := \frac{225}{128}$$

```
> J:='J';  i:='i';
```

$$J := J \qquad i := i$$

```
> for i from 1 to npmax do:
> x:=-xmax+(i-1)*dx;
```

We include the possibility of a source term in the Hamiltonian.

```
> Hmat[i,i]:=Hmat[i,i]+2*dd+J*x:
> if i<npmax then Hmat[i,i+1]:=-dd; Hmat[i+1,i]:=-dd; fi;
> od:
```

We invoke a numerical diagonalization routine which also gives access to the eigenvectors. We are interested in the case of zero source, i.e., $J = 0$, but the algorithm does not make use of the reflection symmetry of the Hamiltonian:

```
> evalf(Eigenvals(subs(J=0,op(Hmat)),vecs));
```

[−5.549769 − 5.549769 − 3.966973 − 3.966951 − 2.586729
−2.586072 − 1.415861 − 1.405658 − .5358296 − .4668494
.1289767 .4169310 .9999764 1.504044 1.538373
1.661803 2.390924 3.244915 4.203118 5.262922
5.833493 5.833493 6.424734 12.82180 12.82180
22.91030 22.91030 36.67344 36.67344 54.88568
54.88568]

We observe that the two lowest energy eigenvalues are nearly degenerate! The eigenvectors contain the discrete representation of the eigenfunction on the discrete mesh of x-values. We are interested in graphing the eigenfunctions associated with the two nearly degenerate eigenenergies. For this purpose we declare a function that returns x_i:

```
> xsi:=i->-xmax+(i-1)*dx;
```

$$xsi := i \rightarrow -\mathrm{xmax} + (i-1)\,dx$$

```
> plot([[xsi(j),-vecs[j,1]] $j=1..npmax],title=`n=0 state`);
```

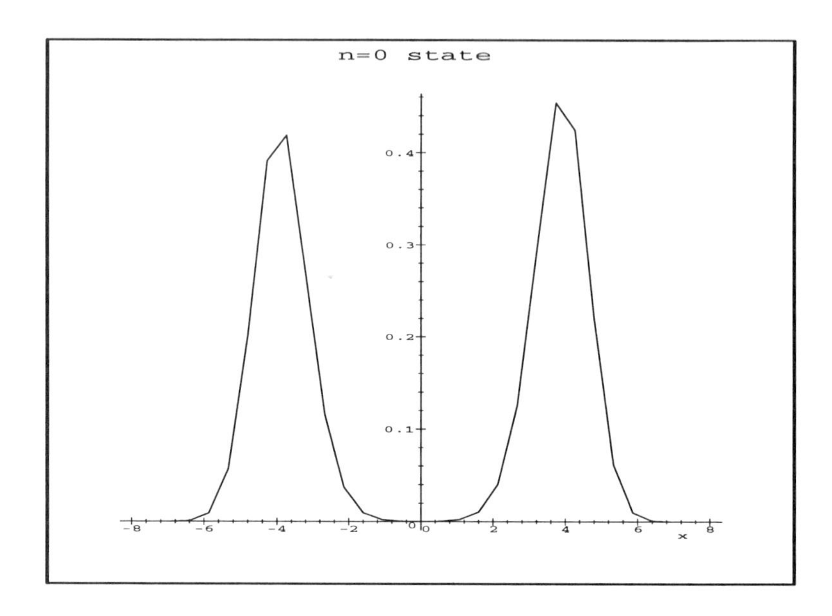

The numerically obtained ground state of the Hamiltonian that commutes with the parity operator is itself at least approximately symmetric. We observe that the humps are located over the potential minima. The imperfect symmetry in the calculated eigenfunction is a consequence of numerical errors occurring in the diagonalization algorithm. These are caused by the near-degeneracy of the $n = 0$ and $n = 1$ levels.

The discretization of x space results in a wavefunction with the information content of a finite-dimensional real vector $[\psi_{-J}, ..., \psi_J]$. To reconstruct a function defined for any x argument one can employ an interpolation scheme of higher order than the piecewise linear interpolation used in the graph. In fact, quadratic interpolation would be consistent with the order of the finite difference method used to replace the second derivative by an elementary arithmetic expression.

To indicate the height of the potential barrier between the two minima we plot the wavefunction together with the potential and shift the baseline of the wavefunction to the position of the ground-state energy level. It becomes evident from the graph that the barrier between the two wells is both very high and appreciably long on the scale of a particle described by a single-hump wavefunction. The tunneling probability for such a particle from the right well into the left and vice versa is very small.

```
> plot({[[VCl[2*k-1],VCl[2*k]]$k=1..ipmax],[[xsi(j),
> -10*vecs[j,1]-5.55] $j=1..npmax]},0..xmax,-10..10,
> title='Vclass and psi(n=0)');
```

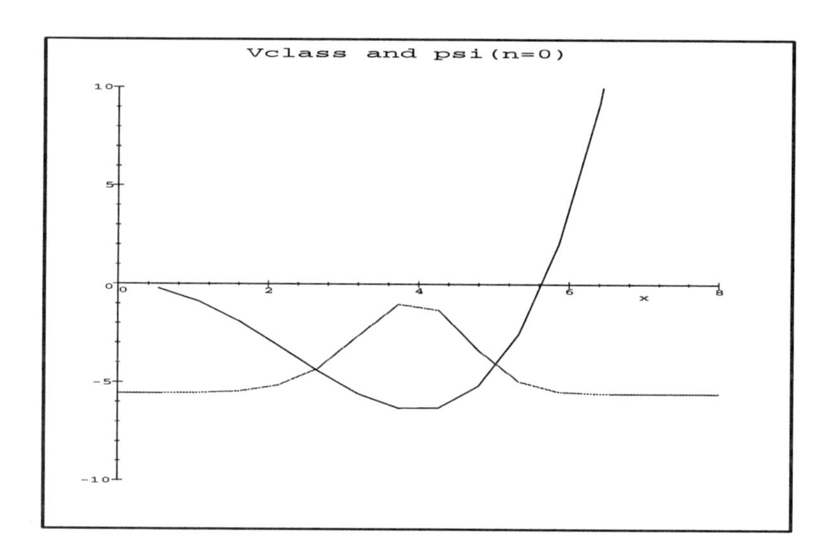

Now we draw the first excited state which is nearly degenerate with ψ_0:

```
> plot([[xsi(j),-vecs[j,2]] $j=1..npmax],title=`n=1 state`);
```

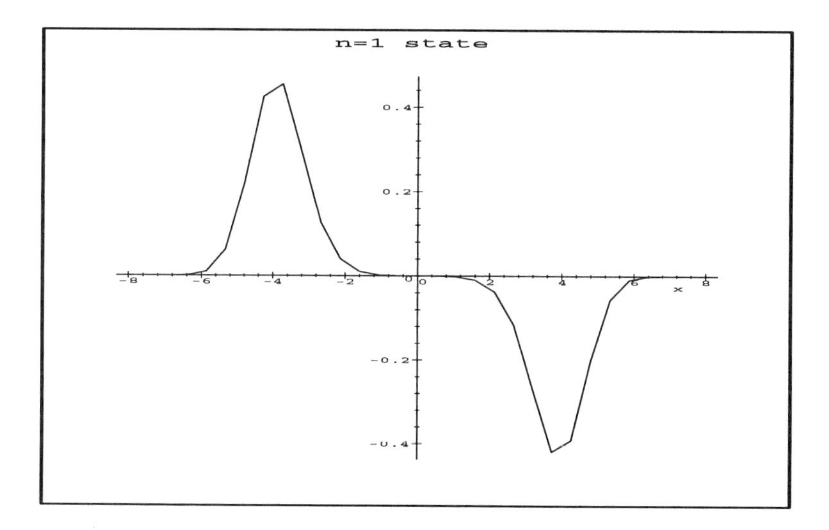

In quantum mechanics the ground state always obeys the symmetry of the Hamiltonian. However, as the double-well potential becomes deeper, and the potential barrier separates the two wells more and more effectively, the $n = 0$ and $n = 1$ states become nearly degenerate. Consequently, they both can be considered approximate ground states. Any linear combination of them also has approximately the same eigenvalue and qualifies as an approximate ground state. In particular, the sum and the difference of the $n = 0$ and $n = 1$

states, which represent single-hump wavefunctions in the left-hand-side and right-hand-side wells become acceptable approximations to the ground state.

One can investigate the time evolution of a single-hump wavefunction in the double-well potential and find that it will develop spatial oscillations from one well to the other with the period of oscillation inversely proportional to the energy spacing $E_1 - E_0$. This tunneling time through the potential barrier separating the two wells is very long if $E_1 \approx E_0$, and it makes good sense to consider the single-hump wavefunction as a ground state. This phenomenon plays an important role in solid-state physics where one considers the ground-state problem for $N \approx 10^{24}$ particles in a potential that contains many wells arranged in a 3D (or sometimes 2D) periodic lattice. For practical purposes one can consider that such a system is highly degenerate, even though mathematically this happens only in the limit as $N \to \infty$.

In quantum field theory, which corresponds to a QM problem of infinitely many particles the ground state does not have to obey the symmetry of the Hamiltonian, but can be spontaneously broken.

Exercise 2.6.1: Perform calculations for the simple AHO $V(x) = x^4/4$ solved accurately in the previous section. Compare the numerical values for the ground and first excited states respectively. Perform a numerical error analysis as a function of the step-size Δx.

Exercise 2.6.2: Use the spatial discretization method to solve for the ground and low-lying excited states of non-symmetric polynomial potentials.

Exercise 2.6.3: Calculate the spectrum of a long-range potential such as $V(x) = -1/\sqrt{1+x^2}$. Note that this potential has $E > 0$ continuum solutions and that the higher levels are affected by the spatial cut-off x_f.

Exercise 2.6.4: Calculate the ground state of a potential that is not bound from below, such as

$$V(x) = -\frac{1}{|x|^{1/3}} \, .$$

This potential requires a discretization that excludes the origin. Perform a convergence analysis, i.e., demonstrate the behaviour of the eigenvalue as a function of the step-size dx. Determine whether quantum zero-point fluctuations are sufficient to prevent the quantum particle from a collapse at $x = 0$. What happens to a classical particle in this potential? What happens in the quantum calculation as the power is increased from the value of 1/3?

3. Scattering in 1D

3.1 Wavepackets

The free-particle solutions to the Schrödinger equation $\exp(\pm ipx/\hbar)$ describe a continuous flux (beam) of particles of momentum p moving to the right or left depending on the sign in the exponent. They are completely delocalized in position space but the momentum is infinitely sharp. Alternatively, one can define free position eigenstates $\psi(x) = \delta(x - x_0)$ that describe a free particle located at x_0, but whose momentum is completely undetermined. Both pictures are in conflict with classical experience and difficult to comprehend. Therefore, one aims for another representation of a free particle that attempts to be in accord with classical experience, namely to have a localization both in position and momentum with the classical position – momentum relationship of a free particle.

In this worksheet we learn how to superimpose free stationary solutions of a fixed sharp energy that are completely delocalized in coordinate space to make up a packet state that has the uncertainty split between momentum and position. We work in units with $\hbar = m = 1$.

As the wavepacket is not stationary in time we have to carry both position and time dependence. This is a price to be paid for the attempted proximity to a classical description of a particle. It is impossible to find a stationary eigenstate of the free Schrödinger equation that satisfies the request of localized distributions both in position and momentum. Note that the interpretation of the wavefunction to be found is not that it describes an individual particle! It just provides the probability amplitude for an ensemble of free particles. If these are prepared at $t = 0$ as a bunch of simultaneously moving particles, or if one produces them sequentially according to the initial distribution, the solution can make predictions for the distribution of particles at later times in position as well as in momentum space.

The free-particle wavepacket is constructed from stationary plane-wave solutions. The latter form a complete set, if one considers both travelling directions and all possible momenta. We can distinguish between both types of basis functions and separately write the right-travelling and left-travelling plane wave solutions with a normalization factor of 1. First we relate the energy and wavenumber k which equals momentum p in our units.

```
> En:=ks->ks^2/2;
```

$$En := ks \to \frac{1}{2}\,ks^2$$

```
> phi_R:=(x,t,ks)->exp(I*ks*x)*exp(-I*En(ks)*t);
```

$$\text{phi\_R} := (\,x, t, ks\,) \to e^{(\,I\,ks\,x\,)}\,e^{(\,-I\,\text{En}(\,ks\,)\,t\,)}$$

```
> phi_L:=(x,t,ks)->exp(-I*ks*x)*exp(-I*En(ks)*t);
```

$$\text{phi\_L} := (\,x, t, ks\,) \to e^{(\,-I\,ks\,x\,)}\,e^{(\,-I\,\text{En}(\,ks\,)\,t\,)}$$

Now we introduce a weight function that superimposes the plane waves into a localized Gaussian wavepacket with a well-defined average momentum. We also have to specify an initial width (uncertainty), which we choose as w_0 in p space.

The spatial distribution is obtained as a result of the above superposition with a width that is inversely proportional to w_0 and carries in addition a factor of 1/2. The weighting coefficient is just a function of momentum, as the free-particle eigenstates are labeled by the momentum variable, and we are working in position space. Parametrically it depends on the center of the requested distribution in p space and, through an optional phase factor, also on the mean position in coordinate space x_0.

We restrict the momenta over which the expansion is summed to a finite range and discuss the limit as $L \to \infty$ in detail, since Maple cannot perform the required integral without assumptions on the allowed range for some variables.

```
> wt:=(p,x0,p0,w0)->exp(-(p-p0)^2/(4*w0^2)-I*p*x0)
> /sqrt(sqrt(Pi)*w0);
```

$$\text{wt} := (\,p, x0, p0, w0\,) \to \frac{e^{\left(-1/4\,\frac{(\,p-\text{p0}\,)^2}{\text{w0}^2}-I\,p\,x0\right)}}{\text{sqrt}(\,\text{sqrt}(\,\pi\,)\,w0\,)}$$

We define a packet composed only of right-travelling plane-wave solutions:

```
> packetR:=(x,t,x0,p0,w0)->int(wt(p,x0,p0,w0)*phi_R(x,t,p),
> p=-L..L);
```

$$\text{packetR} := (\,x, t, x0, p0, w0\,) \to \int_{-L}^{L} \text{wt}(\,p, x0, p0, w0\,)\,\text{phi\_R}(\,x, t, p\,)\,dp$$

We make assumptions about the sign of some parameters and choose a wavepacket with zero initial position and momentum.

```
> assume(L>0);    assume(t>0);    assume(w0>0);
```

```
> packetR(x,t,0,0,w0);
```

$$\frac{\sqrt{\pi}\, e^{\left(-\frac{x^{\sim 2}\, w0^{\sim 2}}{2\, I\, t^{\sim}\, w0^{\sim 2}+1}\right)}\, w0^{\sim}\, \mathrm{erf}\left(\frac{1}{2}\, \frac{2\, I\, L^{\sim}\, t^{\sim}\, w0^{\sim 2} + L^{\sim} - 2\, I\, x^{\sim}\, w0^{\sim 2}}{w0^{\sim}\, \sqrt{2\, I\, t^{\sim}\, w0^{\sim 2}+1}}\right)}{\sqrt{\sqrt{\pi}\, w0^{\sim}}\, \sqrt{2\, I\, t^{\sim}\, w0^{\sim 2}+1}} + \frac{\sqrt{\pi}\, e^{\left(-\frac{x^{\sim 2}\, w0^{\sim 2}}{2\, I\, t^{\sim}\, w0^{\sim 2}+1}\right)}\, w0^{\sim}\, \mathrm{erf}\left(\frac{1}{2}\, \frac{2\, I\, L^{\sim}\, t^{\sim}\, w0^{\sim 2} + L^{\sim} + 2\, I\, x^{\sim}\, w0^{\sim 2}}{w0^{\sim}\, \sqrt{2\, I\, t^{\sim}\, w0^{\sim 2}+1}}\right)}{\sqrt{\sqrt{\pi}\, w0^{\sim}}\, \sqrt{2\, I\, t^{\sim}\, w0^{\sim 2}+1}}$$

```
> limit(",L=infinity):
```

We omit the printed answer, as the limit was not performed. Maple doesn't do the integral between infinite limits, since it finds a problem in evaluating this limit. However, there shouldn't be a problem at all. One can verify this by calculating the expressions for large, yet finite L. The result is a simple Gaussian that spreads in width as a function of time.

We can generate the result with a (somewhat dirty) trick. The difficulty that Maple encounters is related to the complex arguments in the error function erf. So let's cheat a little in intermediate steps, and assume that the final answer comes out correctly. First make sure real variables are treated as real.

```
> assume(x0,real): assume(p0,real): assume(x,real):
```

We can analyze the problem for a packet with fixed parameters, e.g., $x_0 = 0$, $p_0 = 0$, $w_0 = 1$:

```
> packetR(x,t,0,0,1);
```

$$\frac{\pi^{1/4}\, e^{\left(-\frac{x^{\sim 2}}{1+2\, I\, t^{\sim}}\right)}\, \mathrm{erf}\left(\frac{1}{2}\, \frac{L^{\sim} + 2\, I\, L^{\sim}\, t^{\sim} - 2\, I\, x^{\sim}}{\sqrt{1+2\, I\, t^{\sim}}}\right)}{\sqrt{1+2\, I\, t^{\sim}}} + \frac{\pi^{1/4}\, e^{\left(-\frac{x^{\sim 2}}{1+2\, I\, t^{\sim}}\right)}\, \mathrm{erf}\left(\frac{1}{2}\, \frac{L^{\sim} + 2\, I\, L^{\sim}\, t^{\sim} + 2\, I\, x^{\sim}}{\sqrt{1+2\, I\, t^{\sim}}}\right)}{\sqrt{1+2\, I\, t^{\sim}}}$$

```
> res:=factor(");
```

$$\mathrm{res} := \frac{\pi^{1/4}\, e^{\left(\frac{I\, x^{\sim 2}}{-I+2\, t^{\sim}}\right)}}{\sqrt{I\, (-I+2\, t^{\sim})}} \left(\mathrm{erf}\left(\frac{1}{2}\, \frac{L^{\sim} + 2\, I\, L^{\sim}\, t^{\sim} - 2\, I\, x^{\sim}}{\sqrt{I\, (-I+2\, t^{\sim})}}\right) + \mathrm{erf}\left(\frac{1}{2}\, \frac{L^{\sim} + 2\, I\, L^{\sim}\, t^{\sim} + 2\, I\, x^{\sim}}{\sqrt{I\, (-I+2\, t^{\sim})}}\right)\right)$$

We have to isolate the problem part. The **nops** function can be used to find out how many operands there are, some trial-and-error steps help to decompose expressions into pieces.

```
> op(3,res);
```

$$\mathrm{erf}\left(\frac{1}{2}\frac{L\tilde{} + 2\,I\,L\tilde{}\,t\tilde{} - 2\,I\,x\tilde{}}{\sqrt{I\,(-I + 2\,t\tilde{})}}\right) + \mathrm{erf}\left(\frac{1}{2}\frac{L\tilde{} + 2\,I\,L\tilde{}\,t\tilde{} + 2\,I\,x\tilde{}}{\sqrt{I\,(-I + 2\,t\tilde{})}}\right)$$

```
> limit(",L=infinity);
```

$$\lim_{L\tilde{}\to\infty} \mathrm{erf}\left(\frac{1}{2}\frac{L\tilde{} + 2\,I\,L\tilde{}\,t\tilde{} - 2\,I\,x\tilde{}}{\sqrt{I\,(-I + 2\,t\tilde{})}}\right) + \mathrm{erf}\left(\frac{1}{2}\frac{L\tilde{} + 2\,I\,L\tilde{}\,t\tilde{} + 2\,I\,x\tilde{}}{\sqrt{I\,(-I + 2\,t\tilde{})}}\right)$$

```
> subs(t=0,");
```

$$\lim_{L\tilde{}\to\infty} \mathrm{erf}\left(\frac{1}{2}L\tilde{} - I\,x\tilde{}\right) + \mathrm{erf}\left(\frac{1}{2}L\tilde{} + I\,x\tilde{}\right)$$

```
> subs(x=0,");
```

$$2$$

Now we introduce a trick that physicists sometimes use if the results justify the means: replace t by $\tau = it$, obtain your answer and then reverse the substitution back from τ to t. Maple, unlike some other symbolic computing environments, does its mathematics (at least fairly) rigorously and doesn't automatically make assumptions on variables that may seem obvious to the user. (We note that the required integral can be calculated straightforwardly at $t = 0$, i.e., if the imaginary phase factor has no p^2 contribution).

We define a general integral expression required for our purposes.

```
> assume(a>0);   assume(tau>0);
```

```
> GenI:=(a,tau,xi)->int(exp(-(a*p)^2-tau*p^2/2+xi*p),
> p=-infinity..infinity);
```

$$\mathrm{GenI} := (\,a, \tau, \xi\,) \to \int_{-\infty}^{\infty} e^{(\,-a^2\,p^2 - 1/2\,\tau\,p^2 + \xi\,p\,)}\,dp$$

We are interested in an expression of the following type:

```
> GenI(a,I*t,I*x);
```

$$\int_{-\infty}^{\infty} e^{(\,-a\tilde{}^2\,p^2 - 1/2\,I\,p^2\,t\tilde{} + I\,p\,x\tilde{}\,)}\,dp$$

```
> subs(xi=I*x,tau=I*t,GenI(a,tau,xi));
```

$$2 \frac{\sqrt{\pi}\, \mathrm{e}^{\left(-1/2 \frac{x^{\sim 2}}{2\, a^{\sim 2} + I\, t^{\sim}}\right)}}{\sqrt{4\, a^{\sim 2} + 2\, I\, t^{\sim}}}$$

This is the desired form for the answer: a Gaussian packet centered at the origin in x space with zero average translational momentum. However, the width spreads as a function of time. The denominator of the whole expression is a normalization factor, the denominator in the argument of the exponential function represents the width itself, and it is $|\psi|^2$ strictly speaking that is a simple Gaussian, i.e, real valued.

The above result can be generalized to a Gaussian centered on a trajectory with mean position x_0 and mean momentum p_0. The main finding that a free packet increases its width as a function of time remains the same. This can be proven independently of the specific shape of the packet, simply based on the commutation relations for position and momentum. Spreading does not occur, however, if the packet is bound in a potential $V(x)$.

The spreading of the width in position space is in analogy with the classical mechanics result: Given a beam of particles with a position and momentum profile, the position profile will spread in time due to the presence of faster and slower particles. The difference is, however, that in classical mechanics one can begin with a beam with assumed zero width initially in both the momentum distribution and the position distribution. Such a beam will not spread out. In QM this is prohibited by the uncertainty relation.

We apply the trick for an arbitrary Gaussian now and redefine the wavepacket procedure to do the integral over the infinite range for p:

```
> packetR:=(x,t,x0,p0,w0)->int(wt(p,x0,p0,w0)*phi_R(x,t,p),
> p=-infinity..infinity);
```

$$\mathrm{packetR} := (\, x, t, x0, p0, w0\,) \rightarrow \int_{-\infty}^{\infty} \mathrm{wt}(\, p, x0, p0, w0\,)\, \mathrm{phi\_R}(\, x, t, p\,)\, dp$$

```
> packetR(x,t,x0,p0,w0);
```

$$\int_{-\infty}^{\infty} \frac{\mathrm{e}^{\left(-1/4 \frac{(p - p0^{\sim})^2}{w0^{\sim 2}} - I\, p\, x0^{\sim}\right)}\, \mathrm{e}^{(I\, p\, x^{\sim})}\, \mathrm{e}^{(-1/2\, I\, p^2\, t^{\sim})}}{\sqrt{\sqrt{\pi}\, w0^{\sim}}}\, dp$$

This is the required integral. To be able to do it we chop it into pieces and make careful substitutions. First the integrand:

```
> wt(p,x0,p0,w0)*phi_R(x,t,p);
```

$$\frac{\mathrm{e}^{\left(-1/4 \frac{(p - p0^{\sim})^2}{w0^{\sim 2}} - I\, p\, x0^{\sim}\right)}\, \mathrm{e}^{(I\, p\, x^{\sim})}\, \mathrm{e}^{(-1/2\, I\, p^2\, t^{\sim})}}{\sqrt{\sqrt{\pi}\, w0^{\sim}}}$$

We collect the answer into a single exponential:

```
> res:=combine(",exp);
```

$$res := \frac{e^{\left(-1/4\,\frac{(p-p0^{\sim})^2}{w0^{\sim 2}} - I\,p\,x0^{\sim} + I\,p\,x^{\sim} - 1/2\,I\,p^2\,t^{\sim}\right)}}{\sqrt{\sqrt{\pi}\,w0^{\sim}}}$$

We get rid of the normalization factor:

```
> op2res:=op(2,res);
```

$$op2res := e^{\left(-1/4\,\frac{(p-p0^{\sim})^2}{w0^{\sim 2}} - I\,p\,x0^{\sim} + I\,p\,x^{\sim} - 1/2\,I\,p^2\,t^{\sim}\right)}$$

Now we pull out the exponent itself and collect in powers of p:

```
> simplify(ln(op2res));
```

$$-\frac{I}{4w0^{\sim 2}}(-I\,p^2 + 2\,I\,p\,p0^{\sim} - I\,p0^{\sim 2} + 4\,w0^{\sim 2}\,x0^{\sim}\,p - 4\,w0^{\sim 2}\,x^{\sim}\,p + 2\,w0^{\sim 2}\,t^{\sim}\,p^2)$$

```
> res:=collect(",p);
```

$$res := -\frac{1}{4}\,\frac{I\,(-I + 2\,w0^{\sim 2}\,t^{\sim})\,p^2}{w0^{\sim 2}} - \frac{1}{4}\,\frac{I\,(-4\,w0^{\sim 2}\,x^{\sim} + 2\,I\,p0^{\sim} + 4\,w0^{\sim 2}\,x0^{\sim})\,p}{w0^{\sim 2}} - \frac{1}{4}\,\frac{p0^{\sim 2}}{w0^{\sim 2}}$$

Now we assign the coefficients of the different powers in p.

```
> a1:=coeff(res,p,2);
```

$$a1 := -\frac{1}{4}\,\frac{I\,(-I + 2\,w0^{\sim 2}\,t^{\sim})}{w0^{\sim 2}}$$

```
> a2:=coeff(res,p,1);
```

$$a2 := -\frac{1}{4}\,\frac{I\,(-4\,w0^{\sim 2}\,x^{\sim} + 2\,I\,p0^{\sim} + 4\,w0^{\sim 2}\,x0^{\sim})}{w0^{\sim 2}}$$

```
> a3:=coeff(res,p,0);
```

$$a3 := -\frac{1}{4}\,\frac{p0^{\sim 2}}{w0^{\sim 2}}$$

We separate real and imaginary parts:

```
> a1:=evalc(a1);
```

$$a1 := -\frac{1}{4}\,\frac{1}{w0^{\sim 2}} - \frac{1}{2}\,I\,t^{\sim}$$

```
> a1R:=Re(a1);    a1I:=Im(a1);
```

$$a1R := -\frac{1}{4}\frac{1}{w0^{\sim 2}} \qquad a1I := -\frac{1}{2}t^{\sim}$$

```
> a2:=evalc(a2);
```

$$a2 := \frac{1}{2}\frac{p0^{\sim}}{w0^{\sim 2}} + I\,(x^{\sim} - x0^{\sim})$$

```
> a2R:=Re(a2);    a2I:=Im(a2);
```

$$a2R := \frac{1}{2}\frac{p0^{\sim}}{w0^{\sim 2}} \qquad a2I := x^{\sim} - x0^{\sim}$$

Now we define a general integral in which convergence will be ensured by an assumption on the coefficient of p^2:

```
> GenI2:=(a,tau,xi1,xi2)->int(exp(-a*p^2-tau*p^2+xi1*p+xi2*I*p)
> ,p=-infinity..infinity);
```

$$\mathrm{GenI2} := (a, \tau, \xi 1, \xi 2) \to \int_{-\infty}^{\infty} e^{(-a\,p^2 - \tau\,p^2 + \xi 1\,p + I\,\xi 2\,p)}\,dp$$

In the next step we make the appropriate substitutions and define a function with the normalization factor to which the parameters can be supplied.

```
> res:=exp(a3)*subs(xi1=a2R,xi2=-a2I,tau=I*a1I,
> GenI2(-a1R,tau,xi1,xi2));
```

$$\mathrm{res} := 2\,\frac{e^{\left(-1/4\,\frac{p0^{\sim 2}}{w0^{\sim 2}}\right)}\,e^{\left(\frac{w0^{\sim 2}\left(1/2\,\frac{p0^{\sim}}{w0^{\sim 2}} + I\,(-x^{\sim}+x0^{\sim})\right)^2}{1-2\,I\,t^{\sim}\,w0^{\sim 2}}\right)}\,w0^{\sim}\,\sqrt{\pi}}{\sqrt{1-2\,I\,t^{\sim}\,w0^{\sim 2}}}$$

```
> pack:=unapply(res/(Pi^(1/4)*w0^(1/2)),x,t,x0,p0,w0);
```

$$\mathrm{pack} := (`x^{\sim}`, `t^{\sim}`, `x0^{\sim}`, `p0^{\sim}`, `w0^{\sim}`) \to$$

$$2\,\frac{e^{\left(-1/4\,\frac{`p0^{\sim}`^2}{`w0^{\sim}`^2}\right)}\,e^{\left(\frac{`w0^{\sim}`^2\left(1/2\,\frac{`p0^{\sim}`}{`w0^{\sim}`^2} + I\,(-`x^{\sim}`+`x0^{\sim}`)\right)^2}{1-2\,I\,`t^{\sim}`\,`w0^{\sim}`^2}\right)}\,\sqrt{`w0^{\sim}`}\,\pi^{1/4}}{\sqrt{1-2\,I\,`t^{\sim}`\,`w0^{\sim}`^2}}$$

Now we calculate a Gaussian wavepacket for zero initial mean position, $p_0 = 2$, and an initial width $w_0 = 1$:

```
> pack(x,t,0,2,1);
```

$$2\,\frac{e^{(-1)}\,e^{\left(\frac{(1-I\,x^{\sim})^2}{1-2\,I\,t^{\sim}}\right)}\,\pi^{1/4}}{\sqrt{1-2\,I\,t^{\sim}}}$$

The precise form of the packet should be checked against the textbook result due to the tricks applied. The Maple user can display the time evolution using the `animate` feature:

```
> with(plots):
> animate(abs(evalf(pack(x,t,-5,1,1)))^2,x=-10..10,t=0..10,
> numpoints=150);
```

We plot the envelope of the wavepacket at $t = 0$ and some later time ($t = 1$) by changing the second argument in `pack`. We observe how the wavepacket's mean position changes according to $\mathrm{d}x/\mathrm{d}t = p_0/m$, i.e., we obtain the result expected from a classical trajectory. The width spreads in time as discussed above. This spreading depends initially on the value of w_0, but eventually the linear growth takes over. If the packet is initially very narrow, then the linear growth of the width with time sets in very quickly. On the other hand for an initially broad wavepacket it is possible to avoid significant spreading for an appreciable time.

```
> plot(abs(evalf(pack(x,0,0,2,1))),x=-4..6,0..2.7);
```

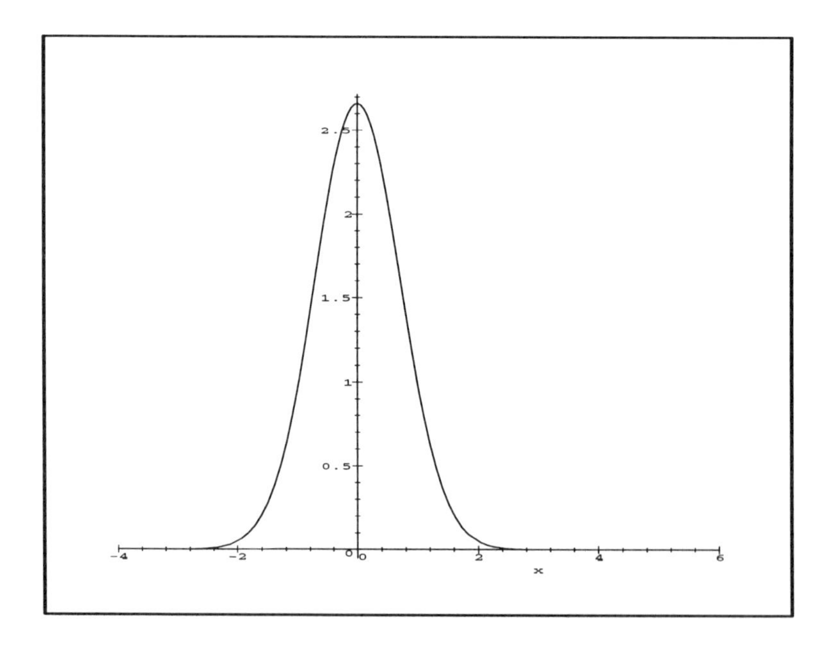


```
> plot(abs(evalf(pack(x,1,0,2,1))),x=-4..6,0..2.7);
```

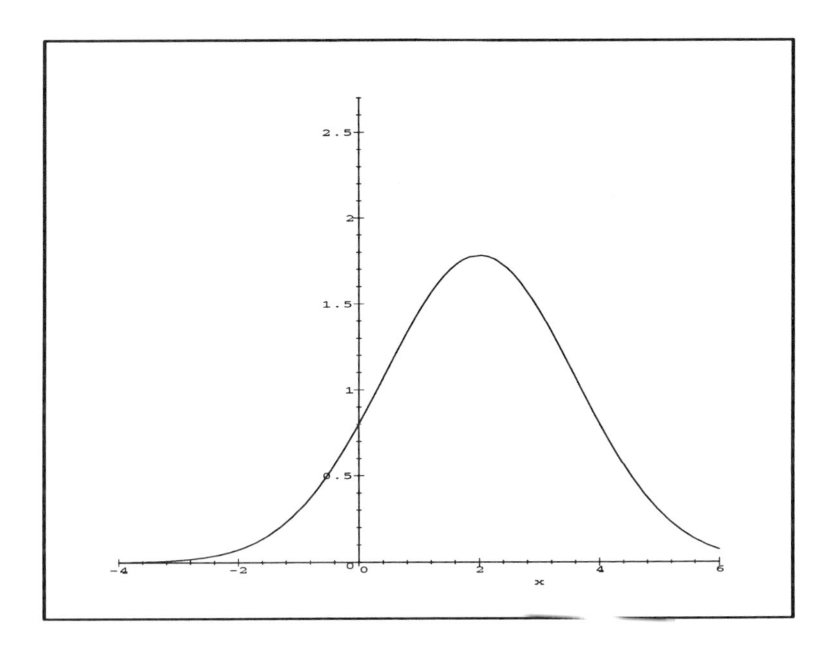

Exercise 3.1.1: Construct left-travelling and right-travelling wavepackets of different initial width. What distribution in coordinate and momentum space do the packets tend to asymptotically?

Exercise 3.1.2: Suppose that one is interested in using a wavepacket that is not spreading appreciably, e.g., less than 10 % over a fixed time, e.g., $\Delta t = 10$ a.u.. What constraint does that put on the initial width w_0?

Exercise 3.1.3: Display the real and imaginary parts of the Gaussian wavepacket for different momenta p_0 and make your observation.

3.2 Potential Well and Barrier

The aim of this section is to discuss the stationary scattering solutions for a potential well or barrier. Although the development is shown for a potential well the results are valid for a barrier by extending the range of the parameter characterizing the well depth to the opposite sign.

The problem of a rectangularly shaped potential is taken up in QM mostly as it is easily solved. The interaction potential is piecewise constant, and in each region the solution of the stationary Schrödinger equation is simply given by the free wave solution discussed in section 1.1, except that the wavenumber takes into account the value of the constant potential for that region. A time-dependent treatment based on the wavepackets discussed in the previous section does not provide exact answers in closed form. However, since it allows to view the processes of tunneling and scattering from potential barriers and wells in a complementary way, it is included in a numerical approach in which space and time are discretized in section 3.5.

Piecewise constant potentials represent an idealization of potential barriers and wells that occur in nature in all branches of physics. We have encountered in section 2.3 an example from solid state physics, namely the periodic lattice problem. In the later section 3.4 the problem of coupling of almost bound states to the scattering continuum is introduced for nuclear alpha decay, and the smoothly varying potential is idealized by a rectangular barrier. For arbitrarily shaped potential barriers or wells an approximate solution of the tunneling problem is based on an incoherent addition of amplitudes describing the scattering from infinitesimal potential steps. This is usually exploited for nuclear alpha decay, but we present it for the tunneling ionization problem for a model atom exposed to a superstrong laser field in section 3.3.

To fully appreciate the interesting results from a quantal treatment of scattering from a potential well or a barrier it is important to understand the motion of a classical particle in such a potential. The barrier of height V_0 (we use below the redefined potential $U_0 = 2mV_0/\hbar^2$) acts as a perfectly reflecting wall for particles of energy $E < V_0$. In reality no such potential exists: over some very short distance scale the potential rises rapidly to the value of V_0. The particle converts kinetic into potential energy over these distances and comes to a halt in order to reverse the energy conversion process subsequently. Over larger distance scales this complicated elastic scattering event is not visible and appears as simple reflection. For energies $E > V_0$, and for attractive potential wells the particle passes over the obstacle with a sudden change in kinetic energy T, such that the overall energy $E = T + V$ remains conserved.

In QM we solve a wave equation and should be prepared for a qualitatively different behaviour. The potential steps can be thought of as obstacles that partly reflect and partly allow the wave to permeate to some extent irrespective of whether the classical force implied by the potential step is attractive or repulsive. Potential steps bring the differences between classical and quantum propagation to a point. The investigation of quantum – classical correspondence, i.e., of Ehrenfest's theorems that show under what circumstances the time evolution of the position expectation value satisfies Hamilton's (or Newton's) equations of motion, shows that the required condition of a slowly varying potential over the extent of a localized wavepacket is violated in a maximal way by a potential step. Nevertheless, it is of interest to scrutinize the physical results for quantum scattering, namely the transmission and reflection probabilities for the correspondence with the classical limit. Some conclusions in that respect are presented at the end of the section.

We consider a potential well of depth U_0 in the range $0 < x < a$. The wavefunction is defined in three regions as psi1, psi2, and psi3. We fix the length to a numerical value of 1, but keep the well depth as a symbolic expression. The energy is fixed to a value of 1/2 and units with $\hbar = m = 1$ are chosen.

```
> a:=1;    U0:='U0';     En:=1/2;     m:=1;
```

$$a := 1 \qquad U0 := U0 \qquad En := \frac{1}{2} \qquad m := 1$$

The wavenumbers in the free and potential regions are:

```
> kf:=sqrt(2*m*En);    kU:=sqrt(2*m*En-U0);
```

$$kf := 1 \qquad kU := \sqrt{1 - U0}$$

Now we find the general solutions in the potential-free region first, and then for the well/barrier region

```
> solfree:=dsolve(diff(psi1(x),x,x)+kf^2*psi1(x)=0,psi1(x));
```

$$\mathrm{solfree} := \psi1(x) = \_C1 \cos(x) + \_C2 \sin(x)$$

```
> solint:=dsolve(diff(psi2(x),x,x)+kU^2*psi2(x)=0,psi2(x));
```

$$\mathrm{solint} := \psi2(x) = \_C1 \cos\left(\sqrt{2}\,x\right) + \_C2 \sin\left(\sqrt{2}\,x\right)$$

The solutions are not expressed as right-travelling and left-travelling plane waves (complex exponentials), which is bad as:

– (a) the interpretation of the coefficients isn't obvious;
– (b) in domain 3 we won't be able to pick just right-travelling waves.

Thus, we convert the solutions to exponential form:

```
> convert(solfree,exp);
```

$$\psi1(x) = \_C1 \left(\frac{1}{2}\,e^{(I\,x)} + \frac{1}{2}\,\frac{1}{e^{(I\,x)}}\right) - \frac{1}{2}\,I\,\_C2 \left(e^{(I\,x)} - \frac{1}{e^{(I\,x)}}\right)$$

```
> evalc(");
```

$$\psi 1(x) = \_C1\left(\frac{1}{2}\cos(x) + \frac{1}{2}\frac{\cos(x)}{\%1}\right) + \frac{1}{2}\_C2\left(\sin(x) + \frac{\sin(x)}{\%1}\right)$$
$$+ I\left(\_C1\left(\frac{1}{2}\sin(x) - \frac{1}{2}\frac{\sin(x)}{\%1}\right) - \frac{1}{2}\_C2\left(\cos(x) - \frac{\cos(x)}{\%1}\right)\right)$$
$$\%1 := \cos(x)^2 + \sin(x)^2$$

This looks like a mess, so let's do it by hand. Presumably a `combine(",trig)` command will help, but also the meaning of the integration constants has to change.

```
> psi1:=x->C1*exp(I*kf*x) + C2*exp(-I*kf*x);
```

$$\psi 1 := x \to C1\, e^{(I\, kf\, x)} + C2\, e^{(-I\, kf\, x)}$$

```
> psi2:=x->C3*exp(I*kU*x) + C4*exp(-I*kU*x);
```

$$\psi 2 := x \to C3\, e^{(I\, kU\, x)} + C4\, e^{(-I\, kU\, x)}$$

```
> psi3:=x->C5*exp(I*kf*x);
```

$$\psi 3 := x \to C5\, e^{(I\, kf\, x)}$$

Now let us match the parts of the wavefunction, which can be verified to satisfy the Schrödinger equation in the different regions by appropriate substitutions. There are four conditions to be satisfied, namely continuity and differentiability of the wavefunction at the boundaries $x = 0$ and $x = a$, but five coefficients. The incoming flux (`C1`) is normalized to unity:

```
> C1:=1: C2:='C2': C3:='C3': C4:='C4': C5:='C5':
```

```
> sol:=solve({psi1(0)=psi2(0),psi2(a)=psi3(a),D(psi1)(0)=
> D(psi2)(0),D(psi2)(a)=D(psi3)(a)},{C2,C3,C4,C5});
```

$$\mathrm{sol} := \left\{ C2 = -\frac{U0\left(\left(e^{(I\sqrt{1-U0})}\right)^2 - 1\right)}{\%1}, \right.$$
$$C3 = 2\frac{\left(-2 - 2\sqrt{1-U0} + U0\right)\left(-1 + \sqrt{1-U0}\right)}{\%1\; U0},$$
$$C4 = 2\frac{\left(e^{(I\sqrt{1-U0})}\right)^2\left(-1 + \sqrt{1-U0}\right)}{\%1},$$

$$C5 = 2\left(-1+\sqrt{1-U0}\right)\mathrm{e}^{\left(I\sqrt{1-U0}\right)}$$
$$\left(-2-2\sqrt{1-U0}+U0+\mathrm{e}^{\left(I\sqrt{1-U0}\right)}\mathrm{e}^{\left(-I\sqrt{1-U0}\right)}U0\right)/(\%1\,U0\,\mathrm{e}^{I})\Big\}$$
$$\%1 := -2\left(\mathrm{e}^{\left(I\sqrt{1-U0}\right)}\right)^2+2\sqrt{1-U0}\left(\mathrm{e}^{\left(I\sqrt{1-U0}\right)}\right)^2$$
$$+\left(\mathrm{e}^{\left(I\sqrt{1-U0}\right)}\right)^2 U0+2+2\sqrt{1-U0}-U0$$

```
> assign(sol);     simplify(C2);
```

$$\frac{\left(-\mathrm{e}^{\left(2I\sqrt{1-U0}\right)}+1\right)U0\left(-2-2\sqrt{1-U0}+U0\right)}{\mathrm{e}^{\left(2I\sqrt{1-U0}\right)}U0^2+8\,U0-8-8\sqrt{1-U0}+4\sqrt{1-U0}\,U0-U0^2}$$

This is indeed a complex number. Let's simplify for a numerical example:

```
> evalf(subs(U0=-1,"));
```

$$-.3261083486+.04853996148\,I$$

```
> simplify(C3);
```

$$\frac{6\,U0-8-8\sqrt{1-U0}+2\sqrt{1-U0}\,U0}{\mathrm{e}^{\left(2I\sqrt{1-U0}\right)}U0^2+8\,U0-8-8\sqrt{1-U0}+4\sqrt{1-U0}\,U0-U0^2}$$

```
> evalf(subs(U0=-1,"));
```

$$.8057959286+.007108512781\,I$$

```
> simplify(C4);
```

$$2\,\frac{\mathrm{e}^{\left(2I\sqrt{1-U0}\right)}U0\left(1+\sqrt{1-U0}\right)}{\mathrm{e}^{\left(2I\sqrt{1-U0}\right)}U0^2+8\,U0-8-8\sqrt{1-U0}+4\sqrt{1-U0}\,U0-U0^2}$$

```
> evalf(subs(U0=-1,"));
```

$$-.1319042772+.04143144868\,I$$

```
> simplify(C5);
```

$$\frac{\mathrm{e}^{\left(I\left(-1+\sqrt{1-U0}\right)\right)}\left(-8\sqrt{1-U0}-8+8\,U0+4\sqrt{1-U0}\,U0\right)}{\mathrm{e}^{\left(2I\sqrt{1-U0}\right)}U0^2+8\,U0-8-8\sqrt{1-U0}+4\sqrt{1-U0}\,U0-U0^2}$$

```
> evalf(subs(U0=-1,"));
```

$$.8608615771+.3875752338\,I$$

We find that the solution is completely determined and that the coefficients are complex numbers. C2 has the interpretation of the amplitude for the backscattered flux. Its magnitude squared represents the reflection coefficient (probability). C3 and C4 determine the wavefunction over the well, while C5 is the transmission amplitude. Let's check unitarity:

```
> conjugate(C2)*C2+conjugate(C5)*C5: simplify(subs(U0=-1,"));
```

$$\frac{\left(2\cos\left(2\sqrt{2}\right)-34\right)\left(-17-12\sqrt{2}\right)}{2\left(-17-12\sqrt{2}\right)\cos\left(2\sqrt{2}\right)+578+408\sqrt{2}}$$

```
> evalf(");
```

$$1.000000000$$

That isn't quite enough for what we should expect from a computer algebra system, but let's accept it for now. We are interested in the position-dependent density $w(x) = \psi^*(x)\psi(x)$. A single expression for the wavefunction can be generated by composing the three pieces with the step function. Strictly speaking, the density $w(x)$ can only be interpreted as a probability density if box normalization is imposed, but qualitatively we obtain the correct features without this noramlization. The lengthy output below is suppressed.

```
> w:=subs(U0=-1,Heaviside(-x)*abs(psi1(x))^2+Heaviside(x)*
> Heaviside(a-x)*abs(psi2(x))^2+Heaviside(x-a)*abs(psi3(x))^2):
```

```
> plot(w,x=-5..5,0..2,title=`probability density:
> E=1/2, U=-1, a=1`);
```

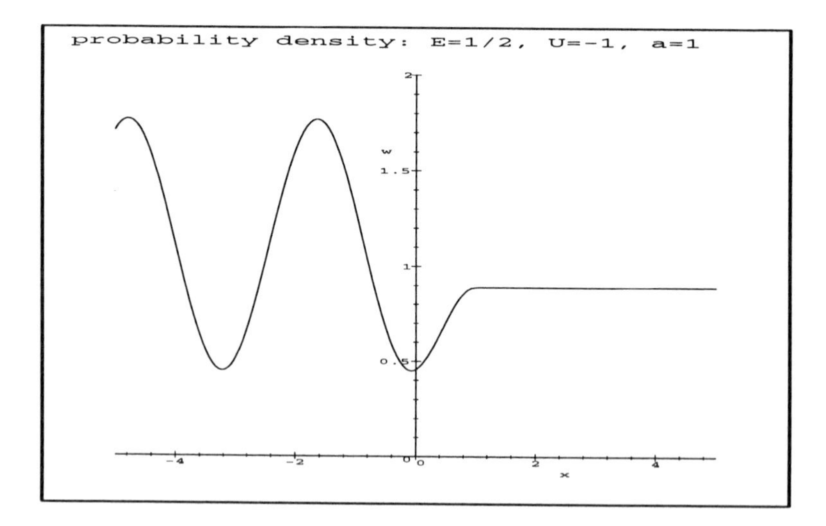

What do we learn from this graph?

– (1) In the region behind the well only a right-travelling wave $T\exp(ik_f x)$ is present. The apparent constant has the value T^*T, i.e., it is equal to the transmission probability t. The complement of t to 1 gives the reflection probability $r = 1 - t$.
– (2) In region 1 we have the sum of an incident and backscattered wave. The reflection coefficient cannot be easily read off. The graph shows a typical standing-wave solution in this range.
– (3) In the region $0 < x < a$ the wavefunction interpolates between the two asymptotic pieces.

The following question arises: can we obtain the coefficients (and wavefunction) as a function of one of the scattering parameters?

We define a function for the probability density and use `animate` to display the probability density for a varying well depth U.

```
> rho:=(x,U)->subs(U0=-U,Heaviside(-x)*abs(psi1(x))^2+
> Heaviside(x)*Heaviside(a-x)*abs(psi2(x))^2+
> Heaviside(x-a)*abs(psi3(x))^2):
```

```
> with(plots):      animate(rho(x,u),x=-5..5,u=0..3,
> title='well depth from 0..3 ');
```

Let us observe how the transmission behaves as a function of increasing attraction in the well. We notice that an attractive well can have a strong reflection and weak transmission which cannot be reconciled with classical mechanics. Observe the large wavenumber inside the well.

```
> plot(rho(x,10),x=-5..5,title='prob. dens.: E=1/2,U=-10,a=1');
```

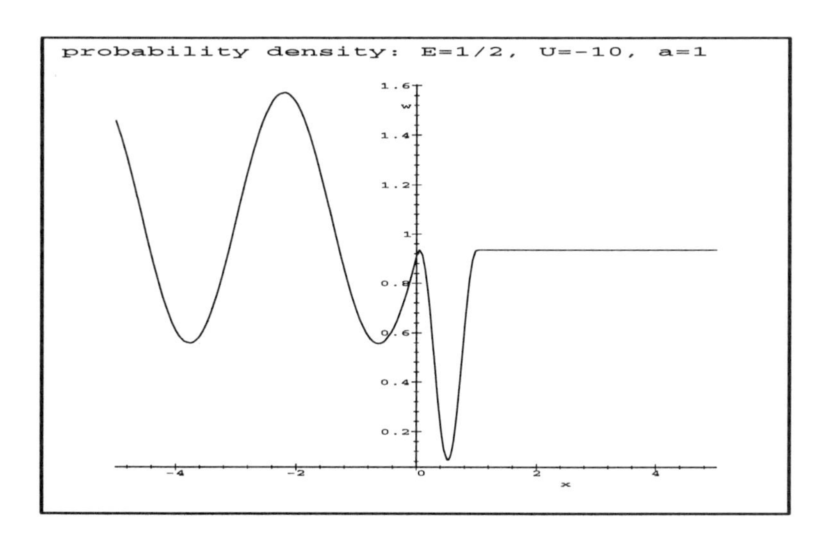

So let's look at the transmission coefficient as a function of well depth:

```
> simplify(conjugate(C5)*C5):  tcof:=unapply(",U0):
```

Unfortunately the evaluated expression keeps a small imaginary part at the level of numerical precision. Before plotting it we have to truncate to real-valued numbers due to numerical inaccuracy in the floating-point arithmetic.

```
> plot(Re(tcof(x)),x=-50..10,t=0..1,title='transmission
> probability for E=1/2 for well/barrier height x');
```

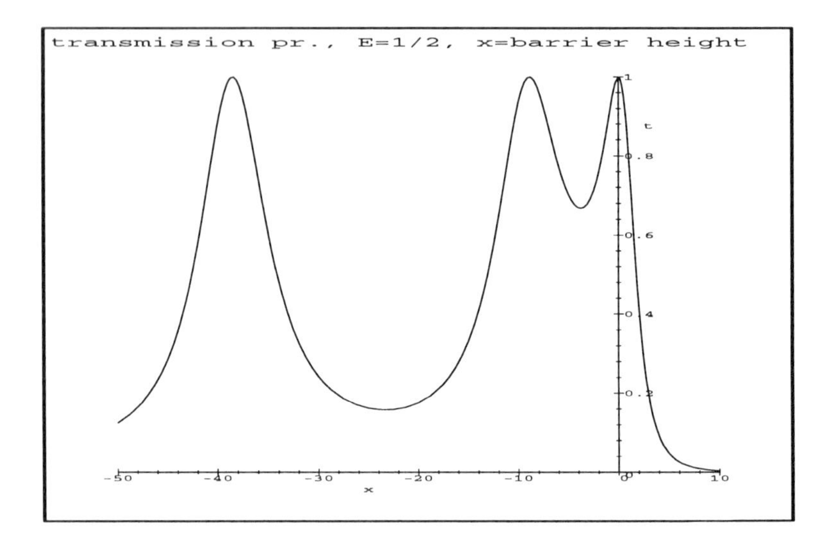

This graph contains a lot of exciting quantum physics:

- (1) $x > 0$: scattering from a potential barrier. Note that a sizeable transmission probability persists beyond the classically allowed region $x = 1/2 = E$! For large barrier height x there is an observable tunneling probability.
- (2) $x < 0$: scattering from a potential well. The transmission dips as the well becomes more attractive. Classically this makes no sense at all, a particle shouldn't be backscattered from an attractive force. Then, as the well depth is increased there are narrow regions for which the well is completely translucent to be followed by wider regions of strong reflectivity.

The strange behaviour for strongly attractive wells can be understood in QM as a wave propagation phenomenon. At $x = 0$ the incident wave matches onto a wave with a high wavenumber (momentum) experiencing both transmission and reflection. The high-momentum wave inside the well scatters at $x = a$ from a high barrier, experiencing mostly reflection. For certain well depths at fixed incident energy E and well length a the transmitted and reflected waves interfere in such a way as to allow perfect transmission.

Exercise 3.2.1: Investigate the transmission probability as a function of the scattering energy for given potential parameters. Explore the spatial probability density in the vicinity of a point where the transmission is perfect, e.g., for fixed scattering energy and width of the well as a function of well depth.

3.3 Tunneling Ionization

We consider a simple 1D model atom described by a potential discussed in Sect. 1.4. The potential has some of the features of the 3D Coulomb potential. The long range causes the spectrum to have a Rydberg series of bound excited states. We immerse the atom in a laser field described in dipole approximation classically by the potential function

$$V(x,t) = E_0 x \sin(\omega t)\,.$$

Here ω is the basic frequency, the pulse is considered to be long enough for us to ignore its shape (turn-on and turn-off effects would be described by $E_0(t)$); E_0 is the peak field strength. The fundamental laser frequency ω is assumed to be small on the scale of atomic energies. This is certainly true at infrared wavelengths and permits us to study an ionization mechanism for which the time evolution of the field strength is irrelevant.

For the tunneling ionization process we consider only those time segments for which $|\sin(\omega t)| \approx 1$. During those times the overall potential forms an ideal scenario for tunneling through a barrier either to the left or to the right of the 1D atom, depending on the sign in the sine function. If the barrier is small enough and the laser frequency low enough, the amount of probability density that can escape through the barrier can be sufficiently large for this to become the dominating photoionization mechanism. Whether the barrier is low enough depends primarily on the laser intensity.

The relationship between the laser intensity I_0 and the electric field strength is given by $I_0 = E_0^2/(8\pi)$. We choose E_0 in atomic units, and are interested in the high-intensity regime, in which dependencies on E_0 show up. In ordinary tabletop lasers the intensity is not strong enough to show dramatic E_0 dependencies in measurable quantities, and tunneling ionization is unimportant. Ionization proceeds through the absorption of sufficiently many photons to reach the continuum part of the atomic spectrum, i.e., electron energies $E > 0$.

As the intensity (field strength) of the laser increases to very high levels, the barrier may disappear altogether for periods of time. In this case one calls the process 'above-the-barrier' ionization.

We define the overall potential during the special phase of the laser pulse:

```
> V:=(x,E0)->-1/sqrt(1+x^2)-E0*x;
```

$$V := (x, E0) \to -\frac{1}{\mathrm{sqrt}(1+x^2)} - E0\,x$$

A plot of the potential together with the approximate ground-state energy level reveals the tunneling problem:

```
> plot({-0.61,V(x,0.05)},x=-5..15);
```

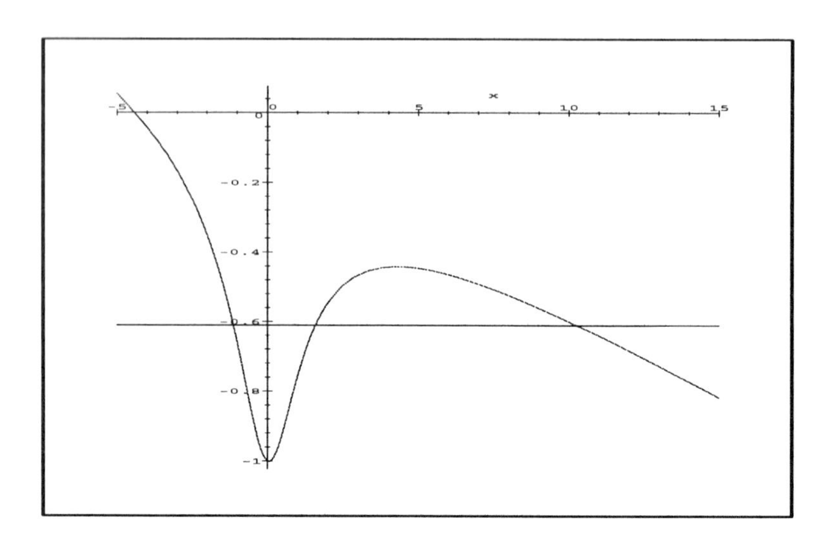

The graph shows the potential at a point in time when $\sin(\omega t)$ takes on the value of -1 and the potential offers to the ground state the possibility for tunneling through a finite barrier. We wish to explore how the tunneling probability depends on the field strength. An approximate treatment of the tunneling mechanism is required as we do not wish to solve the Schrödinger equation numerically.

First we determine the classical turning points for $x > 0$:

```
> eq:=V(x,E0)=-0.61;
```

$$eq := -\frac{1}{\sqrt{1+x^2}} - E0\,x = -.61$$

```
> solve(eq,x):
```

We are out of luck as far as finding a simple general answer. The call `allvalues(")` creates a number of huge expressions and thus we attempt a numerical solution for given $E0$:

```
> fsolve(subs(E0=0.05,eq),x);
```

$$1.599994368$$

There has to be another solution:

```
> fsolve(subs(E0=0.05,eq),x=1.7..50);
```

$$10.25984703$$

Now we program the approximate formula while making use atomic units with $\hbar = m_e = e = 1$:

```
> Tunnel:=(x1,x2,E0,epsilon)->exp(-2*Int(abs(2*V(x,E0)-epsilon)
> ,x=x1..x2));
```

$$\mathrm{Tunnel} := (\, x1 \,, x2 \,, E0 \,, \varepsilon \,) \to \mathrm{e}^{\left(-2 \int_{x1}^{x2} |2\,\mathrm{V}(\, x, E0 \,) - \varepsilon|\, dx\right)}$$

This expression takes into account the shape of the barrier. It can be thought of as arising from a sequence of infinitesimal potential steps summed over in the integral. The formula used originally by Gamow to calculate nuclear α decay ignores, however, in the composition of the tunneling probabilities for the individual steps any interference phenomena. These would arise in an exact treatment from matching the wavefunctions at the infinitely many matching points. We rely on the fact that the formula reflects the exponential decrease of the tunneling probability with increasing barrier height and depth.

For a test, evaluate the case of $E_0 = 0.05$ a.u. with the classical turning points determined above.

```
> evalf(Tunnel(1.6,10.26,0.05,-0.61));
```

$$.001108560773$$

We wish to explore how the answer changes as a function of E_0. We have to calculate the classical turning points each time.

```
> for i from 1 to 7 do:

> E0i:=0.01*i;

> tp1:=fsolve(subs(E0=E0i,eq),x);

> tp2:=fsolve(subs(E0=E0i,eq),x=tp1+0.01..50000);

> Tprob[i]:=evalf(Tunnel(tp1,tp2,E0i,-0.61));

> #print(E0i,");     od;
```

$$E0i := .01 \quad tp1 := 1.345 \quad tp2 := 59.314$$

$$\mathrm{Tprob}_1 := .236\,10^{-11}$$

$$E0i := .02 \quad tp1 := 1.397 \quad tp2 := 28.763$$

$$\mathrm{Tprob}_2 := .59403\,10^{-5}$$

$$E0i := .03 \quad tp1 := 1.455 \quad tp2 := 18.538$$

$$\mathrm{Tprob}_3 := .00008050$$

$$E0i := .04 \quad tp1 := 1.522 \quad tp2 := 13.388$$

$$\mathrm{Tprob}_4 := .00033378$$

$$E0i := .05 \quad tp1 := 1.600 \quad tp2 := 10.260$$

$$\mathrm{Tprob}_5 := .00110876$$

$$E0i := .06 \quad tp1 := 1.694 \quad tp2 := 8.133$$

$$\mathrm{Tprob}_6 := .00328827$$

$$E0i := .07 \quad tp1 := 1.812 \quad tp2 := 6.562$$

$$\mathrm{Tprob}_7 := .00937709$$

Observe how the location of the turning points changes slowly with the field strength. Nevertheless, the increase in the tunneling probability is dramatic. This is a reflection of the exponential dependence of the tunneling probability on the magnitude of the potential barrier. We graph this result to make it obvious. Note that the tunneling probability rises very steeply as the electric field strength approaches less than 10 % of the average strength of the Coulomb field in the model atom.

```
> plot([[0.01*j,(Tprob[j])] $j=1..7],E0=0..0.1);
```

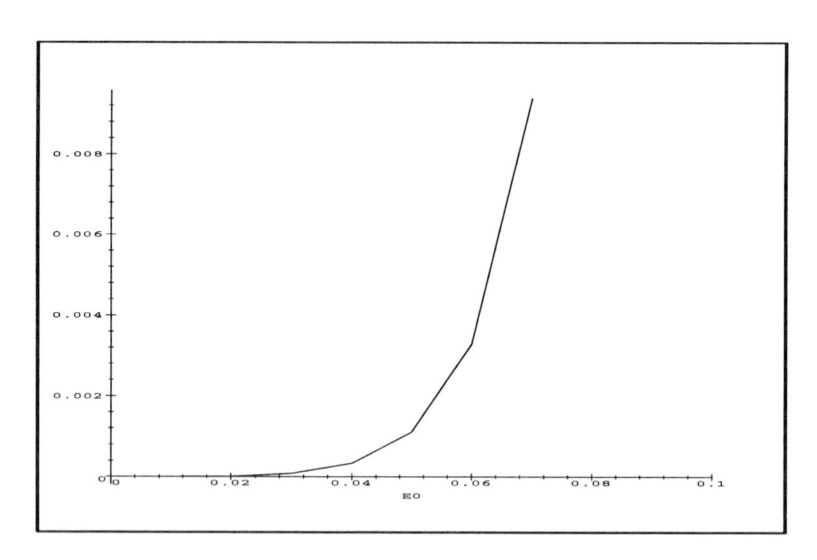

A log-lin plot reveals better the strong variation with E_0:

```
> plot([[0.01*j,log10(Tprob[j])] $j=1..7],E0=0..0.1,style=
> point,title='log of the tunneling probability');
```

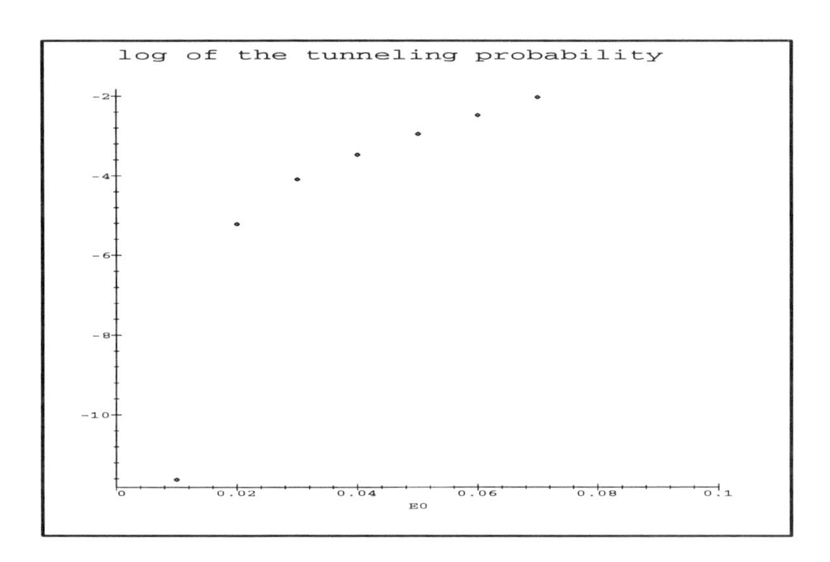

Note that less than one hundred cycles of a laser pulse with a peak strength of $E_0 = 0.07$ are sufficient to convert the full sample of neutral atoms into the singly ionized state.

During each cycle the electron has the possibility to tunnel to the right when $\sin(\omega t) = -1$ and to the left when $\sin(\omega t) = 1$. Due to the sensitivity of the tunneling probability it is not important to consider the times in-between. Also note that the electric field strength has stationary points (as a function of time) at these points in time. This is, in fact, observed experimentally: there are critical values of the peak laser intensity at which a certain charge state appears [AS+89].

Exercise 3.3.1: Observe how the tunneling ionization rate changes as the atomic charge is changed from the hydrogenic value $Z = 1$ to an effective charge $Z_{\text{eff}} = 1.6$ that reflects the binding strength of a He(1s) electron, i.e., consider the atomic model potential $V(x) = -1.6/\sqrt{1+x^2}$. Obtain the approximate ground-state energy variationally (cf. Sect. 1.4) and observe that it does not scale as Z^2, as is true for the 3D hydrogen atom.

Exercise 3.3.2: Multiphoton ionization becomes an effective alternative ionization mechanism for not too short laser pulses, especially when the energy gap $E_1 - E_0$ ($E_{2\text{p}} - E_{1\text{s}}$ in a real hydrogen atom) is tuned into resonance with an integer multiple n of $E_\gamma = \hbar\omega$. Ionization proceeds after the excited state has been reached via n photon absorption either through absorption of an additional photon, or by tunneling from ψ_1. Estimate tunneling ionization rates from the first excited state for the 1D hydrogenic model and compare them to the calculated ionization rates from ψ_0.

3.4 Alpha Decay

Nuclear alpha decay can be calculated approximately as was done for tunneling ionization of a model atom in the preceding section and leads to the Gamow formula. Alternatively it can be calculated accurately, but for a simple potential. We follow the latter approach in this section, i.e, we return to the problem of a simple potential step.

We use units with $\hbar = m = 1$, and consider a potential that is infinite at $x = 0$, zero from $x = 0..a$, V_0 from $x = a..b$, and zero beyond. Such a geometry arises in the radial distance r when a spherically symmetric problem is separated in spherical polar coordinates. The shape of a potential barrier between $x = a$ and $x = b$ represents a gross simplification of the following independent nucleon model: at short distances the nucleons (protons and neutrons) experience a strong attraction. At intermediate to large distances (outside the nucleus) the protons repel each other, which (together with a centrifugal contribution) forms a barrier that is also in effect for simple clusters of nucleons, such as alpha particles.

The alpha particle consists of two protons and two neutrons (a ${}_2\mathrm{He}^4$ nucleus) that can form in the potential well. If that occurs, the nuclear helium binding energy is gained and the cluster winds up in an excited state that is quasi-bound. It is above the particle break-up threshold, but below the top of the Coulomb barrier. It can, however, tunnel through the barrier and the emerging alpha particle is subsequently accelerated by the Coulomb repulsion. Its kinetic energy is a precise indicator of the position of the quasi-bound level.

Alpha decay in this simple model can be viewed as the occurrence of two processes:

- (1) an α cluster forms inside the nucleus and occupies a quasi-bound or resonant level;
- (2) the α cluster tunnels through the Coulomb barrier.

An extremely wide range of lifetimes of alpha emitters can be explained on the basis of this simple model. The lifetimes vary a great deal due to the strong sensitivity of the tunneling probability to the size of the barrier and to the energy of the decaying quasi-bound level. Our interest here is just to investigate precisely a very simple mathematical model that contains such quasi-bound levels. The emphasis is to show how originally bound levels in an isolated potential well become scattering resonances once the well is coupled through a barrier to a continuum of asymptotically free states.

Such a problem appears in different places in modern physics. An obvious example is the problem of positronium states: at the level of simple quantum mechanics they are stable bound states of an electron and a positron. Yet, when coupled to the two-photon continuum (or to three-photon or multiphoton continua) positronium becomes unstable and decays [DHK92]. Similar considerations hold for excited states of atoms [XDH92].

We choose fixed numerical parameters for simplicity:

```
> a:=1;      b:=2;      V0:=4;
```

$$a := 1 \qquad b := 2 \qquad V0 := 4$$

```
> Vpot:=x->V0*(Heaviside(x-a)*Heaviside(b-x)):
```

```
> plot(Vpot(x),x=0..4,V=0..5,title='resonance problem
> in alpha decay');
```

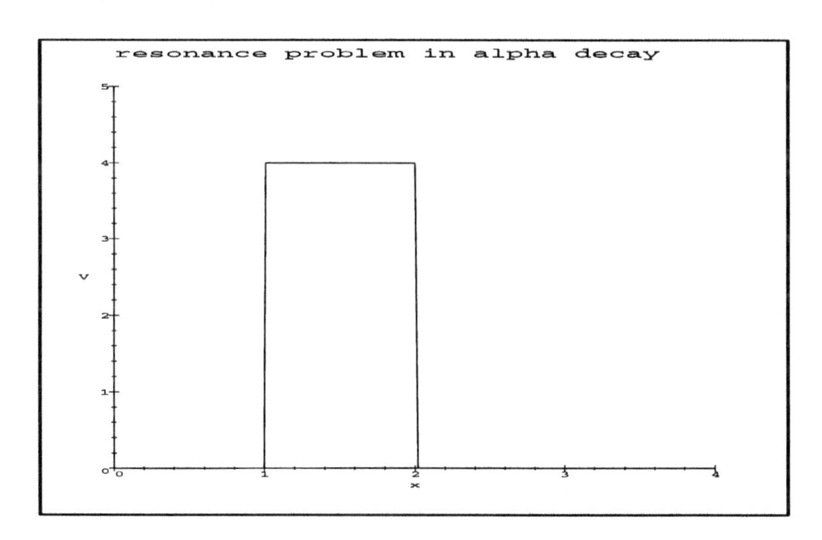

The free wavenumber and the one valid above the constant barrier are given as

```
> ks:=sqrt(2*En);      qs:=sqrt(2*(V0-En));
```

$$ks := \sqrt{2}\,\sqrt{En} \qquad qs := \sqrt{8 - 2\,En}$$

The wavefunction is defined over the three regions from left to right as follows. In region 1 we impose anti-symmetric boundary conditions (a radial orbital that vanishes at $r = 0$):

```
> psi1:=x->A*sin(ks*x);
```

$$\psi 1 := x \to A \sin(\, ks\, x\,)$$

In region 2 we define a function inside the barrier that consists of an exponentially attenuated as well as an increasing part:

```
> psi2:=x->B1*exp(qs*(x-a))+B2*exp(-qs*(x-a));
```

$$\psi 2 := x \to B1\, e^{(\,qs\,(x-a)\,)} + B2\, e^{(\,-qs\,(x-a)\,)}$$

In region 3 we consider incoming and outgoing free waves (left-travelling and right-travelling) with the incoming (left-travelling) flux normalized to unity.

```
> psi3:=x->1*exp(-I*ks*(x-b))+C*exp(I*ks*(x-b));
```

$$\psi 3 := x \to e^{(-I\, ks\, (x-b))} + C\, e^{(I\, ks\, (x-b))}$$

The outgoing flux is also normalized to unity($|C|^2 = 1$), as the wave is fully reflected at $x = 0$. Observe that for an impenetrable barrier of infinite height the inner well has discrete eigenstates determined by the condition $\sin(ka) = 0$. We have four matching conditions to guarantee continuity and differentiability of the overall wavefunction.

```
> A:='A': B1:='B1': B2:='B2': C:='C':
```

```
> eq1:=simplify(psi1(a)-psi2(a));
```

$$\mathrm{eq1} := A \sin\left(\sqrt{2}\,\sqrt{En}\right) - B1 - B2$$

```
> eq2:=simplify(D(psi1)(a)-D(psi2)(a));
```

$$\mathrm{eq2} := A \cos\left(\sqrt{2}\,\sqrt{En}\right)\sqrt{2}\,\sqrt{En} - B1\,\sqrt{8-2\,En} + B2\,\sqrt{8-2\,En}$$

```
> eq3:=simplify(psi2(b)-psi3(b));
```

$$\mathrm{eq3} := B1\, e^{\left(\sqrt{8-2\,En}\right)} + B2\, e^{\left(-\sqrt{8-2\,En}\right)} - 1 - C$$

```
> eq4:=simplify(D(psi2)(b)-D(psi3)(b));
```

$$\mathrm{eq4} := -\frac{1}{2} I \left(I\,\sqrt{2}\,\sqrt{En}\, B1\,\sqrt{8-2\,En}\, e^{\left(\sqrt{8-2\,En}\right)} - I\,\sqrt{2}\,\sqrt{En}\, B2\,\sqrt{8-2\,En}\, e^{\left(-\sqrt{8-2\,En}\right)} - 2\,En + 2\,C\,En\right)\sqrt{2}\,/\sqrt{En}$$

```
> sol:=solve({eq1,eq2,eq3,eq4},{A,B1,B2,C}):
```

The omitted printout looks like a big mess. Nevertheless, we continue in good faith. We assign the solution to the coefficients:

```
> assign(sol);
```

```
> A:=simplify(A):
```

While the (deleted) half-page answer is readable now, it is still not in very useful form. We satisfy ourselves with numerical results below.

```
> evalf(abs(subs(En=1,C))^2);
```

$$1.000000002$$

One should be able to prove for any (real, positive) En that $|C|^2 = 1$, i.e., as much flux comes out as goes in. Of interest to us is the variation of the amplitude of the wavefunction inside the well as a function of energy:

```
> #plot(abs(A),En=0..2*V0);
```

This commented command takes forever to execute, so let's evaluate a sequence of points. First we concentrate on energies below the top of the barrier $E < V_0$.

In anticipation of what's to come we have picked a potential well that, when isolated from the asymptotic continuum, supports a bound state. Remember that the boundary condition $\psi(0) = 0$ corresponds to the antisymmetric solutions in the symmetrical-well problem and that these solutions appear only beyond a certain value of the parameter $V_0 a^2$, namely for

$$2V_0 a^2 \geq \frac{\pi^2}{4}.$$

We have, therefore, chosen $V_0 = 4$, $a = 1$; for a barrier of infinite height the first bound-state energy eigenvalue would be given by $\sin(k_1 a) = 0$, i.e., $k_1 = \pi/a$, or $E_1 = \pi^2/(2a^2) > 4.5$. However, the lower barrier brings the eigenvalue down towards 2.5.

We plot the squared magnitude of the coefficient for the inside part of the wavefunction ψ_1 as a function of free-particle energy in the asymptotic region. Given that the incoming flux was normalized to unity this carries information about the relative probability as a function of energy to form a bound state inside the well. We leave it as an exercise to explore in detail the shape of the function Asq(En) near the critical energy. In 3D scattering the associated phenomenon leads to a resonance which is discussed in some detail in section 6.2.

Resonances play a particularly important role in elementary particle physics, where many new discoveries are observed as unstable resonances in electron – positron scattering at energies in the GeV range. The highly energetic electron and positron (or other particle-antiparticle pairs) convert into photons which can form, e.g., a quark-antiquark resonance, i.e., a meson if the energy fits[PDG92].

```
> for i from 1 to 50 do

> Eni[i]:=0.95*V0/50*i;

> Asq[i]:=abs(evalf(subs(En=Eni[i],A)))^2;

> #print(Eni[i],Asq[i]);

> od:

> plot([seq([Eni[j],Asq[j]],j=1..50)],En=0..V0,Asq=0..150,
> style=point,title=`Relative prob. for particle in well`);
```

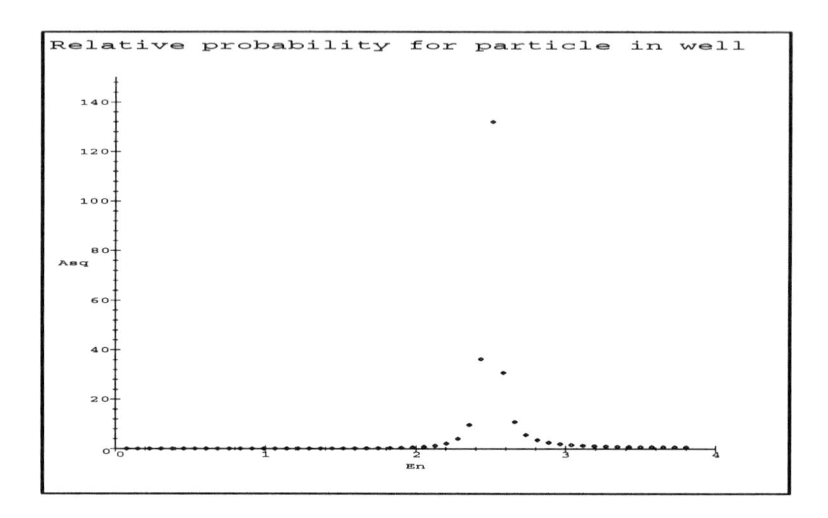

We see that something very curious happens near $En = 2.5$: The squared magnitude of the wavefunction inside the well increases dramatically as the scattering energy approaches the vicinity of the energy level that the well would support if the barrier was of infinite extension. The behaviour can be compared to a damped harmonic oscillator: when excited near its internal frequency it goes into large-amplitude motion. This behaviour is called a resonance.

One can study this problem as a function of the barrier height. Additional resonances appear as the pure bound-state problem, i.e., the problem without coupling to the scattering channel, acquires more eigenvalues. For the scattering problem the sharp structures in the energy spectrum are identified as virtual bound states.

One can find in real 3D settings that scattering is significantly modified by the appearance of resonances with a strong redistribution of flux as a function of the scattering angles. In addition, a delay in the scattering process occurs as a function of time: an incoming particle gets captured, forms a quasi-stationary state, which eventually decays by tunneling through the barrier.

This latter process can be calculated in the following inexact, but intuitive way: initially we pretend that the barrier is infinitely long and a true bound state exists. Given a wave function for this state (which is not an eigenstate of the exact problem with finite barrier) we then estimate the tunneling probability. Tunneling implies a time delay T as the wavepacket requires time to penetrate the barrier.

A decay width Γ, which is an energy, can be defined such that

$$\Gamma T = \hbar .$$

This width can be shown to measure the extent of the energy structure in the scattering problem (see previous figure) in an expression that unlike our first figure has a well-defined maximum, location, and half value that defines the width. Alternatively, the width can be obtained as an imaginary part of the energy eigenvalue in a formulation following the approximate decay time analysis using Gamow states. A detailed presentation can be found in S. Flügge's discussion of problem 15 in [F171].

We can construct an initial packet in the well by a superposition of plane-wave solutions. Each plane-wave component has its own time evolution factor that depends on the energy. Inside the well we have

$$\psi(x,t) = \sum_{n=1}^{\infty} c_n \sin(p_n x) \exp(-\mathrm{i}E_n t/\hbar)\,.$$

At $t = 0$ one should be able to pick an eigenstate of the infinitely long barrier problem, i.e., the sum would be saturated by a single term. Let's look at the wavefunction for $E = 2.5$:

```
> wavef:=x->Heaviside(x)*Heaviside(a-x)*psi1(x)+Heaviside(x-a)
> *Heaviside(b-x)*psi2(x)+Heaviside(x-b)*psi3(x);
```

$$\text{wavef} := x \to \text{Heaviside}(\,x\,)\,\text{Heaviside}(\,a-x\,)\,\psi1(\,x\,) + \text{Heaviside}(\,x-a\,)\,\text{Heaviside}(\,b-x\,)\,\psi2(\,x\,) + \text{Heaviside}(\,x-b\,)\,\psi3(\,x\,)$$

```
> plot(abs(evalf(subs(En=2.5,wavef(x))))^2,x=0..5,rho=0..150);
```

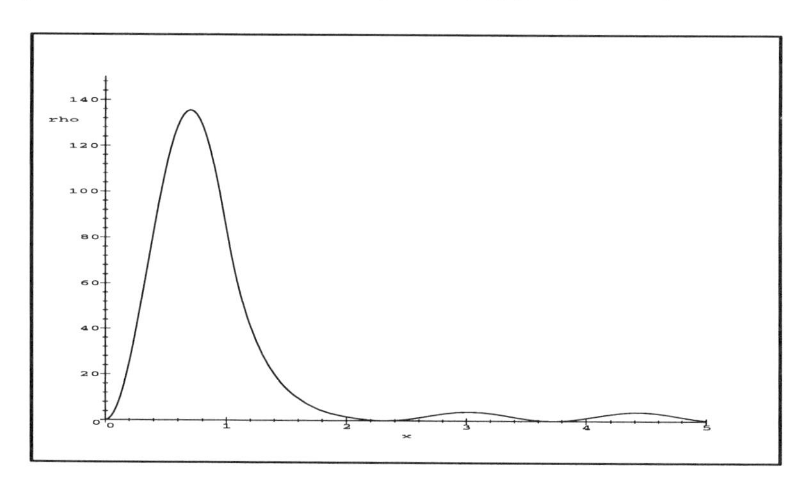

It can be seen that the wavefunction leaks considerably into the barrier, beginning at $x = 1$, and has a much larger amplitude inside the barrier than outside.

We also study what happens at energies above the barrier height. We set up a loop as before to calculate the squared magnitude of the coefficient in ψ_1, i.e., the bound part of the wavefunction.

```
> for i from 1 to 50 do
> Enia[i]:=V0+2*V0/50*i;
> Asqa[i]:=abs(evalf(subs(En=Enia[i],A)))^2;
> #print(Enia[i],Asqa[i]);   od:
> plot([seq([Enia[j],Asqa[j]],j=1..50)],En=V0..3*V0,Asq=0..6
> ,style=point,title=`Relative prob. for particle in well`);
```

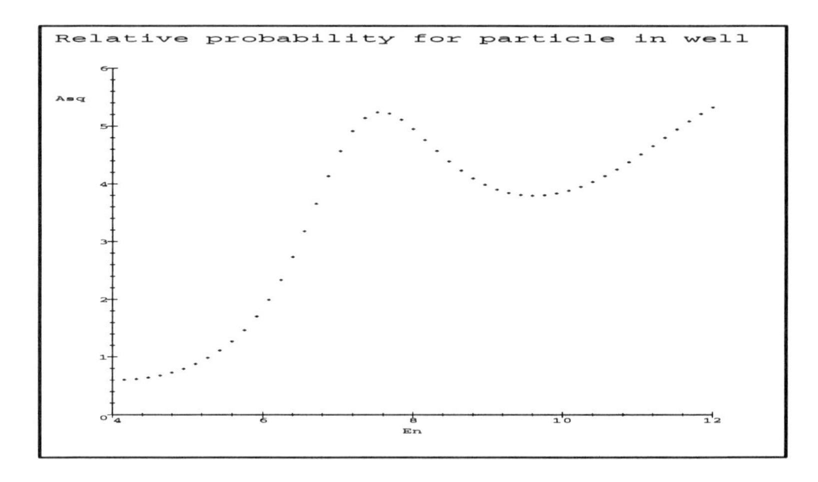

Clearly there are also structures for energies above the barrier height. Let's look at the density at an energy for which a large value of $|A|^2$ occurs:

```
> plot(abs(evalf(subs(En=7.5,wavef(x))))^2,x=0..5,rho=0..10,
> title=`E=7.5`);
```

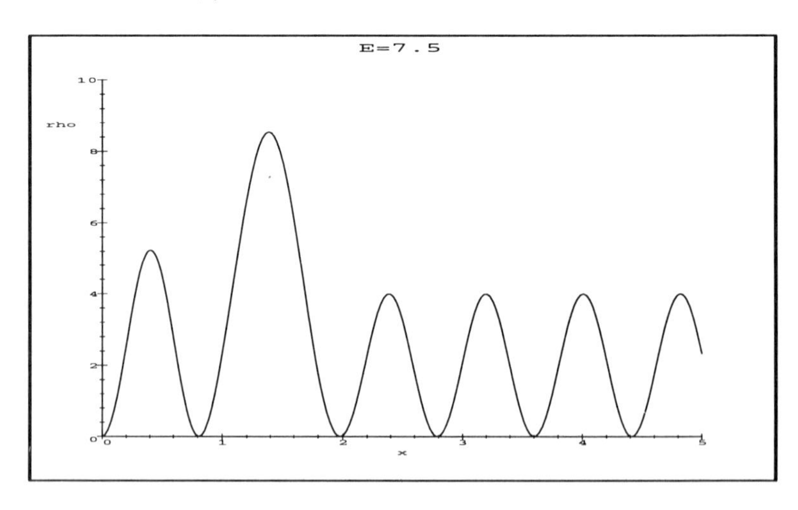

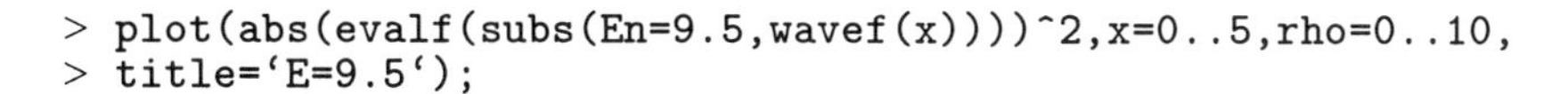

```
> plot(abs(evalf(subs(En=9.5,wavef(x))))^2,x=0..5,rho=0..10,
> title='E=9.5');
```

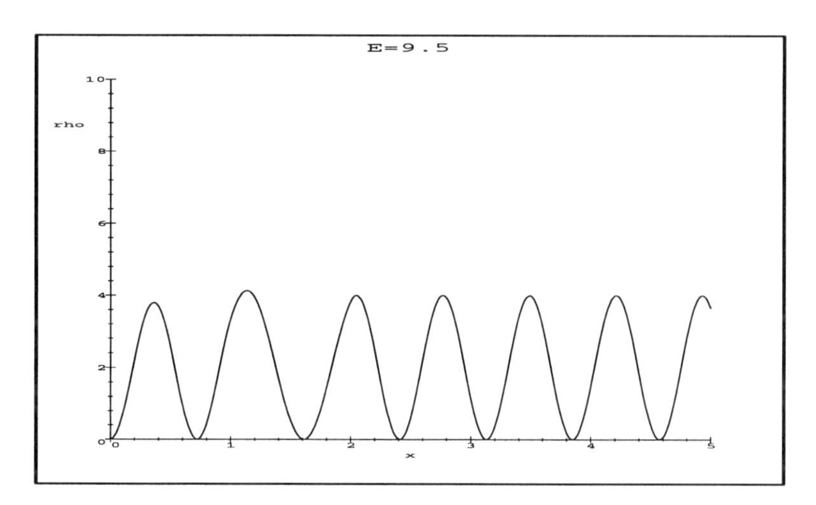

We observe that the wavefunction's amplitude inside the well compared to that outside varies if the scattering energy is changed.

An interesting interpretation of the resonance phenomenon arises in a time-dependent scattering picture based on wavepacket states in the asymptotic region, as opposed to stationary scattering states. In this picture one can interpret resonances as follows. First consider an alpha decay of a parent nucleus to a daughter nucleus and the formation of an alpha particle of fixed energy E_0. Now consider the reverse process: let an alpha particle collide with energy near E_0 with the daughter nucleus. What will happen? There is a good chance that the alpha particle will be gobbled up by the daughter nucleus to form a parent nucleus in the quasi-bound (resonant) state. This state will decay after the decay time determined by the tunneling rate (with a probability distribution). Thus, a considerable time delay occurs in the scattering process with a sticking time inversely proportional to the resonance width Γ.

Exercise 3.4.1: Obtain graphs for the energy-dependent coefficients of the inner-region wavefunction ψ_1, i.e., $|A|^2$, for cases where the barrier is twice as wide, i.e., $a = 2$, and has the same height. Repeat the same for the same width, but double the height. Find an interpretation of your findings, both in terms of the tunneling time, as well as in terms of the width of the energy structure.

3.5 Motion Picture

In this section we consider the numerical solution of the time-dependent Schrödinger equation (TDSE) by a discretization of space and time. The algorithm was designed by P.B. Visscher and is described in [Vi91]. We consider scattering of a wavepacket from an attractive potential well and show the time evolution of the probability density as well as the wavefunction itself.

The TDSE requires a time-evolving wavefunction to be a complex-valued function. It can be rewritten as a pair of differential equations for the real and imaginary parts, respectively, that are of first order in time. The real and imaginary parts of the wavefunction are defined as $\psi = \psi_{\mathrm{R}} + \mathrm{i}\psi_{\mathrm{I}}$ and satisfy

$$\frac{\partial \psi_{\mathrm{R}}}{\partial t} = \hat{H}\psi_{\mathrm{I}}\,, \qquad \frac{\partial \psi_{\mathrm{I}}}{\partial t} = -\hat{H}\psi_{\mathrm{R}} \quad .$$

We consider a stationary Hamiltonian $\hat{H} \neq \hat{H}(t)$. The time dependence in our problem comes from evolving a wavepacket, i.e., an initially free state that is localized both in momentum and in position. Therefore, it is not stationary, but spreads in time as shown in Sect. 3.1.

The time-evolution equations are discretized by a staggered scheme in time. After a choice of stepsize Δt the time axis is marked using intervals of $\Delta t/2$. Since the structure of the evolution equations for ψ_{R} and ψ_{I} is such that one determines the time evolution of the other, it makes sense to employ the following finite difference approximation (known as the leapfrog method for the solution of classical Hamilton's equations of motion):

Start at $t = 0$ with a knowledge of $\psi_{\mathrm{R}}(x, t = 0)$ and $\psi_{\mathrm{I}}(x, t = \Delta t/2)$ as initial conditions. Calculate the real part from

$$\psi_{\mathrm{R}}(x, t + \Delta t) = \psi_{\mathrm{R}}(x, t) + \Delta t H_{op}\psi_{\mathrm{I}}(x, t + \Delta t/2)\,.$$

Having obtained the real part at $t + \Delta t/2$ one can propagate the imaginary part:

$$\psi_{\mathrm{I}}(x, t + (3/2)\Delta t) = \psi_{\mathrm{I}}(x, t + \Delta t/2) - \Delta t H_{op}\psi_{\mathrm{R}}(x, t + \Delta t)\,.$$

This scheme can be iterated to evolve the initial wavefunction to a finite time t_{f}. Note that from the point of view of numerical methods one has discretized a partial differential equation in space and time, i.e., a $1 + 1$ dimensional problem.

The central difference formula for the time derivative is second-order accurate in Δt. The method is stable as long as $\Delta t < (\Delta x)^2$. A disadvantage of the method is that the real and imaginary parts are never known at the same time. However, they can be obtained by interpolation. The same is true for the initial condition, i.e., one has to know the imaginary part of ψ evolved to $t = \Delta t/2$ initially.

For the initial state we choose a Gaussian wavepacket with known mean position x_0, initial width w_0, as well as initial momentum p_0. The expression is lifted from Sect. 3.1. We let it scatter form a potential barrier or well. We begin the scattering process from left to right.

```
> pack:=(x,t,x0,p0,w0)->2*sqrt(w0)*Pi^(1/4)/sqrt(1-2*I*t*w0^2)
> *exp(-p0^2/(2*w0)^2)*exp(w0^2*(I*(x0-x)-p0/(2*w0^2))^2
> /(1-2*I*t*w0^2));
```

$$\text{pack} := (\,x, t, x0, p0, w0\,) \rightarrow 2\, \frac{\text{sqrt}(\,w0\,)\, \pi^{1/4}\, e^{\left(-1/4\, \frac{p0^2}{w0^2}\right)}\, e^{\left(\frac{w0^2\,\left(I\,(\,x0-x\,)-1/2\,\frac{p0}{w0^2}\right)^2}{1-2\,I\,t\,w0^2}\right)}}{\text{sqrt}(\,1-2\,I\,t\,w0^2\,)}$$

We introduce the timestep dt $= \Delta t$ and its half dth :

```
> dt:=1/50; dth:=dt/2;
```

$$dt := \frac{1}{50} \qquad dth := \frac{1}{100}$$

Space is discretized with a resolution dx

```
> dx:=1/5; dxisq:=1/(2*dx^2);
```

$$dx := \frac{1}{5} \qquad dxisq := \frac{25}{2}$$

We consider a spatial mesh that extends from $-x_{\max}$ to $x_{\max}$ and determine the size of the vectors (we use objects called lists in Maple) to hold the discretized wavefunctions:

```
> xmax:=15;    nxm:=2*trunc(xmax/dx)+1;
```

$$xmax := 15 \qquad nxm := 151$$

Define the spatial mesh:

```
> for i from 1 to nxm do:

> xmesh[i]:=-xmax+(i-1)*dx;     od:
>

> xmesh[1];    xmesh[nxm];    xmesh[(nxm-1)/2+1];
```

$$-15 \qquad 15 \qquad 0$$

Now we initialize the real and imaginary parts of the wavefunction at time $t = 0$ and $t = \Delta t/2$, respectively:

```
> w0:=1/2:    x0:=-10:    p0:=2:
```

This choice implies a kinetic energy of $T = p_0^2/(2m)$ which amounts to 2 atomic units.

```
> for i from 1 to nxm do:
```

```
> psiR[i]:=Re(evalf(pack(xmesh[i],0,x0,p0,w0)));
```

```
> psiI[i]:=Im(evalf(pack(xmesh[i],dth,x0,p0,w0)));    od:
```

Now we graph the wavefunction. We show the real part at $t = 0$ and the imaginary part at $t = \Delta t/2$. We have to unassign the dummy variable `i`, if we wish to use it in the `plot` procedure to address elements of the list:

```
> i:='i':
```

```
> plot([[xmesh[i],psiR[i]] $i=1..nxm],title=`Re(psi(x,0))`);
```

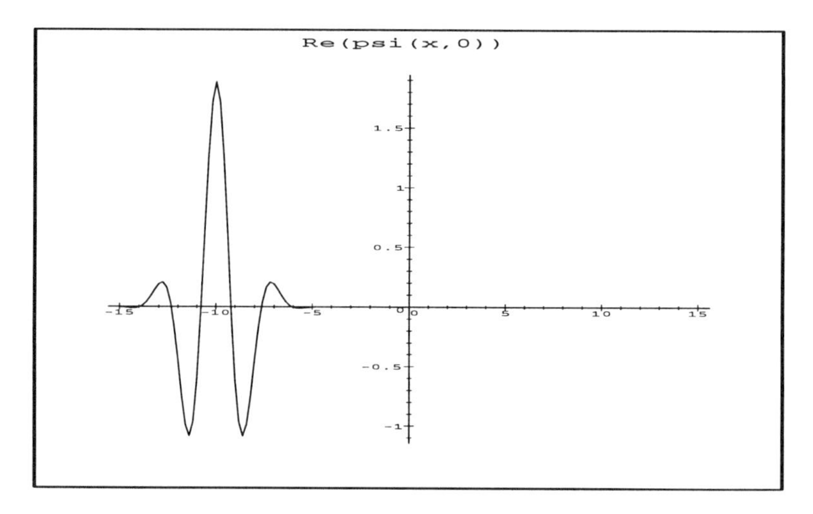

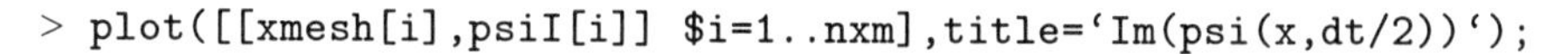

```
> plot([[xmesh[i],psiI[i]] $i=1..nxm],title=`Im(psi(x,dt/2))`);
```

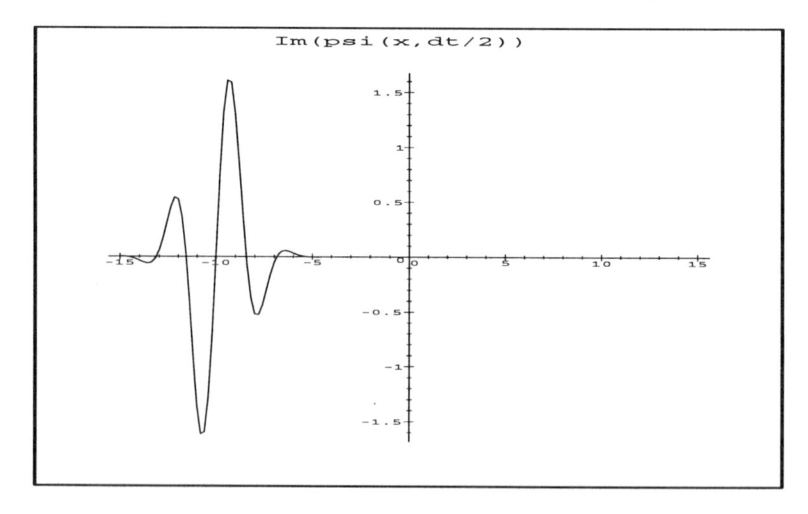

For the density we cheat by ignoring the time difference: the price we pay for this is a deformation. For a smaller value of p_0 the damage would be less visible. In principle, we should interpolate the imaginary part from two bracketing values to the same time level that the real part is defined.

```
> plot([[xmesh[i],psiR[i]^2+psiI[i]^2] $i=1..nxm],
> title=`rho(x,dt/4)`);
```

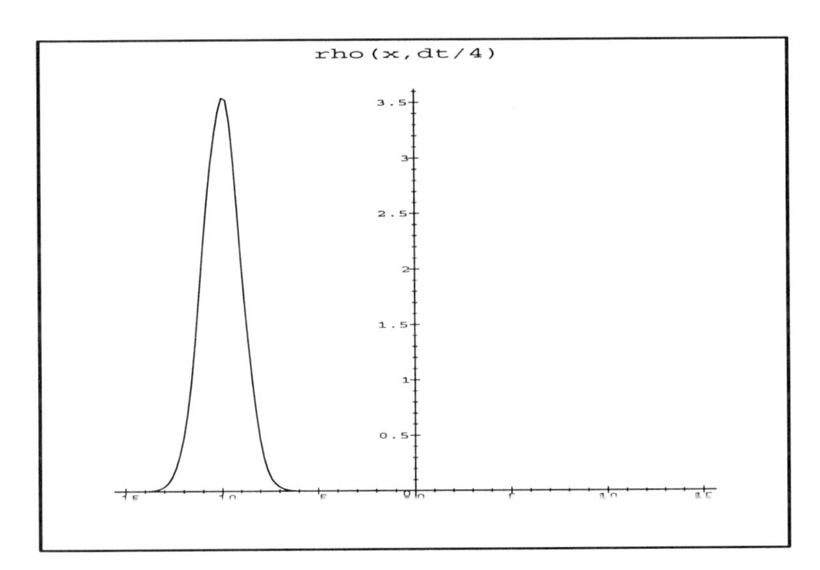

Now we need to define the Hamiltonian on the mesh. The kinetic energy couples nearest-neighbour sites, while the potential energy is local.

We write a procedure that takes as input a list containing the wavefunction (defined on the mesh) to be acted on, and which returns a list with the result. The boundary condition on ψ is incorporated as the Laplacian (kinetic energy) when calculated at the endpoints has to be defined in some way. We impose $\psi(x_{\min} - \Delta x) = \psi(x_{\max} + \Delta x) = 0$, which corresponds to complete reflection. One may want to modify this by adding an absorbing (complex) potential.

First we define the potential; here a simple square well:

```
> for i from 1 to nxm do:

> if abs(xmesh[i]) <= 2 then Vpot[i]:=-4: else Vpot[i]:=0: fi;

> #Vpot[i]:=0;

> od:    i:='i':

> plot([[xmesh[i],Vpot[i]] $i=1..nxm],title=`Vpot`);
```

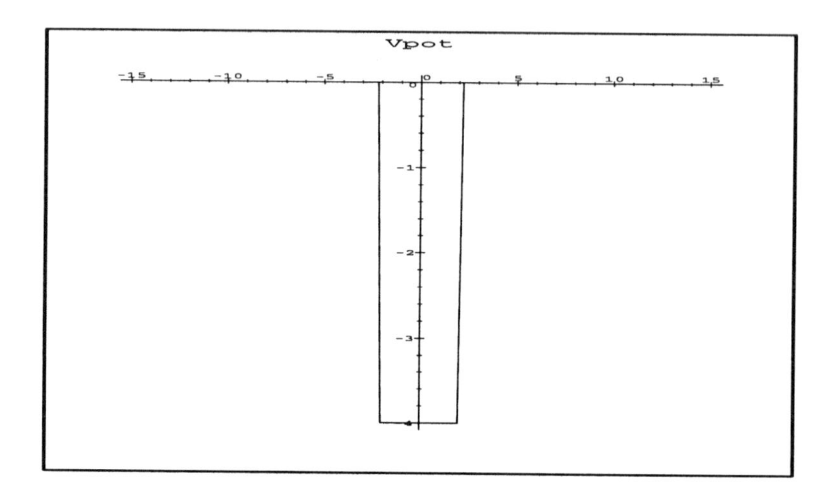

Now we program the discretized kinetic energy:

```
> Ham:=proc() global xmesh,dxisq,nxm,Vpot,Hpsi;

> for i from 2 to nxm-1 do:

> Hpsi[i]:=-dxisq*(args[1][i+1]+args[1][i-1]-2*args[1][i])
> +Vpot[i]*args[1][i];         od;

> Hpsi[1]:=-dxisq*(args[1][2]-2*args[1][1])
> +Vpot[1]*args[1][1];

> Hpsi[nxm]:=-dxisq*(args[1][nxm-1]-2*args[1][nxm])
> +Vpot[nxm]*args[1][nxm];    end;
 Warning, 'i' is implicitly declared local

 Ham :=

 proc()
 local i;
 global xmesh,dxisq,nxm,Vpot,Hpsi;
     for i from 2 to nxm-1 do
         Hpsi[i] := -
             dxisq*(args[1][i+1]+args[1][i-1]-2*args[1][i])+
             Vpot[i]*args[1][i]
     od;
     Hpsi[1] :=
         -dxisq*(args[1][2]-2*args[1][1])+Vpot[1]*args[1][1];
     Hpsi[nxm] := -dxisq*(args[1][nxm-1]-2*args[1][nxm])+
         Vpot[nxm]*args[1][nxm]
 end
```

Next comes the basic time-evolution step for the real and imaginary parts of the wavefunction. It assumes that the imaginary part psiI is a half-step ahead of the real part psiR.

```
> tstep:=proc() global dt,nxm,psiR,psiI,Time;
> Ham(psiI);
> for i from 1 to nxm do:
> psiR[i]:=psiR[i]+dt*Hpsi[i];       od;
> Ham(psiR);
> for i from 1 to nxm do:
> psiI[i]:=psiI[i]-dt*Hpsi[i];       od;
> Time:=Time+dt;
> end;
  Warning, 'i' is implicitly declared local

  tstep :=

      proc()
      local i;
      global dt,nxm,psiR,psiI,Time;
          Ham(psiI);
          for i to nxm do  psiR[i] := psiR[i]+dt*Hpsi[i] od;
          Ham(psiR);
          for i to nxm do  psiI[i] := psiI[i]-dt*Hpsi[i] od;
          Time := Time+dt
      end
```

Now we are ready to set up the time loop. Initially the packet should move freely except for the spreading of the width. Given that we gave it a momentum of $p = 2$, let's evolve for 2 atomic time units and notice the shift in the mean position.

```
> Time:=0:
> for it from 1 to 100 do:
> tstep();
> od:
> i:='i':   Time;
                          2
```

```
> plot([[xmesh[i],psiR[i]^2+psiI[i]^2] $i=1..nxm],
> title='rho(x,t=2)');
```

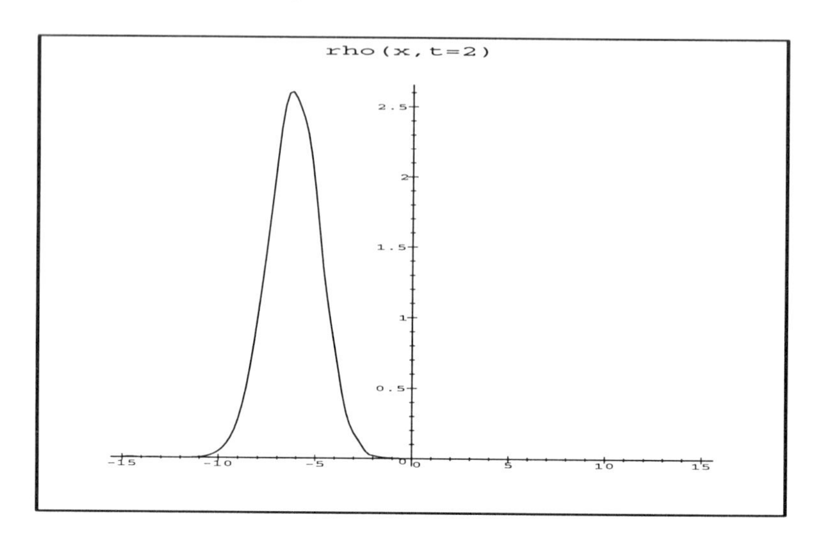

The evolution appears to be correct, i.e., in accord with $\Delta x/\Delta t = p_0/m$.

```
> plot([[xmesh[i],psiR[i]] $i=1..nxm],title='Re(psi(x,t=2))');
```

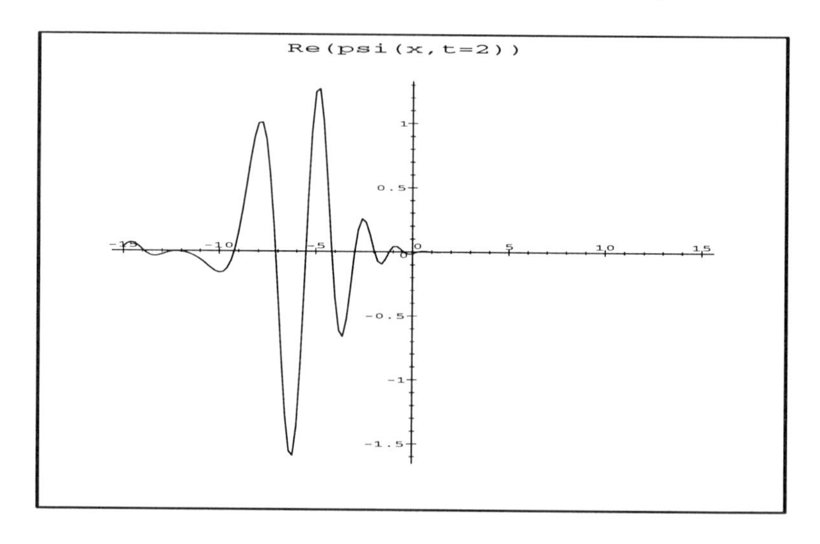


```
> plot([[xmesh[i],psiI[i]] $i=1..nxm],title='Im(psi(x,t=2))');
```

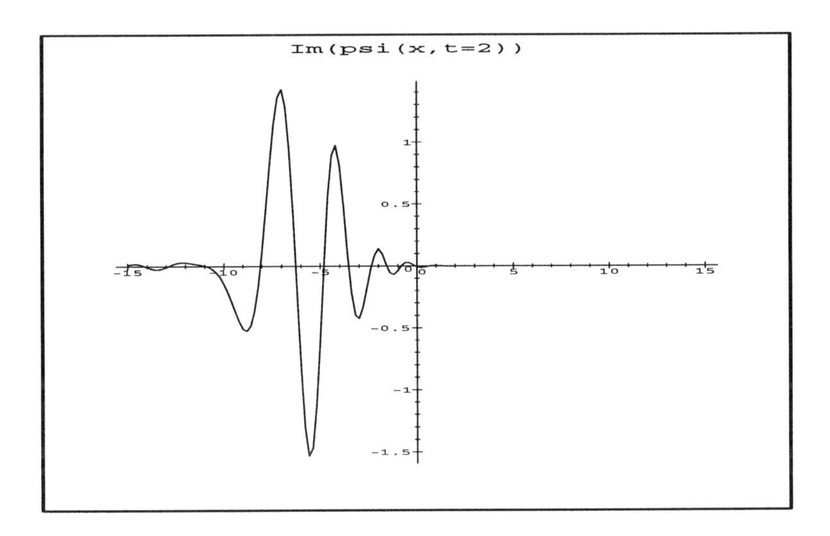

This isn't Fortran on a supercomputer, so it takes a while to execute. But we wish to follow up with what happens as the packet just enters the potential region. While the density shows mostly a shift of the mean position according to $x_0 = (p_0/m)t$, we notice that the real and imaginary parts of ψ have their tails already over the potential well.

Observe how the real and imaginary parts look similar. This is expected since one governs the time evolution of the other. Their oscillatory structures are shifted with respect to each other to make up a hump in the density function that is still smooth. As time progresses and the bulk part of the wavepacket moves over the barrier this changes. Interference patterns appear as parts of the packet are scattered at the points where the potential jumps in value, i.e., at $x = \pm a$. Note that as the packet contains fast (high-k) and slow (low-k) components, these scatter from the points $x = \pm a$ at different times.

```
> for it from 1 to 100 do:
> tstep();
> od:
> i:='i':   Time;
```

$$4$$

```
> plot([[xmesh[i],psiR[i]^2+psiI[i]^2] $i=1..nxm],
> title=`rho(x,t=4)`);
```

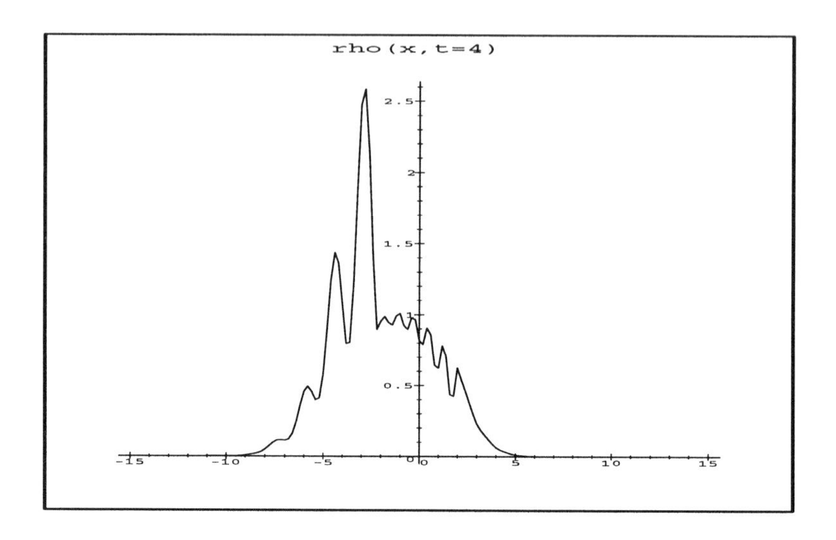


```
> plot([[xmesh[i],psiR[i]] $i=1..nxm],title='Re(psi(x,t=4))');
```

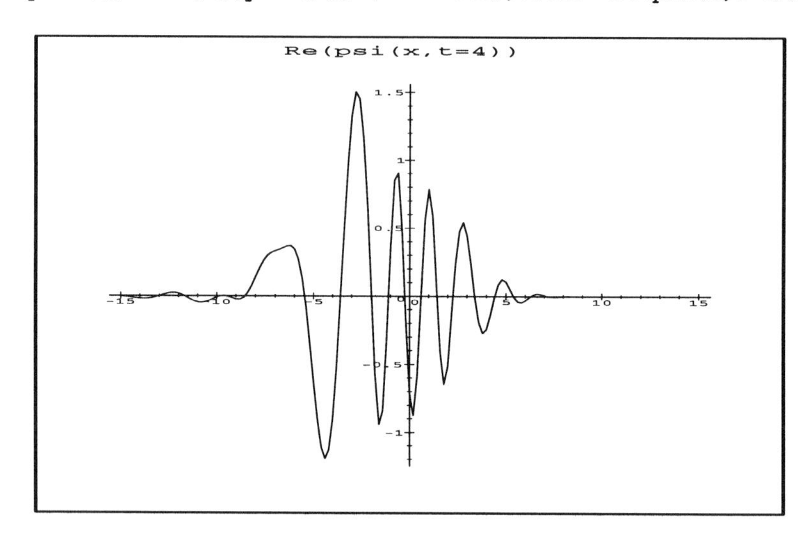


```
> plot([[xmesh[i],psiI[i]] $i=1..nxm],title='Im(psi(x,t=4))');
```

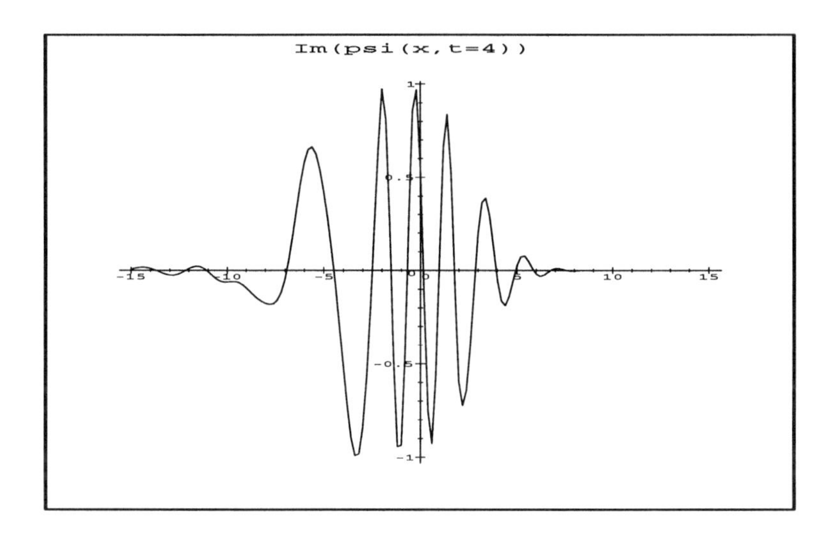


```
> for it from 1 to 100 do:
> tstep();
> od:     i:='i':     Time;
```

6

```
> plot([[xmesh[i],psiR[i]^2+psiI[i]^2] $i=1..nxm],
> title=`rho(x,t=6)`);
```

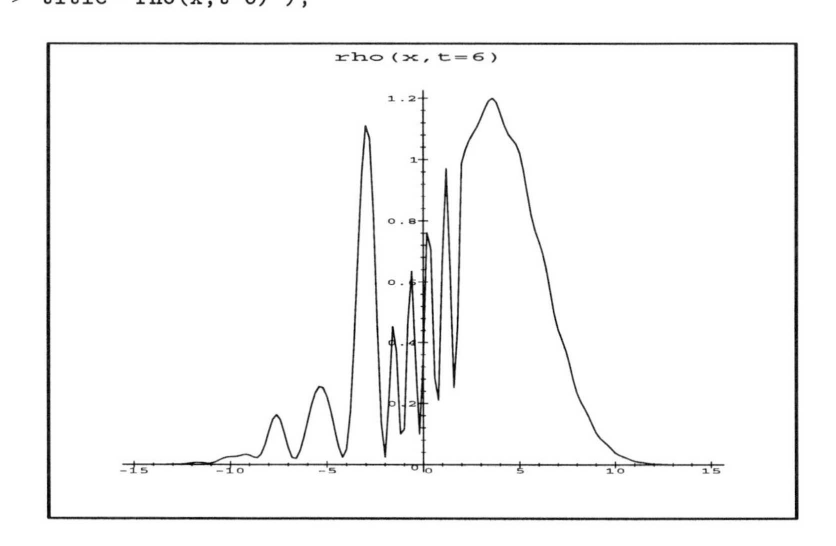


```
> plot([[xmesh[i],psiR[i]] $i=1..nxm],title=`Re(psi(x,t=6))`);
```

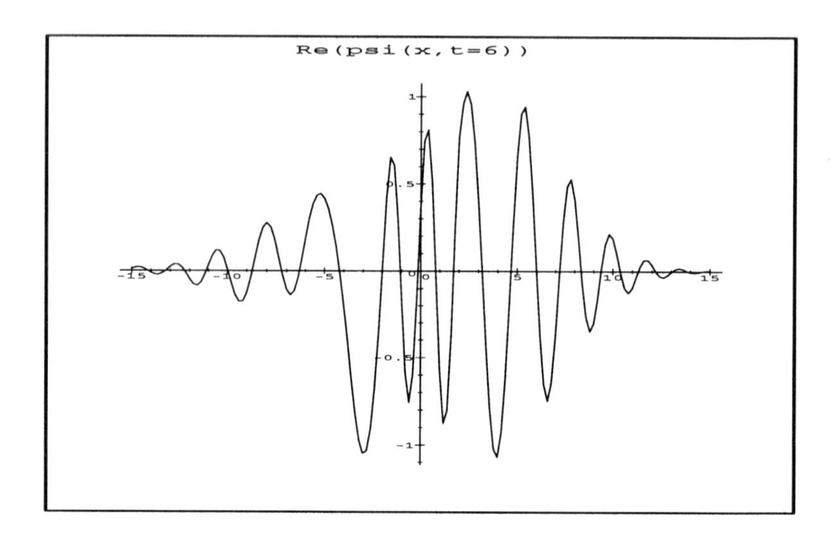


```
> plot([[xmesh[i],psiI[i]] $i=1..nxm],title=`Im(psi(x,t=6))`);
```

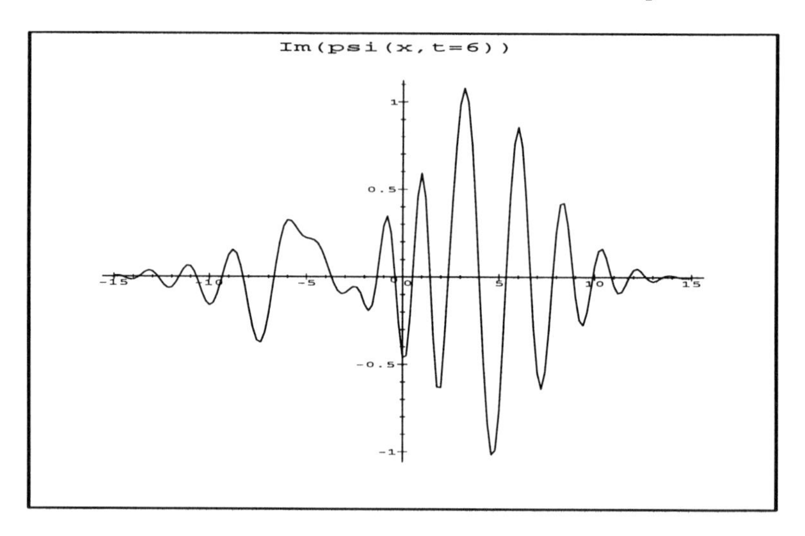


```
> for it from 1 to 100 do:
> tstep();
> od:      i:='i':    Time;
```

8

```
> plot([[xmesh[i],psiR[i]^2+psiI[i]^2] $i=1..nxm],
```

```
> title='rho(x,t=8)');
```

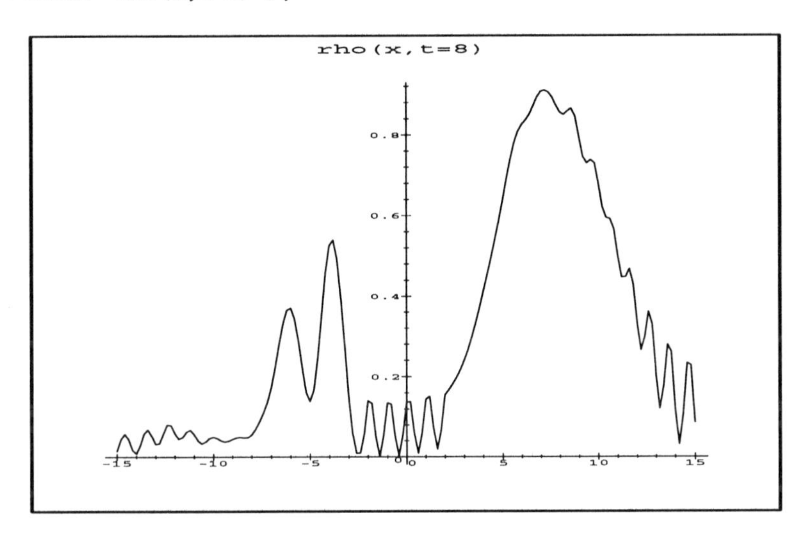


```
> plot([[xmesh[i],psiR[i]] $i=1..nxm],title='Re(psi(x,t=8))');
```

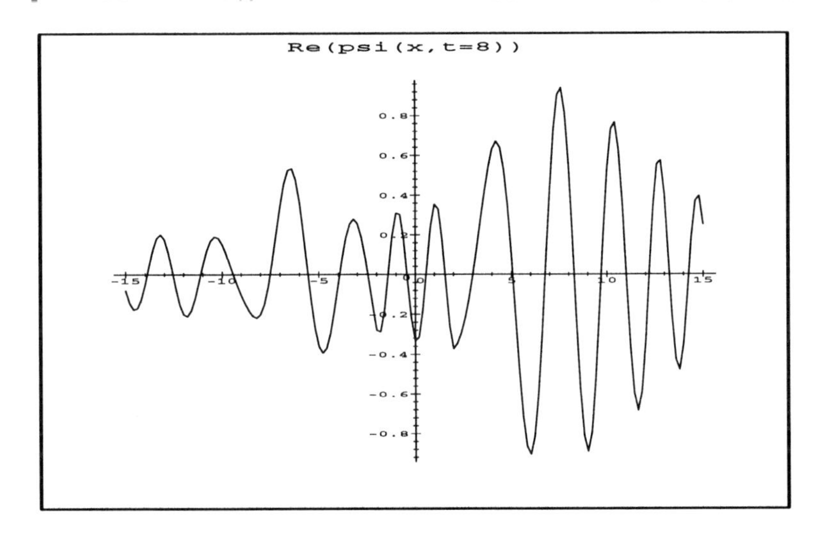


```
> plot([[xmesh[i],psiI[i]] $i=1..nxm],title='Im(psi(x,t=8))');
```

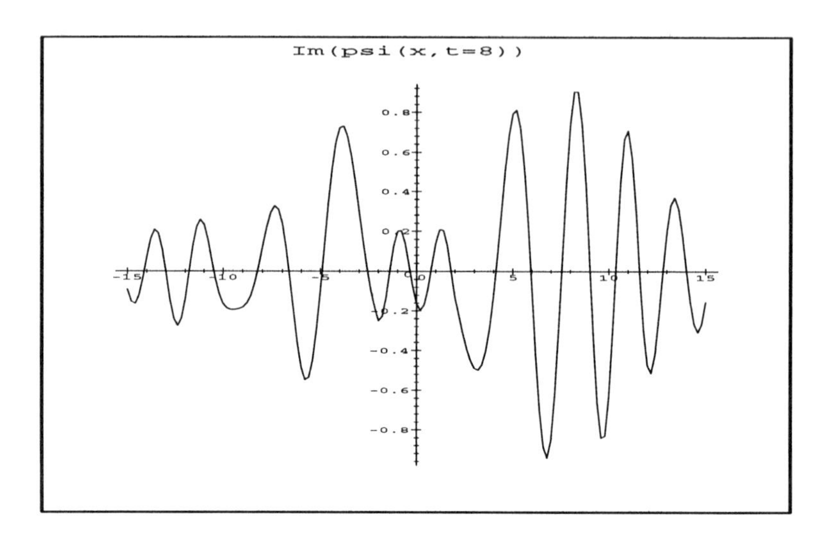

We see that a considerable amount of flux is, in fact, backscattered from an attractive potential step. Also note that the real and imaginary parts of the wavefunction are smoother than the density and give less reason to worry about discretization errors. The density is affected by the fact that we calculated simply $|\psi_R(t)|^2 + |\psi_I(t + \Delta t/2)|^2$, and did not interpolate both parts to the same time level. Further time evolution will be affected by the reflecting boundary condition, and one should use an x mesh that extends to further distances.

Exercise 3.5.1: Explore the time evolution of the wavefunction for barriers and wells of different sizes. Change the kinetic energy of the impinging packet.

Exercise 3.5.2: Replace the square-well potential by a smoothly-shaped well of similar size, e.g., a Gaussian-shaped well. Observe whether the same interference patterns occur in the probability density.

Exercise 3.5.3: Make a careful comparison between the time-dependent wavepacket treatment and the stationary scattering approach used in Sect. 3.2. Verify for some examples that for given potential parameters and scattering energy comparable transmission and reflection probabilities are obtained.

4. Problems in 3D

4.1 Angular Momentum

In this section I introduce the angular momentum operators and their commutation relations (CR). Later on we will find the eigenfunctions and verify the eigenvalues for some examples. Angular momentum plays an important role in QM. In classical mechanics the angular momentum vector $\mathbf{l} = \mathbf{r} \times \mathbf{p}$ of a point particle moving in a central potential $V(r)$ is conserved, which implies that the motion is confined to a plane. In QM the angular momentum operator is obtained by application of the usual quantization rule (cf. Sect. 1.1). The Heisenberg uncertainty principle $\Delta x \Delta p \geq \hbar$ makes it impossible to know precisely all three Cartesian components of angular momentum simultaneously. On the basis of the time evolution equation and the commutation relations between the Hamiltonian and the operators for the square of angular momentum and its components it is possible to show that the conservation law carries over into the statement that the magnitude and one projection is conserved and can be determined exactly.

We begin with a procedure to calculate commutators introduced in Sect. 1.5 and a definition of the angular momentum operator in Cartesian coordinates $\hat{l}_x$, $\hat{l}_y$, $\hat{l}_z$.

```
> CR:=proc();
> (args[1])@(args[2])-(args[2])@(args[1]);
> end:
```

The x component of angular momentum after quantization becomes

$$\hat{l}_x = \mathrm{i}\hbar(z\frac{\mathrm{d}}{\mathrm{d}y} - y\frac{\mathrm{d}}{\mathrm{d}z}) .$$

```
> lx:=proc();
> simplify(I*hbar*(z*diff(args[1],y)-y*diff(args[1],z)));
> end:
```

The other two components are programmed correspondingly as

```
> ly:=proc();
> simplify(I*hbar*(x*diff(args[1],z)-z*diff(args[1],x)));
> end:
> lz:=proc();
> simplify(I*hbar*(y*diff(args[1],x)-x*diff(args[1],y)));
> end:
```

We let $\hat{l}_z$ act on some arbitrary function of the Cartesian coordinates:

```
> lz(f(x,y,z));
```

$$-I\,\mathrm{hbar}\left(-y\left(\frac{\partial}{\partial x}\mathrm{f}(x,y,z)\right)+x\left(\frac{\partial}{\partial y}\mathrm{f}(x,y,z)\right)\right)$$

To verify one of the commutation relations we calculate

```
> CR(lx,ly)(f(x,y,z)):
```

and simplify the lengthy expression to

```
> simplify(");
```

$$-\mathrm{hbar}^2\,y\left(\frac{\partial}{\partial x}\mathrm{f}(x,y,z)\right)+hbar^2\,x\left(\frac{\partial}{\partial y}\mathrm{f}(x,y,z)\right)$$

This indeed equals $i\hbar\hat{l}_z$:

```
> simplify(I*hbar*lz(f(x,y,z)));
```

$$\mathrm{hbar}^2\left(-y\left(\frac{\partial}{\partial x}\mathrm{f}(x,y,z)\right)+x\left(\frac{\partial}{\partial y}\mathrm{f}(x,y,z)\right)\right)$$

I leave it as an exercise to demonstrate the cyclic property

$$[\hat{l}_i,\hat{l}_j]=\mathrm{i}\hbar\hat{l}_k \qquad (i,j,k) = \text{cyclic permutation.}$$

We have demonstrated that one cannot determine simultaneously two (or more) components of angular momentum, since the commutator between them does not vanish. Let us now construct the sum of the three operators squared, i.e., the l^2 operator. Let's try very naively, i.e.,

```
> lsq:=lx^2+ly^2+lz^2;
```

$$lsq := lx^2 + ly^2 + lz^2$$

```
> lsq(f(x,y,z));
```

$$-\mathrm{hbar}^2\left(z\left(\frac{\partial}{\partial y}\mathrm{f}(x,y,z)\right)-y\left(\frac{\partial}{\partial z}\mathrm{f}(x,y,z)\right)\right)^2$$

$$- \mathrm{hbar}^2 \left(x \left(\frac{\partial}{\partial z} \mathrm{f}(x,y,z) \right) - z \left(\frac{\partial}{\partial x} \mathrm{f}(x,y,z) \right) \right)^2$$
$$- \mathrm{hbar}^2 \left(-y \left(\frac{\partial}{\partial x} \mathrm{f}(x,y,z) \right) + x \left(\frac{\partial}{\partial y} \mathrm{f}(x,y,z) \right) \right)^2$$

This didn't work as the operator should be generated first, and then act on the function $f(x, y, z)$. Instead, $\hat{l}_x$, $\hat{l}_y$, $\hat{l}_z$ acted on $f(x, y, z)$ and were then squared, resulting in a non-linear object. How can we fix this? Maple has an operator for the repeated application of functions (note that `CR` is a procedure).

```
> lsq:=lx@@2+ly@@2+lz@@2;
```

$$lsq := lx^{(2)} + ly^{(2)} + lz^{(2)}$$

```
> lsq(f(x,y,z));
```

$$-\mathrm{hbar}^2 \left(z^2 \left(\frac{\partial^2}{\partial y^2} \mathrm{f}(x,y,z) \right) - z \left(\frac{\partial}{\partial z} \mathrm{f}(x,y,z) \right) - 2\, z\, y \left(\frac{\partial^2}{\partial y\, \partial z} \mathrm{f}(x,y,z) \right) \right.$$
$$\left. - y \left(\frac{\partial}{\partial y} \mathrm{f}(x,y,z) \right) + y^2 \left(\frac{\partial^2}{\partial z^2} \mathrm{f}(x,y,z) \right) \right) - \mathrm{hbar}^2 \left(x^2 \left(\frac{\partial^2}{\partial z^2} \mathrm{f}(x,y,z) \right) \right.$$
$$- x \left(\frac{\partial}{\partial x} \mathrm{f}(x,y,z) \right) - 2\, x\, z \left(\frac{\partial^2}{\partial z\, \partial x} \mathrm{f}(x,y,z) \right) - z \left(\frac{\partial}{\partial z} \mathrm{f}(x,y,z) \right)$$
$$\left. + z^2 \left(\frac{\partial^2}{\partial x^2} \mathrm{f}(x,y,z) \right) \right) - \mathrm{hbar}^2 \left(y^2 \left(\frac{\partial^2}{\partial x^2} \mathrm{f}(x,y,z) \right) - y \left(\frac{\partial}{\partial y} \mathrm{f}(x,y,z) \right) \right.$$
$$\left. - 2\, y\, x \left(\frac{\partial^2}{\partial y\, \partial x} \mathrm{f}(x,y,z) \right) - x \left(\frac{\partial}{\partial x} \mathrm{f}(x,y,z) \right) + x^2 \left(\frac{\partial^2}{\partial y^2} \mathrm{f}(x,y,z) \right) \right)$$

Note that this is a linear operator now. We evaluate the commutator $[\hat{l}^2, \hat{l}_x]$ which at first results in a page-long expression. Upon simplification we find that this commutator vanishes.

```
> CR(lx,lsq)(f(x,y,z)):     simplify(");
```

$$0$$

Indeed this CR can be found to hold for any of the components. It implies that any component from the set $\{\hat{l}_x, \hat{l}_y, \hat{l}_z\}$ can be determined simultaneously with $\hat{l}^2$, but only one component, since they don't commute amongst each other. For convenience, particularly to work in spherical polar coordinates, one selects the $\hat{l}_z$ operator. One is, of course, free to choose (once) the direction of the z axis.

We provide the l^2 and l_z operators in spherical polar coordinates. This requires some Calculus operations to perform the change of variables.

```
> with(linalg):
> JM:=jacobian([r*sin(theta)*cos(phi),r*sin(theta)*sin(phi),
> r*cos(theta)],[r,theta,phi]);
```

$$JM := \begin{bmatrix} \sin(\theta)\cos(\phi) & r\cos(\theta)\cos(\phi) & -r\sin(\theta)\sin(\phi) \\ \sin(\theta)\sin(\phi) & r\cos(\theta)\sin(\phi) & r\sin(\theta)\cos(\phi) \\ \cos(\theta) & -r\sin(\theta) & 0 \end{bmatrix}$$

```
> simplify(det(JM));
```

$$\sin(\theta)\, r^2$$

The Jacobian contains the partial derivatives of (x, y, z) expressed as functions of (r, θ, ϕ) with respect to r, θ, ϕ. We convert partial derivatives according to

$$\frac{\mathrm{d}}{\mathrm{d}x} = \frac{\partial r}{\partial x}\frac{\mathrm{d}}{\mathrm{d}r} + \frac{\partial \theta}{\partial x}\frac{\mathrm{d}}{\mathrm{d}\theta} + \frac{\partial \phi}{\partial x}\frac{\mathrm{d}}{\mathrm{d}\phi} \quad .$$

Note that the Jacobian calculated above gives not quite the required information, it contains derivatives of type $\partial x/\partial r$, $\partial x/\partial \theta$, etc.. We cannot simply take inverses of the entries, i.e., it is not correct to state $\partial r/\partial x = (\partial x/\partial r)^{(-1)}$. The reason for this is that the derivatives are partial, and not total. The whole transformation has to be inverted. We can either invert the Jacobian, or calculate the right Jacobian to begin with.

```
> Ji:=inverse(JM);
```

$$Ji := \begin{bmatrix} \dfrac{\sin(\theta)\cos(\phi)}{\%1}, \dfrac{\sin(\theta)\sin(\phi)}{\%1}, \dfrac{\cos(\theta)}{\sin(\theta)^2+\cos(\theta)^2} \\ \dfrac{\cos(\phi)\cos(\theta)}{r\,\%1}, \dfrac{\sin(\phi)\cos(\theta)}{r\,\%1}, -\dfrac{\sin(\theta)}{r\,(\sin(\theta)^2+\cos(\theta)^2)} \\ -\dfrac{\sin(\phi)}{r\,(\cos(\phi)^2+\sin(\phi)^2)\sin(\theta)}, \dfrac{\cos(\phi)}{r\,(\cos(\phi)^2+\sin(\phi)^2)\sin(\theta)}, 0 \end{bmatrix}$$

$$\%1 := \sin(\theta)^2\cos(\phi)^2 + \sin(\theta)^2\sin(\phi)^2 + \cos(\theta)^2\cos(\phi)^2 + \sin(\phi)^2\cos(\theta)^2$$

Unfortunately, the denominator has not been simplified. To correct this in all entries of the matrix Maple requires an mapping of `simplify` with the `map` function:

```
> JMi:=map(simplify,op(Ji));
```

$$JMi := \begin{bmatrix} \sin(\theta)\cos(\phi) & \sin(\theta)\sin(\phi) & \cos(\theta) \\ \frac{\cos(\phi)\cos(\theta)}{r} & \frac{\sin(\phi)\cos(\theta)}{r} & -\frac{\sin(\theta)}{r} \\ -\frac{\sin(\phi)}{r\sin(\theta)} & \frac{\cos(\phi)}{r\sin(\theta)} & 0 \end{bmatrix}$$

For comparison we try to calculate the required Jacobian directly.

```
> JMc:=jacobian([sqrt(x^2+y^2+z^2),arcccos(z/sqrt(x^2+y^2+z^2))
> ,arctan(y/x)],[x,y,z]);
```

$JMc :=$

$$\left[\frac{x}{\sqrt{x^2+y^2+z^2}}, \frac{y}{\sqrt{x^2+y^2+z^2}}, \frac{z}{\sqrt{x^2+y^2+z^2}}\right]$$

[row 2 omitted here to save space]

$$\left[-\frac{y}{x^2\left(1+\frac{y^2}{x^2}\right)}, \frac{1}{x\left(1+\frac{y^2}{x^2}\right)}, 0\right]$$

The result still requires the substitution of $\{x, y, z\}$ in terms of $\{r, \theta, \phi\}$:

```
> JMcsp:=subs(x=r*sin(theta)*cos(phi),y=r*sin(theta)*sin(phi),
> z=r*cos(theta),op(JMc));
```

JMcsp :=

$$\left[\frac{r\sin(\theta)\cos(\phi)}{\sqrt{\%1}}, \frac{r\sin(\theta)\sin(\phi)}{\sqrt{\%1}}, \frac{r\cos(\theta)}{\sqrt{\%1}}\right]$$

[row 2 omitted here to save space]

[row 3 omitted here to save space]

$$\%1 := r^2\sin(\theta)^2\cos(\phi)^2 + r^2\sin(\theta)^2\sin(\phi)^2 + r^2\cos(\theta)^2$$

We note that to simplify matrices or vectors in Maple we have to invoke a call to `map`:

```
> JMcsp:=map(simplify,JMcsp);
```

JMcsp :=

$$[\mathrm{csgn}(r)\sin(\theta)\cos(\phi), \mathrm{csgn}(r)\sin(\theta)\sin(\phi), \%1]$$

$$\left[-\frac{\mathrm{D}(\mathit{arcccos})(\%1)\cos(\theta)\,\mathrm{csgn}(r)\sin(\theta)\cos(\phi)}{r},\right.$$

$$-\frac{\mathrm{D}(\mathit{arcccos})(\%1)\cos(\theta)\,\mathrm{csgn}(r)\sin(\theta)\sin(\phi)}{r},$$

$$\left.\frac{\mathrm{D}(\mathit{arcccos})(\%1)\left(\mathrm{csgn}(r)^2-\cos(\theta)^2\right)\mathrm{csgn}(r)}{r}\right]$$

$$\left[-\frac{\sin(\phi)}{r\sin(\theta)}, \frac{\cos(\phi)}{r\sin(\theta)}, 0\right]$$
$$\%1 := \cos(\theta)\,\mathrm{csgn}(r)$$

This expression can be simplified using the assumption that r is always positive. In Maple the inverse of JM obtained the result somewhat more directly.

The Jacobian contains now in row 1: $[\partial r/\partial x,\ \partial r/\partial y,\ \partial r/\partial z]$, in row 2: $[\partial\theta/\partial x,\ \partial\theta/\partial y,\ \partial\theta/\partial z]$, and in row 3: $[\partial\phi/\partial x,\ \partial\phi/\partial y,\ \partial\phi/\partial z]$. Keeping this in mind we can express the total derivatives now as

$$\frac{\mathrm{d}}{\mathrm{d}x} = \frac{\mathrm{d}r}{\mathrm{d}x}\frac{\mathrm{d}}{\mathrm{d}r} + \frac{\mathrm{d}\theta}{\mathrm{d}x}\frac{\mathrm{d}}{\mathrm{d}\theta} + \frac{\mathrm{d}\phi}{\mathrm{d}x}\frac{\mathrm{d}}{\mathrm{d}\phi} \quad ,$$

and correspondingly $\frac{\mathrm{d}}{\mathrm{d}y}$ and $\frac{\mathrm{d}}{\mathrm{d}z}$. We define:

```
> ddx:=arg->diff(arg,r)*JMi[1,1]+diff(arg,theta)*JMi[2,1]
> +diff(arg,phi)*JMi[3,1];
```

$$ddx := \arg \to \left(\frac{\partial}{\partial r}\arg\right) JMi_{1,1} + \left(\frac{\partial}{\partial\theta}\arg\right) JMi_{2,1} + \left(\frac{\partial}{\partial\phi}\arg\right) JMi_{3,1}$$

```
> ddy:=arg->diff(arg,r)*JMi[1,2]+diff(arg,theta)*JMi[2,2]
> +diff(arg,phi)*JMi[3,2];
```

$$ddy := \arg \to \left(\frac{\partial}{\partial r}\arg\right) JMi_{1,2} + \left(\frac{\partial}{\partial\theta}\arg\right) JMi_{2,2} + \left(\frac{\partial}{\partial\phi}\arg\right) JMi_{3,2}$$

Having defined the derivatives in terms of spherical polar coordinates we are ready to define $\hat{l}_z$ after a corresponding substitution for x and y as well:

```
> lzsp:=arg->simplify(I*hbar*(r*sin(theta)*sin(phi)*ddx(arg)
> -r*sin(theta)*cos(phi)*ddy(arg)));
```

$$lzsp := \arg \to \mathrm{simplify}(I\,\mathrm{hbar}\,(r\sin(\theta)\sin(\phi)\,\mathrm{ddx}(\arg) - r\sin(\theta)\cos(\phi)\,\mathrm{ddy}(\arg)))$$

Now we act with this operator on some function $f(r,\theta,\phi)$:

```
> lzsp(f(r,theta,phi));
```

$$-I\,\mathrm{hbar}\left(\frac{\partial}{\partial\phi}\mathrm{f}(r,\theta,\phi)\right)$$

Thus, in spherical polar coordinates the Cartesian z component of angular momentum, $\hat{l}_z$, simplifies considerably. The other components aren't quite as easy. First, we need the $\frac{\mathrm{d}}{\mathrm{d}z}$ derivative,

```
> ddz:=arg->diff(arg,r)*JMi[1,3]+diff(arg,theta)*JMi[2,3]
> +diff(arg,phi)*JMi[3,3];
```

$$ddz := \arg \to \left(\frac{\partial}{\partial r}\arg\right) JMi_{1,3} + \left(\frac{\partial}{\partial \theta}\arg\right) JMi_{2,3} + \left(\frac{\partial}{\partial \phi}\arg\right) JMi_{3,3}$$

and assemble l_x

```
> lxsp:=arg->simplify(-I*hbar*(r*sin(theta)*sin(phi)*ddz(arg)
> -r*cos(theta)*ddy(arg)));
```

$$lxsp := \arg \to \\ \mathrm{simplify}(-I\,\mathrm{hbar}\,(r\sin(\theta)\sin(\phi)\,\mathrm{ddz}(\arg) - r\cos(\theta)\,\mathrm{ddy}(\arg)))$$

Application of $\hat{l}_x$ to $f(r,\theta,\phi)$ results in an expression that contains still two derivatives.

```
> lxsp(f(r,theta,phi));
```

$$\frac{I\,\mathrm{hbar}}{\sin(\theta)}\left(\sin(\phi)\left(\frac{\partial}{\partial\theta}\mathrm{f}(r,\theta,\phi)\right)\sin(\theta) + \cos(\theta)\left(\frac{\partial}{\partial\phi}\mathrm{f}(r,\theta,\phi)\right)\cos(\phi)\right)$$

We can repeat the same for $\hat{l}_y$ in spherical polar coordinates:

```
> lysp:=arg->simplify(I*hbar*(r*sin(theta)*cos(phi)*ddz(arg)
> -r*cos(theta)*ddx(arg)));
```

$$lysp := \arg \to \\ \mathrm{simplify}(I\,hbar\,(r\sin(\theta)\cos(\phi)\,\mathrm{ddz}(\arg) - r\cos(\theta)\,\mathrm{ddx}(\arg)))$$

```
> lysp(f(r,theta,phi));
```

$$\frac{I\,\mathrm{hbar}}{\sin(\theta)}\left(-\cos(\phi)\left(\frac{\partial}{\partial\theta}\mathrm{f}(r,\theta,\phi)\right)\sin(\theta) + \cos(\theta)\left(\frac{\partial}{\partial\phi}\mathrm{f}(r,\theta,\phi)\right)\sin(\phi)\right)$$

A useful combination of operators is $\hat{l}_x \pm \mathrm{i}\hat{l}_y$ (cf. Exercise 4.1.3):

```
> (lxsp+I*lysp)(f(r,theta,phi)):
> simplify(");
```

$$\frac{\mathrm{hbar}}{\sin(\theta)}\left(I\sin(\phi)\left(\frac{\partial}{\partial\theta}\mathrm{f}(r,\theta,\phi)\right)\sin(\theta) + I\cos(\theta)\left(\frac{\partial}{\partial\phi}\mathrm{f}(r,\theta,\phi)\right)\cos(\phi)\right. \\ \left. + \cos(\phi)\left(\frac{\partial}{\partial\theta}\mathrm{f}(r,\theta,\phi)\right)\sin(\theta) - \cos(\theta)\left(\frac{\partial}{\partial\phi}\mathrm{f}(r,\theta,\phi)\right)\sin(\phi)\right)$$

This allows us to factor out an $\exp(\mathrm{i}\phi)$ and to simplify further. Maple requires massaging to do this and we do not follow up on it. We do, however turn to the calculation of $\hat{l}^2$:

```
> lsqsp:=lxsp@@2+lysp@@2+lzsp@@2;
```

$$lsqsp := lxsp^{(2)} + lysp^{(2)} + lzsp^{(2)}$$

```
> lsqsp(f(r,theta,phi)):
> simplify(");
```

$$\text{hbar}^2\left(\left(\frac{\partial^2}{\partial\theta^2}\text{f}(r,\theta,\phi)\right)-\left(\frac{\partial^2}{\partial\theta^2}\text{f}(r,\theta,\phi)\right)\cos(\theta)^2\right.$$
$$\left.+\cos(\theta)\left(\frac{\partial}{\partial\theta}\text{f}(r,\theta,\phi)\right)\sin(\theta)+\left(\frac{\partial^2}{\partial\phi^2}\text{f}(r,\theta,\phi)\right)\right)\Big/\left(\cos(\theta)^2-1\right)$$

This is close to the desired answer, but why are there these ugly $1-\cos^2\theta$ expressions? A futile attempt is to simplify with respect to trig relations, i.e., the command below just returns the same answer:

```
> simplify(",trig):
```

Given that the unwanted terms won't go away on their own, we make an explicit substitution.

```
> subs(cos(theta)^2=1-sin(theta)^2,");
```

$$-\text{hbar}^2\left(\left(\frac{\partial^2}{\partial\theta^2}\text{f}(r,\theta,\phi)\right)-\left(\frac{\partial^2}{\partial\theta^2}\text{f}(r,\theta,\phi)\right)\left(1-\sin(\theta)^2\right)\right.$$
$$\left.+\cos(\theta)\left(\frac{\partial}{\partial\theta}\text{f}(r,\theta,\phi)\right)\sin(\theta)+\left(\frac{\partial^2}{\partial\phi^2}\text{f}(r,\theta,\phi)\right)\right)\Big/\sin(\theta)^2$$

```
> expand("/hbar^2);
```

$$-\left(\frac{\partial^2}{\partial\theta^2}\text{f}(r,\theta,\phi)\right)-\frac{\cos(\theta)\left(\frac{\partial}{\partial\theta}\text{f}(r,\theta,\phi)\right)}{\sin(\theta)}-\frac{\left(\frac{\partial^2}{\partial\phi^2}\text{f}(r,\theta,\phi)\right)}{\sin(\theta)^2}$$

This is now close enough to the textbook expression.

Next we look at the eigenfunctions that diagonalize $\hat{l}^2$ and $\hat{l}_z$ simultaneously. First the simple case, the eigenfunctions of $\hat{l}_z$. Obviously, an exponential function in ϕ satisfies the eigenvalue equation

$$-\mathrm{i}\hbar\frac{\mathrm{d}}{\mathrm{d}\phi}f(\phi)=\lambda f(\phi),$$

i.e., $f(\phi)=\exp(\mathrm{i}\alpha\phi/\hbar)$ seems to be the solution. The eigenvalue $\lambda=\alpha$ appears to be arbitrary. However, there is a periodicity condition, $f(\phi\pm 2m\pi)=f(\phi)$, for $m=1,2,\ldots$ to be specified to guarantee the uniqueness of the solution. This forces $f(\phi)=\exp(\mathrm{i}m\phi)$ with $m=0,\pm1,\pm2,\ldots$ with eigenvalues $\lambda=m\hbar$. Let's verify this for the operator defined above:

```
> lzsp(exp(I*m*phi))/exp(I*m*phi);
```

$$\text{hbar}\,m$$

Note that the periodicity constraint is only in effect for integer m.

```
> exp(I*m*(phi+2*Pi))-exp(I*m*phi);
```

$$e^{(I\,m\,(\phi+2\,\pi))} - e^{(I\,m\,\phi)}$$

```
> simplify(");
```

$$e^{(I\,m\,(\phi+2\,\pi))} - e^{(I\,m\,\phi)}$$

```
> assume(m,integer);
> exp(I*m*(phi+2*Pi))-exp(I*m*phi);
```

$$e^{(I\,m\tilde{}\,(\phi+2\,\pi))} - e^{(I\,m\tilde{}\,\phi)}$$

```
> simplify(");
```

$$e^{(I\,m\tilde{}\,(\phi+2\,\pi))} - e^{(I\,m\tilde{}\,\phi)}$$

This expression doesn't simplify, because Maple's built-in sine and cosine functions do not recognize periodicity for arbitrary integer m (up to Release 3):

```
> sin(x+2*m*Pi)-sin(x);
```

$$\sin(\,x + 2\,\pi\,m\tilde{}\,) - \sin(\,x\,)$$

This should simplify in some future release of Maple. For the time being we can only obtain an answer for explicit integer values of m, e.g.,

```
> sin(x+2*Pi)-sin(x);
```

$$0$$

The eigenfunctions of $\hat{l}^2$ represent the harder part. We omit a derivation and satisfy ourselves with a demonstration. First we make a call to a package containing orthogonal polynomials.

```
> with(orthopoly):
```

The Legendre polynomials are defined now. Here are some examples:

```
> P(0,x);     P(1,x);     P(2,x);
```

$$1 \qquad x \qquad \frac{3}{2}\,x^2 - \frac{1}{2}$$

We need, however, more than that, namely the associated Legendre polynomials $P(m,n,x)$, or spherical harmonics $Y(l,m,\theta,\phi)$ with $x = \cos(\theta)$. We define them ourselves using the generating expression

$$P_{lm}(\theta) = (-1)^m \sin^m(\theta) \frac{\mathrm{d}^m P_l(x)}{\mathrm{d}x^m} .$$

```
> Plm:=proc(theta,l,m);
> if type(l,integer)=true and type(m,integer)= true then
> x:=cos(theta);
> if m>0 then fak:=subs(y=x,diff(P(l,y),y$m));
> else fak:=subs(y=x,P(l,y)); fi;
> (-1)^m*(sin(theta)^m)*fak;
> fi;      end:
```

To save space we omitted the (structured) printout of the procedure. As an example we calculate $P_{3,2}(\theta, \phi)$:

```
> Plm(theta,3,2);
```

$$15 \sin(\theta)^2 \cos(\theta)$$

The spherical harmonics can be calculated now in a straightforward way:

```
> Ylm:=proc(theta,phi,l,m);
> m1:=abs(m);
> if m1>l then RETURN(`in Ylm l < |m|`,l,m); fi;
> if type(l,integer) <>true and type(m,integer)<>true then
> RETURN(`in Ylm l and or m not integers `,l,m); fi;
> exp(I*m*phi)*Plm(theta,l,m1);        end:
```

Here are some examples to verify the correctness of this procedure:

```
> Ylm(theta,phi,1,0);     Ylm(theta,phi,0,0);
```

$$\cos(\theta) \qquad 1$$

```
> Ylm(theta,phi,0,1);     Ylm(theta,phi,2,-2);
```

$$in\ Ylm\ l < |m|, 0, 1 \qquad 3\, \mathrm{e}^{(-2 I \phi)} \sin(\theta)^2$$

Now we are ready to verify that these are eigenfunctions of $\hat{l}_z$ and $\hat{l}^2$ with the correct eigenvalues $\hbar m$ and $\hbar^2 l(l+1)$ respectively. First a case with $l = 2$ and $m = 1$ for $\hat{l}_z$:

```
> lzsp(Ylm(theta,phi,2,1))/Ylm(theta,phi,2,1);
```

hbar

Now try $\hat{l}_z$ on $Y_{2,-2}$:

```
> lzsp(Ylm(theta,phi,2,-2))/Ylm(theta,phi,2,-2);
```

$$\frac{1}{3}\,\frac{-6\,hbar\,\mathrm{e}^{(-2\,I\,\phi)} + 6\,\mathrm{hbar}\,\mathrm{e}^{(-2\,I\,\phi)}\cos(\theta)^2}{\mathrm{e}^{(-2\,I\,\phi)}\sin(\theta)^2}$$

```
> simplify(");
```

$$-2\,\mathrm{hbar}$$

Next try $\hat{l}^2$ on $Y_{2,1}$:

```
> lsqsp(Ylm(theta,phi,2,1))/Ylm(theta,phi,2,1):
> simplify(");
```

$$6\,\mathrm{hbar}^2$$

Given that $2 \cdot (2+1) = 6$ we should be satisfied. We try one more, namely $l^2 Y_{3,0}$:

```
> lsqsp(Ylm(theta,phi,3,0))/Ylm(theta,phi,3,0);
```

$$\frac{30\,\mathrm{hbar}^2\cos(\theta)^3 - 18\,\mathrm{hbar}^2\cos(\theta)}{\frac{5}{2}\cos(\theta)^3 - \frac{3}{2}\cos(\theta)}$$

```
> simplify(");
```

$$12\,\mathrm{hbar}^2$$

The final thing to worry about in this section is the normalization of these states. We can define an inner product for (θ, ϕ) space:

```
> Ipd:=proc();
> theta1:=args[3]; phi1:=args[4];
> int1:=int(conjugate(args[1])*args[2]*sin(theta1),
> theta1=0..Pi);
> int(int1,phi1=0..2*Pi);      end:
> Ipd(Ylm(theta,phi,0,0),Ylm(theta,phi,0,0),theta,phi);
```

$$4\,\pi$$

The procedure seems to work, our functions Y_{lm} aren't properly normalized yet, but that can always be accounted for later. Let's check orthogonality for an example with $l_1 = 0$, $m_1 = 0$ and $l_2 = 2$, $m_2 = 0$:

```
> Ipd(Ylm(theta,phi,0,0),Ylm(theta,phi,2,0),theta,phi);
```

$$0$$

This orthogonality came from the θ integration, and now we try $l_1 = l_2 = 1$ and $m_1 = 1$, $m_2 = 0$, i.e., a case where the ϕ integration should do the work (this fruitless attempt takes quite a number of CPU cycles, more than 8 minutes on a 486-based PC):

```
> Ipd(Ylm(theta,phi,1,1),Ylm(theta,phi,1,0),theta,phi);
```

$$\int_0^{2\pi}\int_0^{\pi} -\operatorname{conjugate}(e^{(I\phi)}\sin(\theta))\cos(\theta)\sin(\theta)\,d\theta\,d\phi$$

What's going on here? Obviously, the conjugation didn't simplify.

```
> conjugate(exp(-I*phi));
```

$$e^{(I\operatorname{conjugate}(\phi))}$$

```
> assume(phi,real);        conjugate(exp(-I*phi));
```

$$e^{(I\phi\tilde{})}$$

Let's try to incorporate the assumption and make the corresponding assumption about θ, just in case. Before we redefine the procedure we try an example 'by hand'.

```
> assume(theta,real);
```

An example of orthogonality in ϕ space:

```
> conjugate(Ylm(theta,phi,2,2))*Ylm(theta,phi,2,0);
```

$$3\,e^{(-2I\phi\tilde{})}\sin(\theta\tilde{})^2\left(\frac{3}{2}\cos(\theta\tilde{})^2-\frac{1}{2}\right)$$

```
> int("*sin(theta),theta=0..Pi);
```

$$-\frac{4}{5}e^{(-2I\phi\tilde{})}$$

```
> int(",phi=0..2*Pi);
```

$$0$$

Now we provide the corrected procedure (strictly the declaration of real θ and ϕ outside the procedure should be sufficient as long as one uses these variables as arguments).

```
> Ipd:=proc();
```

```
> theta1:=args[3]; phi1:=args[4]; assume(theta1,real);
> assume(phi1,real);
```

```
> int1:=int(conjugate(args[1])*args[2]*sin(theta1),theta1=0..Pi);
```

```
> int(int1,phi1=0..2*Pi);
```

```
> end:
```

We are ready to try now for the inner product $\left(Y_{1,1}, Y_{1,0}\right)$:

```
> Ipd(Ylm(theta,phi,1,1),Ylm(theta,phi,1,0),theta,phi);
```

$$0$$

A not so trivial normalization integral is calculated readily for $\left(Y_{1,1}, Y_{1,1}\right)$

```
> Ipd(Ylm(theta,phi,1,1),Ylm(theta,phi,1,1),theta,phi);
```

$$\frac{8}{3}\pi$$

We finish the section with some graphical illustrations of occupation probabilities in position space. This is relevant for the understanding of solutions to the Schrödinger equation in three dimensions for spherically symmetric, i.e., central, potentials. Such potentials arise in non-relativistic atomic physics as well as in simple models of nuclear structure. For an intuitive understanding of the occupation probability of the particles in three-space one decomposes the problem into the radial dependence and the angular part. The spherical harmonics provide the angular part to be modulated by the radial probability distribution discussed in the subsequent sections.

The visualization problem of the angular part is not trivial and can be done over the angles θ and ϕ. Eventually we need a presentation in Cartesians as we are used to imagine objects in those coordinates. First we define the squared magnitude of the spherical harmonics.

```
> rho:=proc(theta,phi,m,n);
```

```
> Re(abs(evalf(Ylm(theta,phi,m,n)))^2);
```

```
> end:
```

The probability density does not depend on the angle ϕ, and we can graph it purely as a function of θ, such as done here for $Y_{1,1}$:

```
> plot(rho(theta,0,1,1),theta=0..Pi);
```

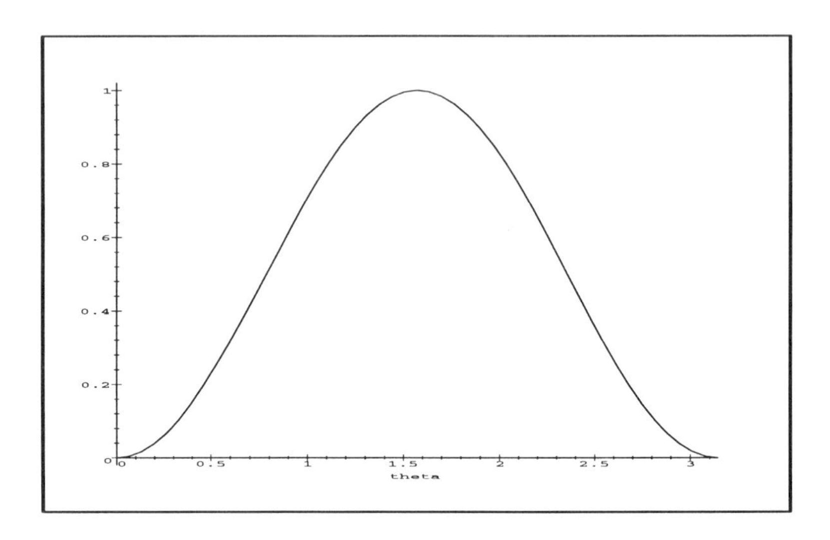

This property is a consequence of the ϕ dependence $\exp(im\phi)$, as demonstrated by the surface plot of the probability for $Y_{2,1}$.

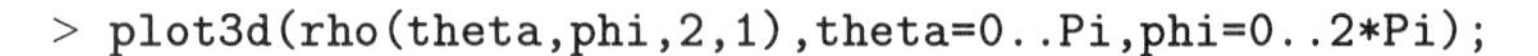

```
> plot3d(rho(theta,phi,2,1),theta=0..Pi,phi=0..2*Pi);
```

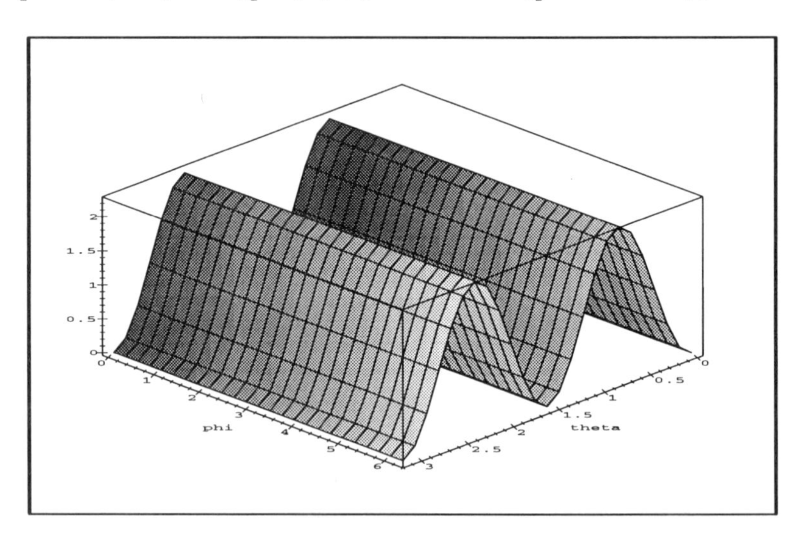

The eigenstates of $\hat{l}_z$ are axially symmetric, i.e., in the azimuthal angle the probability distribution is equal. Of course, one can generate functional dependencies in ϕ by superimposing different eigenstates. This amounts to a Fourier–series decomposition in the angle ϕ.

For further visualization over the Cartesian coordinates it is useful to invoke Maple's `sphereplot` function. It is designed to display objects defined in spherical polar coordinates over (x, y, z). The physicist has to be warned, however, that the assignment of spherical polar coordinates in this function is not in accord with physics tradition. One has to be careful with the order in which the θ and ϕ ranges are supplied. A mix-up results in disaster, i.e., a wrong plot. The routine draws the surface area for constant radius vector $\mathbf{r}$. We choose the previous example, i.e., $l = 2$, $m = 1$. A call to the **plots** package is required.

```
> with(plots):
```

```
> sphereplot(rho(theta,phi,2,1),phi=0..2*Pi,theta=0..Pi);
```

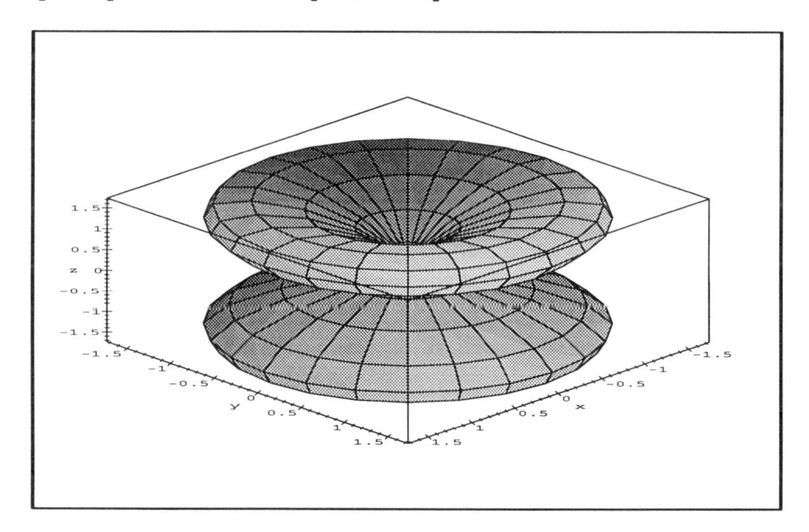

This density is symmetric with respect to the z axis (no dependence on ϕ) as expected.

Exercise 4.1.1: Demonstrate the validity of the CR's $[\hat{l}_i, \hat{l}_j] = i\hbar\hat{l}_k$, where (ijk) is a cyclic permutation of (123). Verify $[\hat{l}^2, \hat{l}_i] = 0$ for $i = 2, 3$.

Exercise 4.1.2: Graph sphereplots of the squared magnitude of the spherical harmonics for $l = 0, 1, 3$ and the possible m sublevels. Develop an intuition as to how an arbitrarily shaped object (begin with an ellipsoid) can be decomposed in spherical harmonics. Draw the density for a linear combination such as $Y_{00} + Y_{10}$.

Exercise 4.1.3: Find the effect of the operators $\hat{l}_\pm = \hat{l}_x \pm i\hat{l}_y$ when acting on a spherical harmonic $Y_{l,m}$ for simple values of l, m.

Exercise 4.1.4: Draw contour diagrams of $\mathrm{e}^{-2r}|Y_{l,m}|^2$ over the $x - z$ plane and compare with the photographic plates given in textbooks for the electron density in an $n = l+1$ state hydrogen atom for different values of l, m. Repeat the same for some radially excited states.

4.2 Radial Equation

The fact that the three-dimensional kinetic energy operator, or basically the Laplacian, permits a separation into a radial kinetic energy plus and angular part that is proportional to the $\hat{l}^2$ operator allows one to treat central potential problems in complete analogy to classical mechanics. The Hamiltonian in the stationary Schrödinger equation (SE) can be written as

$$\hat{H} = \hat{T}_r + V(r) + \frac{\hat{l}^2}{2mr^2}\,.$$

In classical mechanics (e.g., in the Kepler problem [Go80, MT70, Sy71]) one has for central forces the conservation law for angular momentum, and can, therefore, treat l_{cl}^2 as a given constant of the motion specified by the initial conditions. As a result a 1D problem arises in the radial coordinate r with an effective potential:

$$V_{\mathrm{eff}}(r) = V(r) + \frac{l_{\mathrm{cl}}^2}{2mr^2}\,.$$

In QM a similar strategy works. For central potentials the Hamiltonian commutes with the $\hat{l}^2$ operator and the solution to the 3D SE separates into a radial and angular part. The angular part is chosen using eigenstates of $\hat{l}^2$ and $\hat{l}_z$, i.e., spherical harmonics. These were discussed in detail in the previous section.

The radial part is defined through

$$\psi(\mathbf{r}) = R_{n,l}(r)Y_{l,m}(\Omega)\,,$$

where $\Omega = \{\theta, \phi\}$ is the solid angle.

For a given l sector (eigenvalue $\hbar^2 l(l+1)$ for $\hat{l}^2$), we obtain a radial SE with effective potential:

$$V_{\mathrm{eff}}(r) = V(r) + \frac{\hbar^2 l(l+1)}{2mr^2}\,.$$

However, in contrast to classical mechanics a spherical charge distribution corresponds to an eigenvalue of $l = 0$, i.e., no centrifugal potential appears in this case, while circular orbits carry the largest-possible value of angular momentum for a given binding energy.

The analogy with 1D QM problems becomes complete after one rewrites the radial function such that

$$R_{n,l}(r) = \frac{u_{n,l}(r)}{r}\,.$$

As a consequence, the radial kinetic energy simply becomes

$$\hat{T}_r u(r) = -\frac{\hbar^2}{2m}\frac{\partial^2 u}{\partial r^2}\,.$$

The radial distance is restricted to the interval $0 < r < \infty$ and the problem becomes analogous to the one-dimensional SE problem for odd-parity eigenstates. This follows from the constraint that $u(r)$ satisfies the boundary condition $u(0) = 0$, which is required so that the original radial wavefunction $R_{n,l}(r)$ becomes finite at $r = 0$. The first derivative at the origin $u'(r = 0)$ determines the normalization of the wavefunction. We discuss the eigenfunctions for some standard central potentials.

s states of the hydrogen atom

In atomic units we have $\hbar = m_e = e = 1$, where m_e is the electron mass.

```
> Vhyd:=-1/r;    Trad:=arg->-1/2*diff(arg,r$2);
```

$$\mathrm{Vhyd} := -\frac{1}{r} \qquad \mathrm{Trad} := \mathrm{arg} \to -\frac{1}{2}\,\mathrm{diff}(\,\mathrm{arg}, r\,\$\,2\,)$$

```
> Hhyds:=arg->Trad(arg)+Vhyd*arg;
```

$$\mathrm{Hhyds} := \mathrm{arg} \to \mathrm{Trad}(\,\mathrm{arg}\,) + \mathrm{Vhyd}\,\mathrm{arg}$$

We use a radial wavefunction proportional to r at $r = 0$ to verify this eigenvalue/eigenfunction problem:

```
> Hhyds(r*exp(-r))/(r*exp(-r));
```

$$\frac{-1}{2}$$

The ground-state energy of $-1/2$ is obtained in atomic units, and corresponds to $E_{1s} \approx -13.607$ eV.

We proceed with a demonstration that the eigenvalue condition renders a solution to the radial differential equation that is square integrable, i.e., the solution vanishes as $r \to \infty$. For this purpose we use a numerical differential equation solver based on a Runge-Kutta algorithm and solve the problem as an initial value problem at $r = 0$. We use two initial conditions and vary the energy by trial and error:

- (a) $u(0) = 0$ (normalizability);
- (b) $u(0) = 1$ (the numerical value is arbitrary, and fixes the value of the normalization integral if a solution is found);
- (c) the energy eigenvalue is dialed such as to satisfy $u(r = \infty) = 0$.

According to its definition, the function for the radial Hamiltonian `Hhyds` works only if `r` is used as an independent variable. We set up the radial Schrödinger equation:

```
> SE:=Hhyds(u(r))-lambda*u(r);
```

$$\mathrm{SE} := -\frac{1}{2}\left(\frac{\partial^2}{\partial r^2}\,\mathrm{u}(\,r\,)\right) - \frac{\mathrm{u}(\,r\,)}{r} - \lambda\,\mathrm{u}(\,r\,)$$

```
> incond:=u(10^(-Digits))=0,D(u)(10^(-Digits))=1;
```

$$\mathrm{incond} := \mathrm{u}\left(\frac{1}{10000000000}\right) = 0, \mathrm{D}(u)\left(\frac{1}{10000000000}\right) = 1$$

We avoid using the point $r = 0$ due to the singular potential $-1/r$. We verify the ground state:

```
> lambda:=-1/2;      SE;
```

$$\lambda := \frac{-1}{2} \qquad -\frac{1}{2}\left(\frac{\partial^2}{\partial r^2}\mathrm{u}(r)\right) - \frac{\mathrm{u}(r)}{r} + \frac{1}{2}\mathrm{u}(r)$$

```
> sol:=dsolve({SE,incond},{u(r)}, type=numeric,
> output=listprocedure):
```

```
> soln:=subs(sol,u(r)):
```

The resulting procedure allows the solution to be calculated at any point r, e.g., $r = 0.1$:

```
> soln(0.1);
```

$$.09048381919150940$$

We plot by generating an explicit sequence of points:

```
> pseq:=seq([n/20,soln(n/20)],n=1..100):
```

```
> plot([pseq],r=0..5,style=point,title=`s-wave E=-1/2`);
```

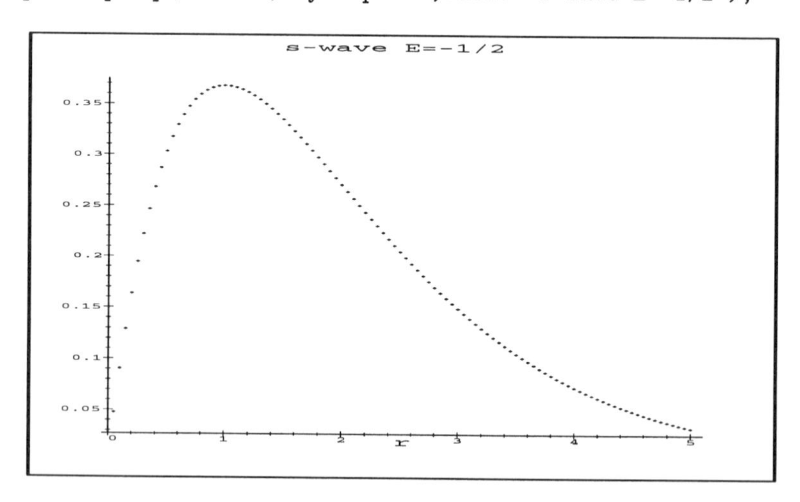

Now we search for the solution that corresponds to the 2s state:

```
> lambda:=-1/8:
```

```
> sol:=dsolve({SE,incond},{u(r)}, type=numeric,
> output=listprocedure):
```

```
> soln:=subs(sol,u(r)):
> pseq:=seq([n/20,soln(n/20)],n=1..400):
> plot([pseq],r=0..20,style=point,title=`s-wave E=-1/8`);
```

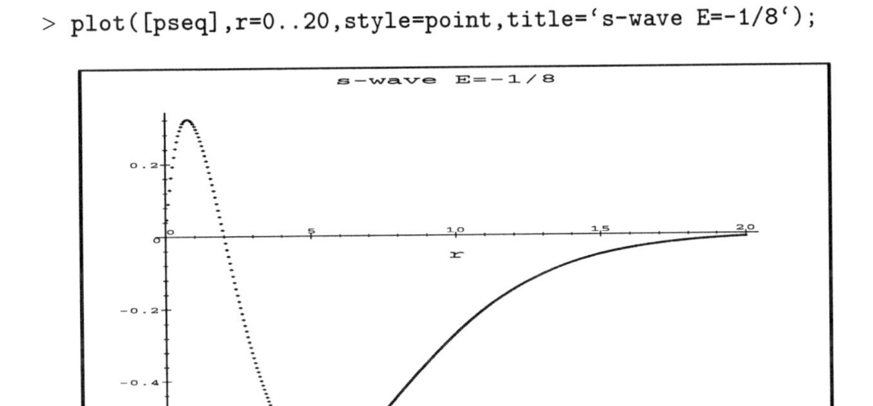

Now for pedagogical purposes we try an energy value in-between:

```
> lambda:=-1/4:
> sol:=dsolve({SE,incond},{u(r)}, type=numeric,
> output=listprocedure):
> soln:=subs(sol,u(r)):
> pseq:=seq([n/20,soln(n/20)],n=1..400):
> plot([pseq],r=0..12,-4..2,style=point,title=`s-wave E=-1/4`);
```

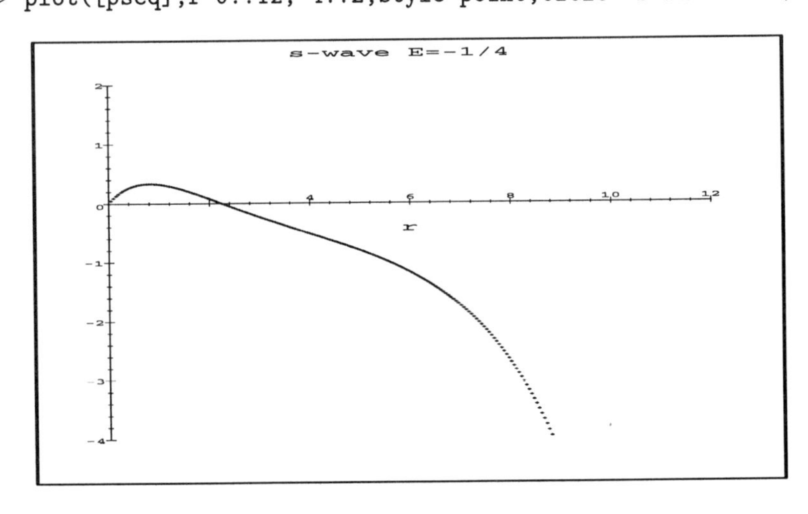

Clearly this solution to the differential equation is not normalizable, and its square cannot be interpretated as a probability. As a result $E = -1/4$ is not an acceptable energy eigenvalue for this Hamiltonian.

p states of the hydrogen atom

Now we demonstrate the accidental degeneracy in the hydrogen atom on the example of the first two degenerate levels 2s and 2p. For $E = -1/8$, i.e. $n = 2$ (according to the Balmer formula: $E_n = -1/(2n^2)$) we have determined previously the first excited state in the s wave sector ($l = 0$), i.e., the 2s state.

We solve the eigenvalue problem in the p-wave sector ($l = 1$) and find that the ground state has a value of $E = -1/8$ in this sector. This is an accidental degeneracy (coming from two different radial eigenvalue equations) that arises as a consequence of a dynamical symmetry. There is a conserved quantity in the problem called the Runge-Lenz vector. For a detailed discussion we refer to classic texts (e.g. [Sch68]) and to a computer algebra based one [Fe94] which shows how to solve the hydrogen atom problem by algebraic methods, i.e., making use of symmetries. Note that with $l(l+1) = 2$ and while in atomic units we have the effective Hamiltonian:

```
> Hhydp:=arg->Hhyds(arg)+arg/r^2:
```

```
> lambda:=-1/8:          SE:=Hhydp(u(r))-lambda*u(r);
```

$$\mathrm{SE} := -\frac{1}{2}\left(\frac{\partial^2}{\partial r^2}\mathrm{u}(r)\right) - \frac{\mathrm{u}(r)}{r} + \frac{\mathrm{u}(r)}{r^2} + \frac{1}{8}\mathrm{u}(r)$$

The effective potential for p states has acquired a centrifugal part:

```
> plot(-1/r+1/r^2,r=0..20,Veff=-0.5..1);
```

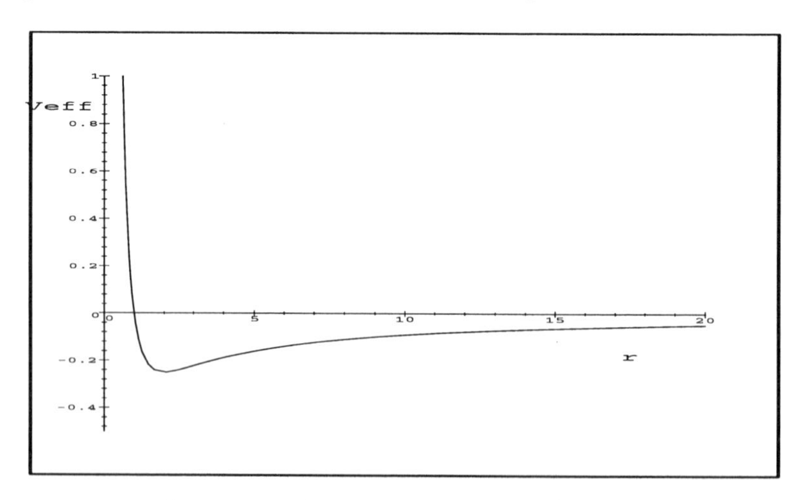

As before, we define a procedure for the solution of the SE:

```
> sol:=dsolve({SE,incond},{u(r)}, type=numeric,
```

```
> output=listprocedure):     soln:=subs(sol,u(r)):
> pseq:=seq([n/20,soln(n/20)],n=1..400):
```

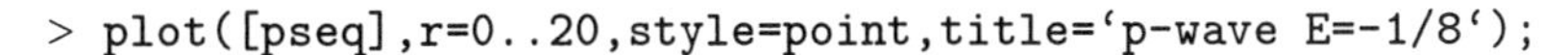

```
> plot([pseq],r=0..20,style=point,title='p-wave E=-1/8');
```

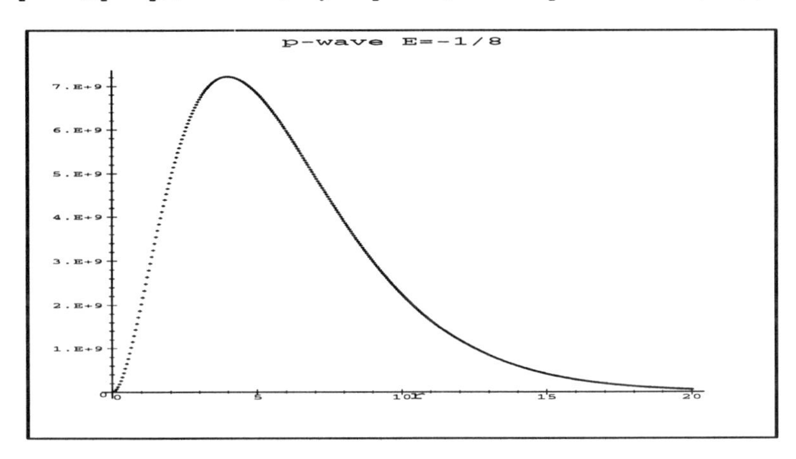

We note the following

- The normalization has gone crazy here, but it is irrelevant. The eigenvalue problem is homogeneous and we can simply rescale the solution.
- The p-wave solution starts with a higher power than linear at $r = 0$, namely quadratically. Thus, our choice of $u'(0) = 1$ is not a reasonable one and causes the mess in the normalization. This can be avoided by an appropriate redefinition of $u(r)$. The slow growth at small r is a consequence of the strong repulsion due to the centrifugal potential.
- This wavefunction is not orthogonal to $u_{1s}(r)$ or $u_{2s}(r)$ (using a 1D inner product in r space), as they were obtained from a different radial equation. Orthogonality in 3D space is guaranteed by the angular part (different l, m in the spherical harmonics).
- This radial function is to be combined with three different angular eigenfunctions to form the distinct states $2p_0$ $(m = 0)$, $2p_1$ $(m = 1)$, and $2p_{-1}$ $(m = -1)$. Thus, it is triply degenerate. This degeneracy arises as a consequence of the spherical symmetry in the potential.
- The accidental 2s-2p symmetry (and infinitely many higher ones, the lowest d state, i.e., 3d is also degenerate with excitations in the s-state and p-state sectors, namely 3s and 3p, etc.) is related to the Kepler problem itself, i.e., the $-1/r$ potential. This symmetry (degeneracy) is lifted in central field approximations to atomic structure. For example in a lithium atom with a $(1s)^2$ - $(2s)$ configuration in the ground state, the (2p) level is not degenerate with (2s), as the effective potential is not of simple $-Z/r$ type. Close to the nucleus it varies as $-Z/r$, while at large distances it behaves like $-1/r$. The latter is obvious, as it is the effective potential for one of the neutral atom's electrons and it experiences a nuclear attraction that

is screened by the two inner electrons. Close to the nucleus screening is (almost) irrelevant.

Exercise 4.2.1: Verify the predictions of the Balmer formula for $n = 1, 2, 3$ in all relevant l sectors. Store the numerically obtained wavefunctions on a mesh as used for plotting, normalize them and graph the radial probability densities.

Exercise 4.2.2: Following the advice of the preceding exercise calculate the expectation values for kinetic and potential energies for the numerical wavefunctions defined on a mesh. Use the discretized kinetic energy operator. Verify the virial theorem.

Exercise 4.2.3: Find the bound-state eigenvalues and eigenfunctions for other sphericially symmetric problems, such as a Gaussian well $V(r) = -V_0 \exp(-\beta^2 r^2)$. How many bound states can be found for given V_0 and β?

Exercise 4.2.4: Find the eigenfunctions for a spherical shell potential that could be used in a model to describe spherically symmetric carbon clusters (called fullerenes or buckyballs), such as C_{60}, namely $V(r) = -V_0 \exp(-\beta^2 (r - r_0)^2)$. In the case of C_{60} one can assume a sphere radius of $r_0 = 7$ a.u. and a width of $\beta^{-1} = 2$ a.u. Given that the 1s electrons remain bound by the individual nuclei one can assume that the bottom of the potential well should reach about $V_0 = -4$ atomic units.

Exercise 4.2.5: Consider a simple model for the description of the valence electron of a lithium atom. Suppose the two core electrons in the $1s^2 2s$ ground-state configuration are described by hydrogenic orbitals resulting in a density $\rho_{1s}(r) = \alpha^3 \exp(-\alpha r)/(4\pi)$ with $\alpha \approx 5.38$ a.u. [CR74]. According to Poisson's equation this results in an overall electrostatic potential experienced by the valence electron of the form

$$V(r) = -\frac{3}{r} + \frac{2}{r}[1 - (1 + \frac{\alpha r}{2} \exp(-\alpha r)] ,$$

where the first contribution describes the attraction to the nucleus, while the second represents the repulsion from the core. Using the radial part of the Laplacian verify, using Poisson's equation, that the second part is indeed consistent with the charge density given above, and normalized properly such that the density represents a charge of $N = 2$ electrons. Graph the potential together with the asymptotic limits $-3/r$ (for small r), and $-1/r$ (for large r). Calculate the 2s and 2p levels to demonstrate the breaking of the dynamical 2s-2p degeneracy present in a pure Coulomb potential. How does the first excitation threshold, as calculated here, compare to a model potential result of 1.84 eV [Pe78, Pe82, WH+89]?

4.3 Atomic Model

We present a simple central-field approximation that can be used to discuss the structure of spherically symmetric closed-shell atoms without a self-consistent field scheme. It represents a very naive independent particle model (IPM). The mathematical foundation for such an approximation scheme to the N particle problem is discussed in Sect. 7.1 and 7.2.

For the reader with some background in the many-electron problem I make the following remarks. Exchange effects are treated in a very approximate fashion: the Hartree-Fock-Slater local-density approximation requires an asymptotic correction to cancel the self-energy contribution present in the direct interaction term. In exact Hartree-Fock theory the self-energy contributions from the direct and exchange contributions cancel perfectly. The Slater approximation to the exchange integral necessicates an ad-hoc correction in the effective interaction potential. We incorporate this so-called Latter correction in the choice of the effective potential.

We assume that a simple effective central potential can be found that interpolates between the purely nuclear attraction at small distances $-Z/r$ and the asymptotically required behaviour of $-1/r$. An elaborate potential model that is competitive with self-consistent field calculations can be found in [GS+71].

The screening function depends on a single parameter that determines at which distances the changeover occurs between the asymptotic values of the charge experienced by an electron in the model atom. One can try to adjust the parameter to observe some of the main features of a self-consistent field calculation. The effective single-particle potential is defined via an effective charge function (screening function) $Z_{\mathrm{S}}(r)$ as

$$V(r) = -\frac{Z_{\mathrm{S}}(r)}{r},$$

where $Z_{\mathrm{S}}(r)$ approaches the nuclear charge Z as $r \to 0$ and $Z_{\mathrm{S}}(r \to \infty) \to 1$.

One can easily extend this to treat ionic structure by requesting a different asymptotic charge value, such as $Z(\infty) = q + 1$, where $q = Z - N$ is the residual positive charge of the ion, and N is the number of bound electrons. The adjustable parameter to set the scale where the screening changes is b_0.

```
> Zeff:=(Z,b0,r)->Z*(1+(b0/Z)*r)/(1+b0*r);
```

$$\mathrm{Zeff} := (Z, b0, r) \to \frac{Z\left(1 + \frac{b0\, r}{Z}\right)}{1 + b0\, r}$$

```
> plot({Zeff(10,1,r),Zeff(10,2,r),Zeff(10,3,r)},r=0..5,
> Zef=0..10);
```

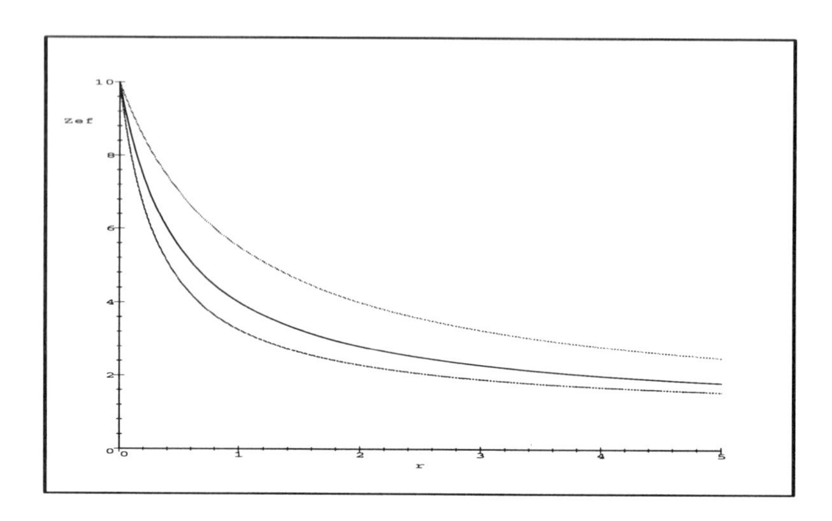

For better results one should probably use a multi-parameter form that allows more flexibility in the interpolation between the known asymptotic limits.

In an IPM calculation for the neon atom the 1s level is occupied twice (spin degeneracy), the 2s level twice, and the 2p level six times. We cannot determine these orbitals exactly. Rather than making use of the numerical technique of the previous section we use the variational method with hydrogenic orbitals. In order to just calculate radial integrals we normalize the radial functions to unity and assume that a spherical harmonic $Y_{00} = 1/\sqrt{4\pi}$ is attached to them, such as to cancel the 4π from the $d\Omega$ integration in the normalization integral. We will, in fact, simply normalize the energy expression.

We use orbitals that are multiplied by r, namely $R_{nl}(r) = u_{nl}(r)/r$, in order to have a simple radial kinetic energy. As a result the radial integrals in the inner products do not carry an r^2 factor.

For a first attempt we choose $b_0 = 3$ in anticipation that the neon atom is relatively compact. We define an atomic radial Hamiltonian for s states:

```
> Hatoms:=arg->int(arg*(-1/2*diff(arg,r,r)-Zeff(10,3,r)/r*arg)
> ,r=0..infinity)/int(arg^2,r=0..infinity);
```

$$\text{Hatoms} := \text{arg} \rightarrow \frac{\displaystyle\int_0^\infty \text{arg}\left(-\frac{1}{2}\left(\frac{\partial^2}{\partial r^2}\,\text{arg}\right) - \frac{\text{Zeff}(10,3,r)\,\text{arg}}{r}\right) dr}{\displaystyle\int_0^\infty \text{arg}^2\, dr}$$

For p states ($l = 1$) we have to add the correct centrifugal term:

```
> Hatomp:=arg->int(arg*(-1/2*diff(arg,r,r)-Zeff(10,3,r)/r*arg
> +1/r^2*arg),r=0..infinity)/int(arg^2,r=0..infinity);
```

$$\text{Hatomp} := \text{arg} \to \frac{\displaystyle\int_0^\infty \text{arg}\left(-\frac{1}{2}\left(\frac{\partial^2}{\partial r^2}\text{arg}\right) - \frac{\text{Zeff}(10,3,r)\,\text{arg}}{r} + \frac{\text{arg}}{r^2}\right) dr}{\displaystyle\int_0^\infty \text{arg}^2\, dr}$$

The hydrogenic variational wavefunctions are defined for the 1s, 2s, and 2p levels, respectively:

```
> psi10:=(r,g)->r*exp(-g*r);
```

$$\psi 10 := (r, g) \to r\, e^{(-r\, g)}$$

```
> psi20:=(r,g)->r*(1-g*r)*exp(-g*r/2);
```

$$\psi 20 := (r, g) \to r\,(1 - r\, g)\, e^{(-1/2\, r\, g)}$$

```
> psi21:=(r,g)->r*r*exp(-g*r/2);
```

$$\psi 21 := (r, g) \to r^2\, e^{(-1/2\, r\, g)}$$

We keep in mind that according to the Rayleigh-Ritz theorem we are allowed to optimize only the lowest state's energy within each symmetry sector. We are not permitted to optimize E_{2s} as a function of the wavefunction parameter g as we have no guarantee of being orthogonal to the true (exact) ground state. Also orthogonality to the approximate ground state is guaranteed only if the same g value is used as for the 1s state.

The ground-state energy for the given model potential is a function of the variational parameter g:

```
> assume(g>0);      E1s:=Hatoms(psi10(r,g));
```

$$E1s := \frac{1}{2}\left(8\, g^{\sim\,-2}\, e^{(2/3\, g^\sim)}\, \text{Ei}\left(1, \frac{2}{3}\, g^\sim\right) - 11\, g^\sim - 2\right) g^\sim$$

Our first guess at the wavefunction parameter is based on the hydrogen result $g = Z$:

```
> evalf(subs(g=10,E1s));      evalf("*27.12*eV);
```

$$-30.8403232 \qquad -836.3895652\, eV$$

We graph the calculated ground-state energy as a function of the variational parameter g to find the minimum value.

```
> plot(E1s,g=7..12,E1=-32..-28);
```

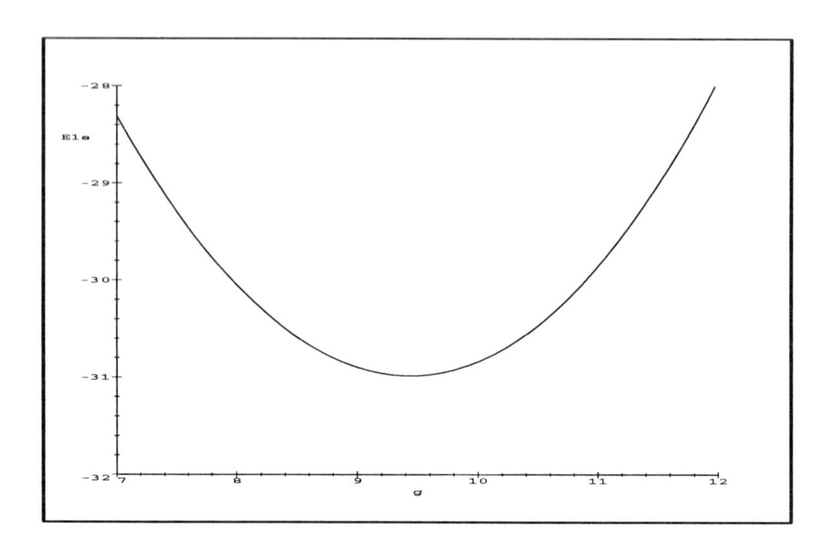

It is evident from the variational calculation that there are sizeable outer-screening effects. The no-screening approximation would have given for the wave function $g = Z$ and for the energy $E = -Z^2/2 = -50$ a.u. Our model potential overestimates somewhat inner screening: in a Hartree-Fock (HF) calculation [CR74] one obtains an orbital binding energy of about -32.8 a.u. per 1s orbital. From the graph we read off the approximate optimum value $g_{\text{opt}} \approx 9.5$ and evaluate the orbital energy:

```
> evalf(subs(g=9.5,E1s));        evalf("*27.12*eV);
```

$$-30.9850483 \qquad -840.3145099\, eV$$

We can minimize the energy of the 2p state:

```
> E2p:=Hatomp(psi21(r,g));
```

$$E2p := \frac{1}{216}\left(-54 - 3\,g^{\sim 3} - 27\,g^{\sim} + 9\,g^{\sim 2} + g^{\sim 4}\,e^{(1/3\,g^{\sim})}\,\mathrm{Ei}\left(1, \frac{1}{3}\,g^{\sim}\right)\right)\,g^{\sim}$$

The answer involves an exponential integral. Maple has no difficulty to evaluate this special function numerically. An interesting observation can be made about the variational method from the following graph. The value of g for which the 2p-state energy is minimized is different from the g value that minimizes the 1s state.

```
> plot(E2p,g=4.5..8);
```

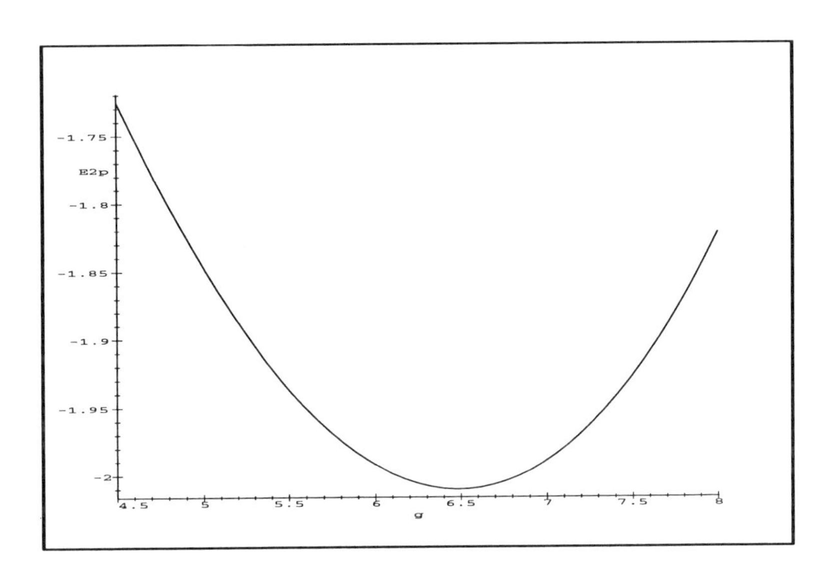

Again, we lift the optimum value $g_{opt} \approx 6.5$ from the figure and substitute it into the expression for the 2p orbital energy.

```
> evalf(subs(g=6.5,E2p));     evalf("*27.12*eV);
```

$$-2.01034321 \qquad -54.52050786\, eV$$

This does not compare well with the HF result of $E_{2p} \approx -0.85$ a.u. [CR74]. The reason for the overbinding of the higher levels in our model can be found in the screening function which decreases too slowly towards unity as r increases. Finally, we just calculate the 2s energy

```
> E2s:=Hatoms(psi20(r,g));
```

$$E2s := \frac{1}{504}\left(-108 + 24\, g^{\sim 3}\, e^{(1/3\, g^{\sim})}\, \mathrm{Ei}(1, \frac{1}{3}\, g^{\sim}) + 4\, g^{\sim 4}\, e^{(1/3\, g^{\sim})}\, \mathrm{Ei}(1, \frac{1}{3}\, g^{\sim}) \right.$$
$$\left. -81\, g^{\sim} - 36\, g^{\sim 2} - 12\, g^{\sim 3} + 36\, g^{\sim 2}\, e^{(1/3\, g^{\sim})}\, \mathrm{Ei}(1, \frac{1}{3}\, g^{\sim})\right) g^{\sim}$$

```
> evalf(subs(g=9.5,E2s));     evalf("*27.12*eV);
```

$$-5.1241800 \qquad -138.9677616\, eV$$

Again, this level is overbound, a more realistic value is given by the HF result of $E_{2s} \approx -1.93$ a.u. [CR74].

```
> plot(E2s,g=5..30,E20=-8..0);
```

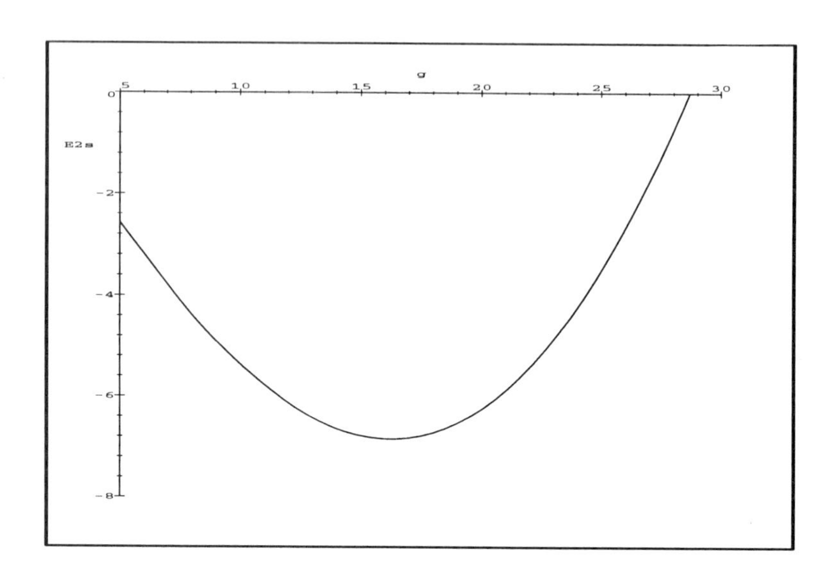

It is certainly incorrect to minimize the 2s energy as a function of g! The graph shows that the state maximizes its overlap with the true ground state at some very large value of $g \approx 16$ and gives a misleading result. Instead we use $g_{1s} \approx 9.5$ determined before which provides an orbital energy $E_{2s} \approx -139$ eV. We note the strong splitting of 2s and 2p due to the fact that the effective charge Z_{eff} varies as a function of r.

The estimates obtained for the orbital energies are:

$E_{1s} = -2 \times 840$ eV (the factor of 2 comes from the spin degeneracy, i.e., the orbital can be occupied by a spin-up as well as a spin-down electron);

$E_{2s} = -2 \times 139$ eV;

$E_{2p} = -6 \times 54.5$ eV (spin and l degeneracy).

This can be summed to provide a total energy of

```
> 2*(840+139+3*54.5)*eV;
```

$$2285.0\, eV$$

In a more sophisticated IPM, such as the HF approximation [CO80] to the many body problem the proper treatment of so-called exchange effects leads to the result that the total energy is given by the sum over the orbital energies only in a lowest-order approximation [GR86, GRH91, CO80]. In this approximation the total energy of the neon atom comes out as $E \approx -3485$ eV, which is significantly below the sum over the HF orbital energies $E^{(0)} \approx -2020$ eV [CR74].

To calculate the experimentally easily accessible ionization energy one would have to repeat the calculation for Ne^+ and subtract the total energies. It is not quite correct to simply subtract the energy for the last occupied level in the neutral atom calculation. For Ne the observed ionization potential is about 21.5 eV.

The ionization potential for subsequent electron removal is considerably larger, even though it is from the same 2p shell. The independent particle model 2p shell for neutral Ne, therefore, cannot be simply considered to provide the energy and wavefunction for the last electron in the Ne atom. The level is representative for all Ne(2p) electrons in the neutral atom, which may not be realistic, but is in accord with the simplistic nature of an IPM.

To indicate the artificial nature of occupied orbitals in an IPM one gives them the name of quasi-particle levels. For atoms with valence electrons (e.g., alkalis) the difficulty is not pronounced and one has also the possibility to discuss excited states within an IPM meaningfully.

An interesting problem is the determination of the total energy of atoms across the periodic table, i.e., as a function of Z. We note the large contribution of the K shell (or inner shells in general) towards the total energy.

We conclude with some general remarks about atomic structure, i.e., the problem of predicting the electronic shell structure throughout the periodic table of elements. Within the framework of an IPM it is straightforward to consider, e.g., how the first nontrivial row between the elements beryllium and neon is built up by filling the up to six possible m sublevels of the 2p shell (including the spin degeneracy). A closer inspection of the problem reveals, however, that the answer is more complicated. The incomplete population of the 2p shell results in an electrostatic potential that is no longer spherically symmetric (cf. exercise 4.3.2). However, for a non-central potential the Schrödinger equation is not solved by a simple product of a radial orbital times a spherical harmonic, but requires an expansion in these. Thus, the treatment of open-shell systems requires more work[CO80] and the coupling of orbital angular momenta (cf. Sect. 7.4), in particular. Different ground states with conserved total angular momentum can be calculated for atoms like boron, or nitrogen[CR74].

Exercise 4.3.1: Perform an analogous calculation for the argon atom ($Z = 18$) and compare your results with the Hartree-Fock orbital energies in atomic units $E_{1s} = -118.6$, $E_{2s} = -12.32$, $E_{3s} = -1.28$, $E_{2p} = -9.57$, $E_{3p} = -0.59$ [CR74].

Exercise 4.3.2: Using the spherical harmonics provided in Sect. 4.1 show that the probability density in a closed-shell atom such as neon is spherically symmetric. Use Poisson's equation in spherical polar coordinates to calculate the electrostatic potential produced by the Ne($1s^2$, $2s^2$, $2p^6$) configuration.

Exercise 4.3.3: Estimate the ionization energy in neon by performing an analogous model calculation for the singly ionized neon atom.

4.4 Zeeman Splitting

In this short section we discuss the behaviour of the hydrogen atom in a homogeneous magnetic field following the discussion of problem 65 in [Fl71]. The treatment of magnetic fields in nonrelativistic QM requires either

- the Pauli equation (includes spin), or
- an extended Schrödinger equation in which the momentum operator has been replaced by canonical momentum in the presence of a vector potential.

The vector potential for a homogeneous magnetic field of strength B in the z direction, $\mathbf{B} = (0, 0, B)$, is given in Cartesian coordinates as

$$\mathbf{A} = \left(-\frac{By}{2}, \frac{Bx}{2}, 0\right).$$

This is verified in Maple from $\mathbf{B} = \nabla \times \mathbf{A}$:

```
> with(linalg):     curl([-B*y/2,B*x/2,0],[x,y,z]);
```

$$[0\ 0\ B]$$

The extended Schrödinger equation, ignoring terms quadratic in B, is given by:

$$(T + V)\psi + \frac{\mathrm{i}\hbar e}{\mu c}(\mathbf{A} \cdot \nabla\psi) = E\psi \quad ,$$

where μ is the particle (electron) mass.

The additional term due to the vector potential can be written as

$$\frac{\mathrm{i}\hbar e B}{2\mu c}\frac{\partial\psi(r, \theta, \phi)}{\partial\phi} \quad .$$

This operator is, however, proportional to the $\hat{l}_z$ angular momentum operator ($\hat{l}_z = -\mathrm{i}\hbar\partial/\partial\phi$). Thus, the Schrödinger equation is separated again by spherical harmonics. Substitution of the hydrogen atom eigenstates $R_{nl}(r)Y_{lm}(\Omega)$ shows that the $\hat{l}_z$ operator can be replaced by its eigenvalue ($\hbar m$) while acting on the eigenstate. Therefore, we observe an m-dependent shift of the hydrogenic energy levels

$$E_{nl} = -\frac{1}{2n^2} \to E'_{nlm} = -\frac{1}{2n^2} + \frac{B}{2c}m \quad .$$

This modification of the energy level gave the m quantum number the name magnetic quantum number. In atomic units the speed of light equals the inverse of the fine structure constant $c = \alpha^{-1} \approx 137$, and, therefore, the energy levels are given approximately by

```
> Enlm:=(n,l,m)->-1/(2*n^2)-B/(2*137)*m;
```

$$Enlm := (n, l, m) \to -\frac{1}{2}\frac{1}{n^2} - \frac{1}{274}B\,m$$

We display graphically how some of the degeneracies in the $n = 2$ levels are broken. The $2p_0$ level doesn't shift and remains degenerate with 2s, while $2p_1$ and $2p_{-1}$ split linearly as a function of B.

```
> plot({Enlm(2,0,0),Enlm(2,1,0),Enlm(2,1,1),Enlm(2,1,-1)},
> B=0..0.1);
```

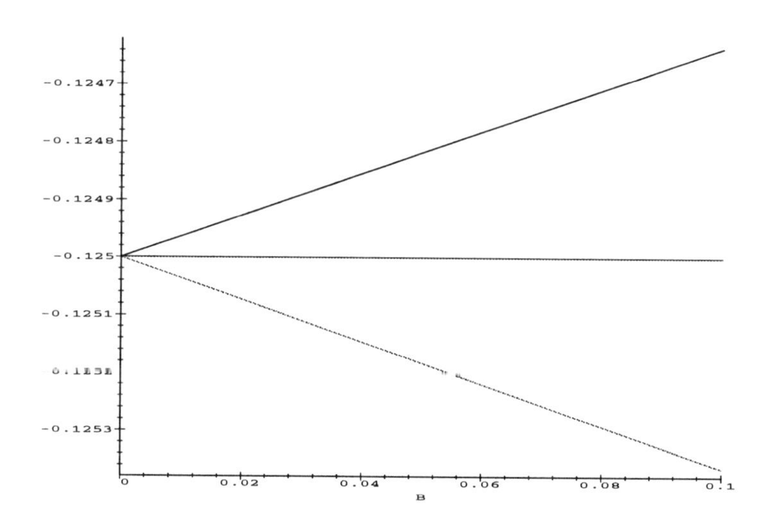

The considerations can be generalized to include the electronic spin degree of freedom. The additional interaction term in the Pauli Hamiltonian

$$\boldsymbol{\sigma} \cdot \mathbf{B} = \left(\sigma_x, \sigma_y, \sigma_z\right) \cdot \left(B_x, B_y, B_z\right)$$

becomes, due to the choice of axes, $[e\hbar B/(2\mu c)]\sigma_z$, which again is a diagonal operator for spin orbitals.

Therefore, the level splitting is proportional to $m_j = m_l + m_s$. As a consequence there is a splitting of the 1s ground state in a magnetic field $(m_j = m_s)$ into a doublet and also the 2p levels split according to $m_s = \pm 1/2$ and into two triplets $(m_l = 1, 0, -1)$. Spontaneous decay occurs with sizeable probability, however, only between states of the same spin projection and the $|\Delta m| = 1$ selection rule allows only transitions for the $m_l = \pm 1$ states to the ground state.

The Zeeman effect considered here for a simple Schrödinger-Pauli model atom forms the basis for a deeper understanding of atomic structure. Orbital angular momentum with a non-zero magnetic quantum number gives rise to an electric current around the z axis. This current induces a magnetic field that interacts with the spin of the particle. This results in an additional coupling of type

$$\mathbf{l} \cdot \mathbf{s} = (l_x, l_y, l_z) \cdot (s_x, s_y, s_z) .$$

A proper treatment is given in a relativistic framework (Dirac equation) in which the spin-orbit couplings emerge from first principles.

Exercise 4.4.1: Using the linearized treatment of the magnetic field, estimate at which strength of the magnetic field the energy level for the $2p_1$ state crosses that of the $3d_{-2}$ state. Note that the estimate is unrealistic, as it ignores effects quadratic in B.

Exercise 4.4.2: Consider a model that describes single-electron excitations of helium atoms of type He(1s,nlm) by assuming that one electron forms a core represented by the hydrogen-like He^+(1s) state, while the other (excited) electron forms a valence-type orbital around the core. Make an educated guess, on the basis of the relevant orbital probability densities, from which value of n on the valence orbital should be very similar to the hydrogen atom orbitals. Calculate and graph the energy levels in the He(1s,4f) shell as a function of the magnetic field strength B.

Exercise 4.4.3: A nuclear shell model is deemed successful, if it is capable to explain within the framework of an independent particle model the binding energies of stable nuclei for different proton and neutron number. This is analogous to the atomic structure problem, where the strong binding of the rare gas atoms is explained by the formation of completely filled electron shells. Particularly strong binding occurs in nuclei with filled neutron and proton shells (magic numbers 2, 8, 20, 28, etc.)[RS80, chapter 2]. For the low-lying levels it is sufficient to consider either an isotropic harmonic oscillator or a square well potential in order to obtain the single-particle energy levels. These potentials have a different degeneracy structure. The harmonic oscillator model is defined by

$$V(r) = \frac{m}{2}\omega^2(r^2 - R_0^2) ,$$

where the nuclear radius is given as $R_0 = 1.2A^{1/3}$ [fm], A is the nuclear mass number, and $\hbar\omega \approx 41A^{-1/3}$ [MeV]. Obtain solutions for the three lowest energy levels by separating in spherical polar coordinates and compare with the solution in Cartesian coordinates. The energy spectrum is given by

$$E_N = \hbar\omega\left(N + \frac{3}{2}\right) - \frac{m}{2}\omega^2 R_0^2 ,$$

with $N = 2(n-1) + l$, where the radial quantum number $n = 1, 2, ...$, and the angular momentum quantum number $l = 0, 1, 2, ...$. At $N = 2$ a degeneracy occurs between the $l = 0$ radial $n = 2$ excited state and the $l = 2$, $n = 1$ angular momentum excited state.

Based on the developments in Sect. 7.4 one can show that this degeneracy is lifted by a spin-orbit coupling term by calculating its expectation value using $2\mathbf{l} \cdot \mathbf{s} = \mathbf{j}^2 - \mathbf{l}^2 - \mathbf{s}^2$, where $\mathbf{j} = \mathbf{l} + \mathbf{s}$.

4.5 DC Stark Effect

Consider an atom exposed to an external electric DC field acting along a fixed direction (e.g., the z axis), and investigate what influence the field has on atomic structure. This problem is considered in this section for two kinds of atoms with one 'active' electron:

- (a) alkali atoms (no 2s-2p degeneracy);
- (b) hydrogen(-like) atoms.

Obviously, this is a case for perturbation theory (PT), introduced in Sect. 1.5, as the external macroscopic field is small compared to the nuclear Coulomb attraction. A warning is, however, in order: the asymptotic behaviour of the potential for a constant electric field is very different from the Coulomb potential, which vanishes asymptotically. For strong enough fields the atomic eigenstates are no longer stationary, but decay with an appreciable tunneling ionization rate as discussed in Sect. 3.3. The appearance of this phenomenon is not obvious from the application of perturbation theory, as long as one ignores couplings to states which have an appreciable amplitude in the asymptotic domain.

In the case of no degeneracy in the 2s-2p levels, perturbation theory can be used straightforwardly. The total Hamiltonian is given by

$$H = H_0 + e|E|z = T + V(r) + e|E|z \, ,$$

where E is the strength of the electric field. We denote atomic energy eigenvalues as e_{nlm} in this section.

The perturbing matrix element calls for the calculation of the dipole operator, in particular, the matrix element $\langle nlm|z|nlm\rangle$. For the non-degenerate (s) states it is the only required matrix element in first-order perturbation theory and it is obvious that it vanishes as demonstrated below.

It should be possible to obtain the unperturbed hydrogen atom solution from the confluent hypergeometric function. This we found difficult to accomplish with Maple's **hypergeom** function (cf. Sect. 8.4), and, thus, we use the Laguerre polynomials instead:

```
> with(orthopoly);
```

$$[G, H, L, P, T, U]$$

```
> Rnl:=proc();

> n:=args[1]; l:=args[2];

> if type(n,integer) <> true then RETURN(` 1^st argument
> integer please`,n) fi;

> if type(l,integer) <> true then RETURN(` 2^nd argument
> integer please`,l) fi;
```

```
> if l>n then RETURN(`l cannot be greater than n. n,l = `,n,l)
> fi;
> n1:=n-l-1; l1:=2*l+1;
> n1!*l1!/(n+l)!*L(n1,l1,2/n*args[3])*r^l*exp(-args[3]/n);
> end:
```

For example the radial 3s wavefunction is calculated as:

```
> Rnl(3,0,r);
```

$$\frac{1}{3}\left(3-2\,r+\frac{2}{9}\,r^2\right)\,e^{(-1/3\,r)}$$

Now calculate the normalization integral for the 1s state.

```
> int(r^2*Rnl(1,0,r)^2,r=0..infinity);
```

$$\frac{1}{4}$$

Clearly the radial functions are not normalized properly. Thus, we divide each matrix element by the norm integral to correct for this. We proceed to define the angular parts by making use of the Legendre polynomials. The procedures are lifted from Sect. 4.1.

```
> Plm:=proc(theta,l,m);
> if type(l,integer)=true and type(m,integer)= true then
> x:=cos(theta);
> if m>0 then fak:=subs(y=x,diff(P(l,y),y$m));
> else fak:=subs(y=x,P(l,y)); fi;
> (-1)^m*(sin(theta)^m)*fak;
> fi;
> end:
```

We try the procedure for $P_{3,2}(\theta)$:

```
> Plm(theta,3,2);
```

$$15\sin(\theta)^2\cos(\theta)$$

and define the spherical harmonics.

```
> Ylm:=proc(theta,phi,l,m);
> m1:=abs(m);
> if m1>l then RETURN(`in Ylm l < |m|`,l,m); fi;
```

```
> if type(l,integer) <>true and type(m,integer)<>true then
> RETURN('in Ylm l and or m not integers ',l,m); fi;

> exp(I*m*phi)*Plm(theta,l,m1);

> end:
```

Thus, $Y_{1,0}(\theta, \phi)$ is calculated as

```
> Ylm(theta,phi,1,0);
```

$$\cos(\theta)$$

Now we calculate the complete state specified by the quantum numbers $n = 2$, $l = 1$, $m = 1$, i.e., the yet unnormalized $2p_1$ state:

```
> Rnl(2,1,r)*Ylm(theta,phi,1,1);
```

$$-r\,e^{(-1/2\,r)}\,e^{(I\,\phi)}\sin(\theta)$$

We assume now that the 2s and 2p levels are not degenerate, as is true for alkali atoms and other atoms with an effective single-electron potential more complicated than $1/r$. The dipole operator does not induce a linear correction to the 2s level alone, since $z = r\cos(\theta)$, and the matrix element vansishes due to $P_0(\mu) = 1$, $P_1 = \mu = cos(\theta)$:

$$\langle P_0|P_1|P_0\rangle = \langle P_1\rangle = \int_{-1}^{1} \mu \mathrm{d}\mu = 0\,.$$

There is, however, a second-order contribution, e.g., from the inclusion of an intermediate $2p_0$ state

$$\Delta e_{2s} = e^2|E|^2 \frac{\langle 2s|r\cos(\theta)|2p_0\rangle\langle 2p_0|r\cos(\theta)|2s\rangle}{e_{2s} - e_{2p0}}\,,$$

where e_{2s} and e_{2p0} are the unperturbed energy levels of the 2s and $2p_0$ states, respectively.

The complete energy-level correction includes an infinite sum over intermediate states in the above expression instead of just the $2p_0$ contribution. Often this infinite sum is dominated by the closely lying states due to the energy denominator. It is evident that the second-order expression cannot be used if a degeneracy is present, i.e., $e_{2s} = e_{2p0}$.

The second-order expression displays a characteristic dependence on the external field strength, i.e., it can be recognized by the $|E|^2$ dependence. A typical matrix element such as $\langle 2p_0|z|2s\rangle$ is calculated in spherical polar coordinates with the integration measure $r^2 \mathrm{d}r \sin\theta \mathrm{d}\theta \mathrm{d}\phi$ with the three integrations done successively. First we perform the ϕ integration:

```
> Iphi:=int(r^2*sin(theta)*r*cos(theta)*Rnl(2,1,r)*Ylm
> (theta,phi,1,0)*Rnl(2,0,r)*Ylm(theta,phi,0,0),phi=0..2*Pi);
```

$$\text{Iphi} := -r^4 \sin(\theta) \cos(\theta)^2 (\mathrm{e}^{(-1/2\,r)})^2 (-2+r)\,\pi$$

next the θ integration

```
> Itheta:=int(Iphi,theta=0..Pi);
```

$$\text{Itheta} := -\frac{2}{3} r^4 (\mathrm{e}^{(-1/2\,r)})^2 (-2+r)\,\pi$$

and finally the radial integral:

```
> Ir:=int(Itheta,r=0..infinity);
```

$$\text{Ir} := -48\,\pi$$

As we haven't normalized our states beforehand, we should do so now. We request the calculation of the norm integral for the 2p state with a single command as

```
> In1:=int(int(int(r^2*sin(theta)*(Rnl(2,1,r)*Ylm(theta,phi,
> 1,0))^2,phi=0..2*Pi),theta=0..Pi),r=0..infinity);
```

$$\text{In1} := 32\,\pi$$

Similarly, for the 2s state we obtain

```
> In2:=int(int(int(r^2*sin(theta)*(Rnl(2,0,r)*Ylm(theta,phi,
> 0,0))^2,phi=0..2*Pi),theta=0..Pi),r=0..infinity);
```

$$\text{In2} := 8\,\pi$$

The relevant dipole matrix element including normalization gives a number of order unity:

```
> Ir/sqrt(In1*In2);
```

$$-3$$

Now we wish to demonstrate the case of an existing degeneracy between 2s and $2p_0$. Degeneracies can never occur in ground states in quantum mechanics, although they do play a role in many-particle systems, and near-degeneracies are possible in molecules as one increases the inter-nuclear separation.

For the degenerate case, the first-order PT has to be modified in that the unperturbed levels cannot simply be used to calculate the energy correction. This is the case as we can assume that the degeneracy will be lifted by the perturbation.

An existing degeneracy between two levels implies that any linear combination of the two eigenfunctions associated with the degenerate energy is also an eigenstate with the same energy. With switched on perturbing interaction, the directions of the eigenvectors (functions) will be uniquely specified, no matter how small the perturbation. This breaking of symmetry requires that the two eigenstates be prediagonalized before the perturbation is calculated.

For the DC Stark effect in hydrogen we consider the four energetically degenerate states: 2s, $2p_0$, $2p_1$, $2p_{-1}$. We label these states as $\nu = 1, 2, 3, 4$. We have to make the assumption for first-order PT that even the zeroth-order wavefunction is a linear combination of these levels

$$\phi = \sum_{\nu=1}^{4} c_\nu \psi_\nu \, .$$

This induces non-vanishing first-order matrix elements of $e|E|z$. The expansion coefficients c_ν and the energy corrections for the four states are obtained from the linear algebra eigenvalue problem:

$$[e^{(0)}_{n=2} - e + W_{\nu,\nu}]c_\nu + \sum_{\mu=1,\mu\neq\nu}^{4} W_{\mu,\nu} c_\mu = 0 \quad , \quad \nu = 1..4 \quad .$$

Here $e^{(0)}$ is the unperturbed and e the corrected $n = 2$ energy value, and $W_{\mu,\nu}$ is the potential matrix

$$\langle \mu | W | \nu \rangle = \langle nlm | \quad e|E|z \quad | n'l'm' \rangle \quad .$$

This matrix has a simple structure as the dipole operator matrix elements vanish within the l-degenerate multiplet, which is obvious from the ϕ integration. The only non-vanishing matrix element is $W_{12} = W_{21}$, calculated above.

```
> W12:=-3*Eabs;
```

$$\text{W12} := -3\,\text{Eabs}$$

With the unperturbed energy of the $n = 2$ levels in atomic units given as $e_{n=2} = -1/(2n^2) = -1/8$, we can assemble the relevant 2-by-2 submatrix:

```
> eq1:=(-1/8-en)*c1+W12*c2=0;
```

$$\text{eq1} := \left(-\frac{1}{8} - en\right) c1 - 3\,\text{Eabs}\, c2 = 0$$

```
> eq2:=(-1/8-en)*c2+W12*c1=0;
```

$$\text{eq2} := \left(-\frac{1}{8} - en\right) c2 - 3\,\text{Eabs}\, c1 = 0$$

```
> solve({eq1,eq2},{c1,en});
```

$$\left\{en = -\frac{1}{8} - 3\,\text{Eabs},\, c1 = c2\right\}, \left\{en = -\frac{1}{8} + 3\,\text{Eabs},\, c1 = -c2\right\}$$

We have solved the problem as an algebraic eigenvalue problem. The two possible solutions represent an equal superposition of the 2s and $2p_0$ levels independent of the field strength magnitude `Eabs`. The coefficient pairs (c_1, c_2) are determined completely for each of the two states by the normalization condition.

Thus, to first order (!) we find that in hydrogen the $n = 2$ multiplet splits into three:

- the unshifted $2p_1$ and $2p_{-1}$ levels that form a doublet at $e^{(0)}_{n=2} = -1/8$ a.u.;
- the two possible linear combinations of 2s and $2p_0$, of which one moves up linearly(!) with `Eabs`, while the other comes down by the same amount.

Note that these splittings are small for macroscopic electric fields: field strengths of order $10\,\mathrm{kVcm}^{-1}$ translate into fractions of a meV, while the $(n = 1) - (n = 2)$ splitting is of the order of 10 eV.

Exercise 4.5.1: The valence electrons in the alkali atoms can be described approximately using an effective electrostatic potential for the inner electrons as demonstrated in Exercise 4.2.5 for the lithium atom. Use the potential given in that exercise and approximate the 2s and 2p levels variationally with hydrogen-like wavefunctions. Using these 2s and 2p wavefunctions estimate the shift of the 2s state in an external DC field due to the second-order PT correction using only the 2p level as an intermediate state.

Exercise 4.5.2: Calculate the first-order DC field level shift of the 2s level in the singly ionized helium atom.

5. Spin and Time-Dependent Processes

5.1 Pauli Equation

The Pauli equation represents an enhanced Schrödinger equation in which, in addition to a full incorporation of external electromagnetic fields, the spin degree of freedom is added for spin-$\frac{1}{2}$ particles. A fully satisfying description of fermions is only possible in the context of relativistic QM. Nevertheless, if one accepts the construction of the Pauli equation, many relevant results can be calculated with it.

One can take the experimentally motivated point of view that spin has to be postulated as an internal property of the electron, as the Schrödinger description of electrons with zero orbital angular momentum passing through a strong magnetic field is in conflict with experimental findings. The Stern-Gerlach experiment and atomic spectroscopy can only be explained properly if electrons are postulated to possess a magnetic moment. Due to the interaction of this magnetic moment with the magnetic field, a beam splits into two halves. Such a property is not allowed for a classical point-particle.

In QM this internal degree of freedom that shows the properties of an angular momentum variable with two exactly known projections with respect to one Cartesian axis aligned with the magnetic field and unknown projections with respect to the other two axes can be added to the Schrödinger wavefunction. In principle, one could write two separate wavefunctions for the electrons with the two different spin projections. However, if one makes the assumption that the coordinate space degrees of freedom are unaffected by the spin degrees of freedom, one incorporates spin by multiplying the coordinate space wavefunction by a spinor, i.e., a simple 2-component vector with only two degrees of freedom labeled as 'up' and 'down'.

In order to add this additional degree of freedom to the wavefunction to distinguish between two possible projections of the fermion's spin one has to supplement the Hamiltonian and other operators with a 2-by-2 algebra. This algebra is conveniently represented by the Pauli spin matrices σ_i with $i = 1, 2, 3$ labeling Cartesian components.

We demonstrate first the commutation and anti-commutation properties of the Pauli matrices. We label $\sigma_x = \sigma_1, \sigma_y = \sigma_2, \sigma_z = \sigma_3$. The spin matrices s_x, s_y, s_z differ from these by a factor of $\hbar/2$. We choose a representation in which σ_z is diagonal. The spin operators (and the Pauli matrices) satisfy the same commutation relations (CRs) as the angular momentum operators, i.e.,

$$[\hat{s}_x, \hat{s}_y] = i\hbar\hat{s}_z \quad .$$

As before, in the CR above $\{x, y, z\}$ can be replaced by any cyclic permutation thereof. Many representations of this algebra are possible, even after the constraint that $\hat{s}_z$ be diagonal. Instead of deriving all possible representations we chose the conventional one and demonstrate that it works.

```
> with(linalg):
> sigma3:=matrix([[1,0],[0,-1]]);  evalm(sigma3 &* sigma3);
```

$$\sigma 3 := \begin{bmatrix} 1 & 0 \\ 0 & -1 \end{bmatrix} \quad \begin{bmatrix} 1 & 0 \\ 0 & 1 \end{bmatrix}$$

Obviously the square of σ_3 gives a unit matrix. Now we find another matrix with the property that its square equals unity:

```
> sigma1:=matrix([[0,1],[1,0]]);  evalm(sigma1 &* sigma1);
```

$$\sigma 1 := \begin{bmatrix} 0 & 1 \\ 1 & 0 \end{bmatrix} \quad \begin{bmatrix} 1 & 0 \\ 0 & 1 \end{bmatrix}$$

```
> eigenvals(sigma1);
```

$$1, -1$$

The property $\sigma_x^2 = (1)$ forces the matrix to have the same eigenvalues, but note that $\sigma_1 = \sigma_x$ is not diagonal in this representation. We make use of the commutation relation to determine $\sigma_2 = \sigma_y$:

```
> CR:=proc();
> evalm(args[1]&*args[2]-args[2]&*args[1]);  end:
> CR(sigma3,sigma1);
```

$$\begin{bmatrix} 0 & 2 \\ -2 & 0 \end{bmatrix}$$

Note that we used a cyclic permutation of $\{1, 2, 3\}$, i.e., $(3, 1, 2)$ and that we have to split off a factor of $2i$:

```
> sigma2:=matrix([[0,-I],[I,0]]);
```

$$\sigma 2 := \begin{bmatrix} 0 & -I \\ I & 0 \end{bmatrix}$$

We can now complete the demonstration of the CR:

```
> CR(sigma1,sigma2);        evalm("/(2*I));
```

$$\begin{bmatrix} 2I & 0 \\ 0 & -2I \end{bmatrix} \quad \begin{bmatrix} 1 & 0 \\ 0 & -1 \end{bmatrix}$$

```
> CR(sigma2,sigma3);      evalm("/(2*I));
```

$$\begin{bmatrix} 0 & 2I \\ 2I & 0 \end{bmatrix} \quad \begin{bmatrix} 0 & 1 \\ 1 & 0 \end{bmatrix}$$

Next we show that some anti-commutation relations are satisfied:

```
> ACR:=proc();
> evalm(args[1]&*args[2]+args[2]&*args[1]);  end:
> ACR(sigma1,sigma2);    ACR(sigma1,sigma3);
```

$$\begin{bmatrix} 0 & 0 \\ 0 & 0 \end{bmatrix} \quad \begin{bmatrix} 0 & 0 \\ 0 & 0 \end{bmatrix}$$

```
> ACR(sigma2,sigma3);     eigenvals(sigma2);
```

$$\begin{bmatrix} 0 & 0 \\ 0 & 0 \end{bmatrix} \quad 1, -1$$

The hermeticity of σ_y is also obvious. Note that the eigenfunctions common to $\sigma^2 = \sigma_x^2 + \sigma_y^2 + \sigma_z^2$ and to σ_z are trivial to determine, since σ^2 is proportional to the identity matrix. Now we determine the (trivial) eigenvectors of the diagonal matrix σ_3:

```
> eigenvects(sigma3);
```

$$[1, 1, \{[1\ \ 0]\}], [-1, 1, \{[0\ \ 1]\}]$$

For the eigenvalue $+1$ we have multiplicity 1 (no degeneracy) and the eigenvector $[1, 0]$, while for $\sigma_z = -1$ the eigenvector is $[0, 1]$. We note that for reasons of a compact output, column vectors are displayed as row vectors.

The spin's z projection eigenvalues are, therefore, $+\hbar/2$ and $-\hbar/2$. The eigenvectors are the same as for σ_z. That they are also eigenvectors of s^2 is trivial. The eigenvalues of s^2 are obtained via those of σ^2:

```
> sigmasq:=evalm(sigma1^2+sigma2^2+sigma3^2);
```

$$\text{sigmasq} := \begin{bmatrix} 3 & 0 \\ 0 & 3 \end{bmatrix}$$

This result is to be multiplied by $(\hbar/2)^2$, and we thus have a doubly degenerate eigenvalue of $(3/4)\hbar^2$, which is in accord with $s(s+1)\hbar^2$.

We consider now the Pauli equation for a pure spin wavefunction [Gr93, p. 242]). The electron (or any charged spin-$\frac{1}{2}$ particle) is taken to be at rest, or in simple rectilinear motion, such that we can ignore the spatial part of the wavefunction. We are interested in the precession of the spin in a constant magnetic field.

Classically, a charged particle orbits around the direction of a magnetic field of strength B with a frequency determined by equating the Lorentz and centrifugal forces:

$$-\frac{eBv}{c} = mr\omega^2 ,$$

where v is the tangential velocity which for a circular orbit equals the circular frequency ω times the radius r.

For the QM discussion we begin with a spin wavefunction that describes a linear superposition of 'up' and 'down'.

```
> up:=vector([1,0]);      down:=vector([0,1]);
```

$$\text{up} := [1\ \ 0] \qquad \text{down} := [0\ \ 1]$$

```
> a0:='a0': b0:='b0':      chi:=evalm(a0*up+b0*down);
```

$$\chi := [\text{a0}\ \ \text{b0}]$$

To make the normalization of the wavefunction obvious we reparametrize the two complex valued coefficients a_0, b_0 in terms of real parameters g, d and θ with an explicit incorporation of the normalization condition.

```
> a0s:=exp(I*g)*cos(theta/2);     b0s:=exp(I*d)*sin(theta/2);
```

$$\text{a0s} := e^{(I\,g)} \cos\left(\frac{1}{2}\theta\right) \qquad \text{b0s} := e^{(I\,d)} \sin\left(\frac{1}{2}\theta\right)$$

```
> norm:=simplify(abs(a0s)^2+abs(b0s)^2);
```

$$\text{norm} := e^{(-2\,\Im(g))} \left|\cos\left(\frac{1}{2}\theta\right)\right|^2 + e^{(-2\,\Im(d))} \left|\sin\left(\frac{1}{2}\theta\right)\right|^2$$

Maple fails to simplify further due to the magnitude signs and θ, d, g should all be assumed as real variables.

```
> assume(g,real); assume(d,real); assume(theta,real);
> simplify(norm);
```

$$\left|\cos\left(\frac{1}{2}\theta\tilde{}\right)\right|^2 + \left|\sin\left(\frac{1}{2}\theta\tilde{}\right)\right|^2$$

```
> combine(",trig);
```

The failure to return the value of 1 is caused by the magnitude signs. It is disappointing that Maple fails to recognize that the answer is identical to the one given below:

```
> simplify(cos(theta/2)^2+sin(theta/2)^2);
```

$$1$$

Note that in a_0, b_0 there are four real degrees of freedom reduced by the normalization constraint by 1. Consequently there are three real parameters g, d, and θ. The Pauli equation for the problem without dynamics in coordinate space is given as

$$\mathrm{i}\hbar\frac{\partial\chi}{\partial t} = \mu(\boldsymbol{\sigma}\cdot\mathbf{B})\chi(t),$$

where μ denotes the magnetic moment of the particle. If we fix the coordinate system such that the z axis coincides with the direction of the magnetic field, $\mathbf{B} = (0, 0, B)$, the equation becomes

$$\mathrm{i}\hbar\frac{\partial\chi}{\partial t} = -\frac{e\hbar B}{2mc}\sigma_z\chi(t),$$

as we have made use of the electron's magnetic moment. With the abbreviation for the Larmor frequency $\omega_L = -eB/(2mc)$ it simplifies to

$$\mathrm{i}\hbar\frac{\partial\chi}{\partial t} = \hbar\omega_L\sigma_z\chi(t).$$

We define the Pauli equation making use of the spinor declared previously. Note how Maple automatically transfers the time dependence to $a_0(t)$ and $b_0(t)$.

```
> PauliEq:=evalm(I*map(diff,chi(t),t))=evalm(omega*sigma3*chi);
```

$$\text{PauliEq} := \left[I\left(\frac{\partial}{\partial t}\,\mathrm{a0}(t)\right) \quad I\left(\frac{\partial}{\partial t}\,\mathrm{b0}(t)\right)\right] = [\omega\,\mathrm{a0} \quad -\omega\,\mathrm{b0}]$$

It is more convenient to work with the spinor components, i.e.,

```
> eq1:=lhs(PauliEq)[1]=rhs(PauliEq)[1];
```

$$\text{eq1} := I\left(\frac{\partial}{\partial t}\,\mathrm{a0}(t)\right) = \omega\,\mathrm{a0}$$

```
> eq2:=lhs(PauliEq)[2]=rhs(PauliEq)[2];
```

$$\text{eq2} := I\left(\frac{\partial}{\partial t}\,\mathrm{b0}(t)\right) = -\omega\,\mathrm{b0}$$

The general solution is obtained as

```
> dsolve({eq1},a0(t));
```

$$\mathrm{a0}(t) = e^{(-I\,\omega\,t)}\,\_C1$$

Alternatively, we find a solution with specified initial condition in terms of the previously declared value `a0s`.

```
> sol1:=dsolve({eq1,a0(0)=a0s},a0(t));
```

$$\mathrm{sol1} := \mathrm{a0}(t) = \cos(\omega\,t)\left(\cos(g)\cos\left(\frac{1}{2}\theta\right) + I\sin(g)\cos\left(\frac{1}{2}\theta\right)\right)$$
$$- I\sin(\omega\,t)\left(\cos(g)\cos\left(\frac{1}{2}\theta\right) + I\sin(g)\cos\left(\frac{1}{2}\theta\right)\right)$$

Similarly the general solution to the second equation can be obtained.

```
> dsolve({eq2},b0(t));
```

$$\mathrm{b0}(t) = e^{(I\,\omega\,t)}\,\_C1$$

```
> sol2:=dsolve({eq2,b0(0)=b0s},b0(t));
```

Nevertheless, we prefer to use the special solution given by

$$\mathrm{sol2} := \mathrm{b0}(t) = \cos(\omega\,t)\left(\cos(d)\sin\left(\frac{1}{2}\theta\right) + I\sin(d)\sin\left(\frac{1}{2}\theta\right)\right)$$
$$+ I\sin(\omega\,t)\left(\cos(d)\sin\left(\frac{1}{2}\theta\right) + I\sin(d)\sin\left(\frac{1}{2}\theta\right)\right)$$

The time evolution of the spinor wavefunction determined uniquely through a choice of initial conditions (fixing of g, d, θ) is then given as

```
> chioft:=evalm(rhs(sol1)*up+rhs(sol2)*down);
```

$$\mathrm{chioft} := \left[\cos(\omega\,t)\left(\cos(g)\cos\left(\frac{1}{2}\theta\right) + I\sin(g)\cos\left(\frac{1}{2}\theta\right)\right)\right.$$
$$- I\sin(\omega\,t)\left(\cos(g)\cos\left(\frac{1}{2}\theta\right) + I\sin(g)\cos\left(\frac{1}{2}\theta\right)\right)$$
$$\cos(\omega\,t)\left(\cos(d)\sin\left(\frac{1}{2}\theta\right) + I\sin(d)\sin\left(\frac{1}{2}\theta\right)\right)$$
$$\left.+ I\sin(\omega\,t)\left(\cos(d)\sin\left(\frac{1}{2}\theta\right) + I\sin(d)\sin\left(\frac{1}{2}\theta\right)\right)\right]$$

We can calculate now the expectation values for the components of the spin operator, which up to the factor of $\hbar/2$ is identical with the Pauli matrices $\sigma_x = \sigma_1$, $\sigma_y = \sigma_2$, $\sigma_z = \sigma_3$.

We begin with the conserved z projection and demonstrate the calculation of an expectation value of some operator $\hat{O}$ for pure spinors: $\langle\chi|\hat{O}|\chi\rangle$. We need to form the hermitean adjoint of χ and make use of matrix multiplication:

```
> chiconj:=evalm(map(conjugate,chioft)):
```

```
> chidag:=transpose(chiconj):
```

Now we evaluate $\langle\chi|\sigma_3|\chi\rangle$ as a specific example and display the simplified result:

```
> evalm(chidag &* sigma3 &* chioft):
```

```
> combine(evalc(simplify(")),trig);
```

$$\cos(\theta)$$

Given that θ is a parameter fixed by the initial condition we see that the z projection of spin remains conserved for a particle interacting with a magnetic field aligned along the z axis. Now we try for the other components, namely first the x component of spin (up to $\hbar/2$). We calculate the expectation value $\langle\chi|\sigma_1|\chi\rangle$ and show the simplified answer:

```
> evalm(chidag &* sigma1 &* chioft):
```

```
> sx:=combine(evalc(simplify(")),trig);
```

$$sx := \frac{1}{2}\sin(d+\theta+2\,\omega\,t-g) - \frac{1}{2}\sin(d-\theta+2\,\omega\,t-g)$$

We define a procedure to which the parameters can be passed:

```
> sx1t:=unapply(sx,t,omega,d,g,theta);
```

$$\mathrm{sx1t} := (t,\omega,d,g,\theta) \to \frac{1}{2}\sin(d+\theta+2\,\omega\,t-g) - \frac{1}{2}\sin(d-\theta+2\,\omega\,t-g)$$

We are interested in a mixed state 'up' and 'down', say $\theta = \pi/4$,

```
> sx1t(t,1,0,0,Pi/4);
```

$$\frac{1}{2}\sin\left(\frac{1}{4}\pi+2\,t\right) + \frac{1}{2}\cos\left(\frac{1}{4}\pi+2\,t\right)$$

The expectation value of the x component of spin $\hat{s}_x$, which is represented by σ_1 times $\hbar/2$, oscillates as a function of time, i.e., $\hat{s}_x$ is not a conserved quantity once we have chosen a superposition of states with known z projection. The oscillatory behaviour of $\langle s_x\rangle(t)$ is demonstrated in the graph below.

```
> plot(sx1t(t,1,0,0,Pi/4),t=0..2*Pi);
```

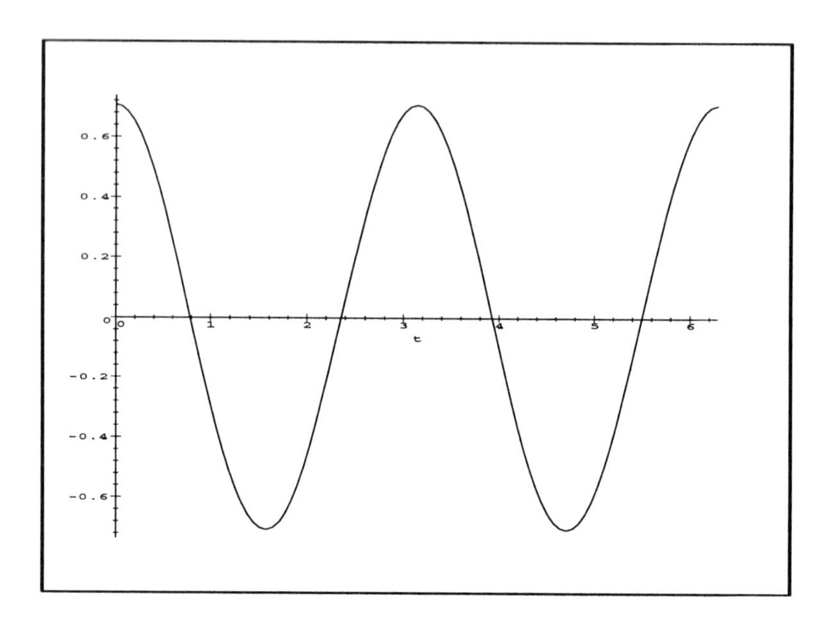

Similarly we obtain the expectation value for the y component of spin, i.e., $\langle\chi|\sigma_2|\chi\rangle$, again suppressing the intermediate output.

```
> evalm(chidag &* sigma2 &* chioft):
```

```
> sy:=combine(evalc(simplify(")),trig);
```

$$sy := -\frac{1}{2}\cos(\,d+\theta+2\,\omega\,t-g\,)+\frac{1}{2}\cos(\,d-\theta+2\,\omega\,t-g\,)$$

```
> sy1t:=unapply(sy,t,omega,d,g,theta);
```

$$\mathrm{sy1t} := (\,t,\omega,d,g,\theta\,)\rightarrow -\frac{1}{2}\cos(\,d+\theta+2\,\omega\,t-g\,)+\frac{1}{2}\cos(\,d-\theta+2\,\omega\,t-g\,)$$

The result of a conserved z projection and out-of-phase constant-amplitude oscillations of the x and y projections of the spin vector corresponds to a precessing motion of the spin vector around the z axis. I leave it to the reader to visualize, on the basis of the next graph, how the projection of the spin vector onto the $x-y$ plane executes a circular motion. This behaviour of the spin vector is in complete analogy with the motion of the orbital angular momentum vector. In that case we argued that the incomplete information arose as a consequence of the Heisenberg uncertainty principle, which also manifested itself in the commutation relations. In the case of spin we derived the precession of the vector from the non-commutativity of its Cartesian components.

```
> plot({sx1t(t,1,0,0,Pi/4),sy1t(t,1,0,0,Pi/4)},t=0..2*Pi);
```

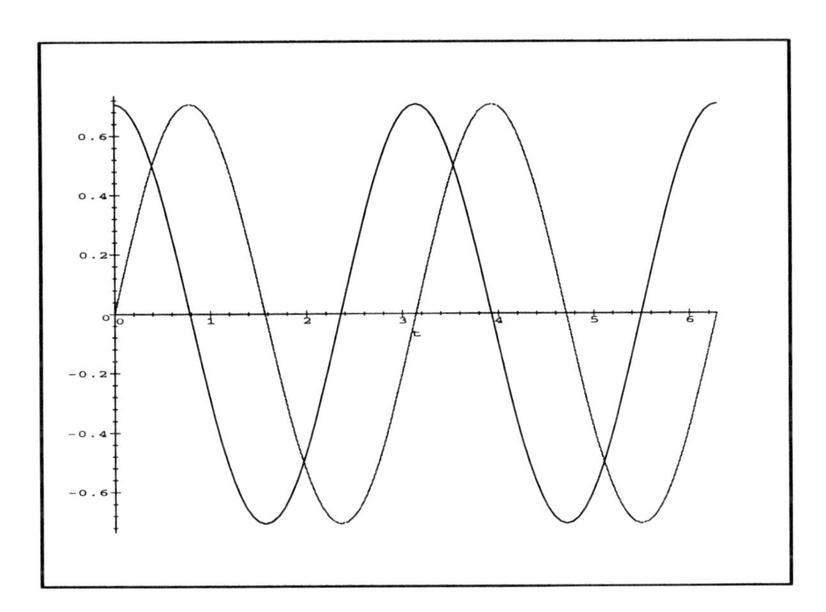

Exercise 5.1.1: Consider the spin operators for a vector particle (cf. [Sch68, p. 198]), namely

$$[S_x, S_y, S_z] = \mathrm{i}\hbar \left[\begin{pmatrix} 0 & 0 & 0 \\ 0 & 0 & -1 \\ 0 & 1 & 0 \end{pmatrix}, \begin{pmatrix} 0 & 0 & 1 \\ 0 & 0 & 0 \\ -1 & 0 & 0 \end{pmatrix}, \begin{pmatrix} 0 & -1 & 0 \\ 1 & 0 & 0 \\ 0 & 0 & 0 \end{pmatrix} \right]$$

Show that these matrices satisfy the commutation relations for an angular momentum operator in QM. Also show that $\mathbf{S^2} = 2\hbar^2$. What are the eigenvalues of the components of $\mathbf{S}$)?

5.2 Magnetic Moments

This section contains a discussion of the differential equation for the time evolution of spin in a time-dependent magnetic field as relevant for the Rabi experiment to measure the magnetic moment of the proton and neutron. A pure spin wavefunction for a spin-$\frac{1}{2}$ particle has two degrees of freedom

$$S(t) = [S1(t), S2(t)],$$

which really is a column vector. The wavefunction satisfies a Schrödinger (or Pauli) equation with a single term in the Hamiltonian that describes the interaction of spin with a magnetic field:

$$-\mu(\mathbf{B}\cdot\boldsymbol{\sigma}),$$

where μ is the particle's internal magnetic moment. Such an interaction arises even for a QM point particle as the internal angular momentum variable spin can be thought of as implying a simple electric current which interacts with the magnetic field. This results in a simple magnetic moment – the electron case was presented in the previous section. If the particle has a complex internal structure with induced currents, it will interact differently with the magnetic field. This manifests itself in a different value of the magnetic moment M. Thus, the study of the behaviour of particles in magnetic fields serves as an important tool for probing their internal structure.

$S_1(t)$ is the expansion coefficient for the $m_s = +\frac{1}{2}$ spin projection (wavefunction $[1,0]$), and $S_2(t)$ for the $m_s = -\frac{1}{2}$ state $[0,1]$. The particle's magnetic moment M is represented by the operator $\mu\boldsymbol{\sigma}$. The spin operator is represented by the Pauli matrices introduced in the previous section:

```
> with(linalg):
> sigma1:=matrix([[0,1],[1,0]]):sigma2:=matrix([[0,-I],[I,0]]):
> sigma3:=matrix([[1,0],[0,-1]]): Svec:=vector([S1(t),S2(t)]);
```

$$\mathrm{Svec} := [\,\mathrm{S1}(t)\;\;\mathrm{S2}(t)\,]$$

The Schrödinger-Pauli equation for a particle with trivial dynamics, i.e., at most rectilinear spatial motion (with $\hbar = 1$) is given by:

```
> Bx:='Bx':  By:='By':  Bz:='Bz':  assume(omega>0);
> SPeq:=evalm(I*map(diff,op(Svec),t))=evalm(-mu*(Bx*sigma1+
> By*sigma2+Bz*sigma3)*Svec);
```

$$\mathrm{SPeq} := \left[I\left(\frac{\partial}{\partial t}\mathrm{S1}(t)\right)\;\; I\left(\frac{\partial}{\partial t}\mathrm{S2}(t)\right)\right] =$$
$$[-\mu\,(Bz\,\mathrm{S1}(t) + \mathrm{S2}(t)\,Bx - I\,\mathrm{S2}(t)\,By)$$
$$-\mu\,(\mathrm{S1}(t)\,Bx + I\,\mathrm{S1}(t)\,By - Bz\,\mathrm{S2}(t))]$$

Now we specify a circularly polarized magnetic radiation field:

```
> Bx:=B1*cos(omega*t); By:=B1*sin(omega*t); Bz:=B0;
```

$$Bx := B1\cos(\omega\tilde{}\,t) \qquad By := B1\sin(\omega\tilde{}\,t) \qquad Bz := B0$$

```
> eval(SPeq);
```

$$\left[I\left(\frac{\partial}{\partial t}\mathrm{S1}(t)\right)\ I\left(\frac{\partial}{\partial t}\mathrm{S2}(t)\right)\right] = \left[-\mu\,(Bz\,\mathrm{S1}(t)+\mathrm{S2}(t)\,Bx - I\,\mathrm{S2}(t)\,By)\quad -\mu\,(\mathrm{S1}(t)\,Bx + I\,\mathrm{S1}(t)\,By - Bz\,\mathrm{S2}(t))\right]$$

We extract the equations for further simplification:

```
> eq1:=lhs(SPeq)[1]=rhs(SPeq)[1];
```

$$\mathrm{eq1} := I\left(\frac{\partial}{\partial t}\mathrm{S1}(t)\right) = -\mu(B0\,\mathrm{S1}(t)+\mathrm{S2}(t)\,B1\cos(\omega\tilde{}\,t) - I\,\mathrm{S2}(t)\,B1\sin(\omega\tilde{}\,t))$$

```
> eq2:=lhs(SPeq)[2]=rhs(SPeq)[2];
```

$$\mathrm{eq2} := I\left(\frac{\partial}{\partial t}\mathrm{S2}(t)\right) = -\mu(\mathrm{S1}(t)\,B1\cos(\omega\tilde{}\,t) + I\,\mathrm{S1}(t)\,B1\sin(\omega\tilde{}\,t) - B0\,\mathrm{S2}(t))$$

We wish to solve the pair of ordinary differential equations subject to the initial condition that the spin projection is 'up'. However, first we need to convert to an equation with constant coefficients, since the direct attempt to solve them fails:

```
> sol:=dsolve({eq1,eq2},{S1(t),S2(t)});
```

$$\mathrm{sol} :=$$

First we convert to exponentials:

```
> eq1:=simplify(convert(eq1,exp));
```

$$\mathrm{eq1} := I\left(\frac{\partial}{\partial t}\mathrm{S1}(t)\right) = -\mu(B0\,\mathrm{S1}(t)+\mathrm{S2}(t)\,B1\cos(\omega\tilde{}\,t) - I\,\mathrm{S2}(t)\,B1\sin(\omega\tilde{}\,t))$$

```
> eq2:=simplify(convert(eq2,exp));
```

$$\mathrm{eq2} := I\left(\frac{\partial}{\partial t}\,\mathrm{S2}(t)\right) = -\mu(\mathrm{S1}(t)\,B1\cos(\omega\tilde{}\;t) + I\,\mathrm{S1}(t)\,B1\sin(\omega\tilde{}\;t) - B0\,\mathrm{S2}(t))$$

```
> sol:=dsolve({eq1,eq2},{S1(t),S2(t)});
```

$$\mathrm{sol} :=$$

The time-dependent coefficients still prevent **dsolve** from providing the answer. The equations have to be simplified by hand. We introduce two new constants ν and D that absorb B_1, B_0 and μ

$$D = \frac{B_1}{2B_0}, \quad \nu = 2\mu B_0 \quad ,$$

and perform the substitutions in the equations:

```
> eq1:=subs(B1=2*B0*D,eq1);
```

$$\mathrm{eq1} := I\left(\frac{\partial}{\partial t}\,\mathrm{S1}(t)\right) = -\mu(B0\,\mathrm{S1}(t) + 2\,\mathrm{S2}(t)\,B0\,D\cos(\omega\tilde{}\;t) - 2\,I\,\mathrm{S2}(t)\,B0\,D\sin(\omega\tilde{}\;t))$$

```
> eq2:=subs(B1=2*B0*D,eq2);
```

$$\mathrm{eq2} := I\left(\frac{\partial}{\partial t}\,\mathrm{S2}(t)\right) = -\mu(2\,\mathrm{S1}(t)\,B0\,D\cos(\omega\tilde{}\;t) + 2\,I\,\mathrm{S1}(t)\,B0\,D\sin(\omega\tilde{}\;t) - B0\,\mathrm{S2}(t))$$

```
> eq1:=simplify(subs(B0=nu/(2*mu),eq1/I));
```

$$\mathrm{eq1} := \frac{\partial}{\partial t}\,\mathrm{S1}(t) = \frac{1}{2}\,I\,\nu\,\mathrm{S1}(t) + I\,\nu\,\mathrm{S2}(t)\,D\cos(\omega\tilde{}\;t) + \nu\,\mathrm{S2}(t)\,D\sin(\omega\tilde{}\;t)$$

```
> eq2:=simplify(subs(B0=nu/(2*mu),eq2/I));
```

$$\mathrm{eq2} := \frac{\partial}{\partial t}\,\mathrm{S2}(t) = I\,\nu\,\mathrm{S1}(t)\,D\cos(\omega\tilde{}\;t) - \nu\,\mathrm{S1}(t)\,D\sin(\omega\tilde{}\;t) - \frac{1}{2}\,I\,\nu\,\mathrm{S2}(t)$$

Now we differentiate the second equation and eliminate $S_1(t)$:

```
> eq2p:=diff(eq2,t);
```

$$\text{eq2p} := \frac{\partial^2}{\partial t^2}\,\mathrm{S2}(t) = I\,\nu\left(\frac{\partial}{\partial t}\,\mathrm{S1}(t)\right)D\cos(\omega\tilde{}\,t) - I\,\nu\,\mathrm{S1}(t)\,D\sin(\omega\tilde{}\,t)\,\omega\tilde{} - \nu\left(\frac{\partial}{\partial t}\,\mathrm{S1}(t)\right)D\sin(\omega\tilde{}\,t) - \nu\,\mathrm{S1}(t)\,D\cos(\omega\tilde{}\,t)\,\omega\tilde{} - \frac{1}{2}\,I\,\nu\left(\frac{\partial}{\partial t}\,\mathrm{S2}(t)\right)$$

Next we substitute for the derivative $S_1'(t)$ using eq1:

```
> eq2p:=simplify(subs(diff(S1(t),t)=rhs(eq1),eq2p));
```

$$\text{eq2p} := \frac{\partial^2}{\partial t^2}\,\mathrm{S2}(t) = -\frac{1}{2}\,\nu^2\,D\cos(\omega\tilde{}\,t)\,\mathrm{S1}(t) - I\,\nu\,\mathrm{S1}(t)\,D\sin(\omega\tilde{}\,t)\,\omega\tilde{} - \frac{1}{2}\,I\,\nu^2\,D\sin(\omega\tilde{}\,t)\,\mathrm{S1}(t) - \nu^2\,D^2\,\mathrm{S2}(t) - \nu\,\mathrm{S1}(t)\,D\cos(\omega\tilde{}\,t)\,\omega\tilde{} - \frac{1}{2}\,I\,\nu\left(\frac{\partial}{\partial t}\,\mathrm{S2}(t)\right)$$

Now we prepare the ground for a substitution of $S_2'(t)$ without affecting the second derivative $S_2''(t)$. We protect S_2'' from the substitution by giving it a distinct name:

```
> eq2p:=subs(diff(S2(t),t$2)=S2pp,eq2p);
```

$$\text{eq2p} := \mathrm{S2pp} = -\frac{1}{2}\,\nu^2\,D\cos(\omega\tilde{}\,t)\,\mathrm{S1}(t) - I\,\nu\,\mathrm{S1}(t)\,D\sin(\omega\tilde{}\,t)\,\omega\tilde{} - \frac{1}{2}\,I\,\nu^2\,D\sin(\omega\tilde{}\,t)\,\mathrm{S1}(t) - \nu^2\,D^2\,\mathrm{S2}(t) - \nu\,\mathrm{S1}(t)\,D\cos(\omega\tilde{}\,t)\,\omega\tilde{} - \frac{1}{2}\,I\,\nu\left(\frac{\partial}{\partial t}\,\mathrm{S2}(t)\right)$$

```
> eq2p:=simplify(subs(diff(S2(t),t)=rhs(eq2),eq2p));
```

$$\text{eq2p} := \mathrm{S2pp} = -I\,\nu\,\mathrm{S1}(t)\,D\sin(\omega\tilde{}\,t)\,\omega\tilde{} - \nu^2\,D^2\,\mathrm{S2}(t) - \nu\,\mathrm{S1}(t)\,D\cos(\omega\tilde{}\,t)\,\omega\tilde{} - \frac{1}{4}\,\nu^2\,\mathrm{S2}(t)$$

We are almost done, and recognize that the remaining dependence on S_1 can be replaced by the S_2 equation:

```
> subs(S1(t)=solve(eq2,S1(t)),eq2p);
```

$$\mathrm{S2pp} = I\,\nu\,\left(\left(\frac{\partial}{\partial t}\,\mathrm{S2}(t)\right) + \frac{1}{2}\,I\,\nu\,\mathrm{S2}(t)\right)\,D\sin(\omega\tilde{}\,t)\,\omega\tilde{}\,/$$
$$(-I\,\nu\,D\cos(\omega\tilde{}\,t) + \nu\,D\sin(\omega\tilde{}\,t)) - \nu^2\,D^2\,\mathrm{S2}(t)$$
$$+\frac{\nu\,\left(\left(\frac{\partial}{\partial t}\,\mathrm{S2}(t)\right) + \frac{1}{2}\,I\,\nu\,\mathrm{S2}(t)\right)\,D\cos(\omega\tilde{}\,t)\,\omega\tilde{}}{-I\,\nu\,D\cos(\omega\tilde{}\,t) + \nu\,D\sin(\omega\tilde{}\,t)}$$
$$-\frac{1}{4}\,\nu^2\,\mathrm{S2}(t)$$

Finally we are ready to revert the substitution of $S_2''(t)$:

```
> simplify(convert(",exp));
```

$$\mathrm{S2pp} = -\frac{1}{4}\left(4\,I\sin(\omega\tilde{}\,t)\,\omega\tilde{}\,\left(\frac{\partial}{\partial t}\,\mathrm{S2}(t)\right)\right.$$
$$-2\sin(\omega\tilde{}\,t)\,\omega\tilde{}\,\nu\,\mathrm{S2}(t) + 4\,I\,\nu^2\,D^2\,\mathrm{S2}(t)\cos(\omega\tilde{}\,t)$$
$$-4\,\nu^2\,D^2\,\mathrm{S2}(t)\sin(\omega\tilde{}\,t)$$
$$+4\cos(\omega\tilde{}\,t)\,\omega\tilde{}\,\left(\frac{\partial}{\partial t}\,\mathrm{S2}(t)\right)$$
$$+2\,I\cos(\omega\tilde{}\,t)\,\omega\tilde{}\,\nu\,\mathrm{S2}(t) + I\,\nu^2\,\mathrm{S2}(t)\cos(\omega\tilde{}\,t)$$
$$\left.-\nu^2\,\mathrm{S2}(t)\sin(\omega\tilde{}\,t)\right)/(I\cos(\omega\tilde{}\,t) - \sin(\omega\tilde{}\,t))$$

```
> eq2p:=simplify(subs(S2pp=diff(S2(t),t$2),"));
```

$$\mathrm{eq2p} := \frac{\partial^2}{\partial t^2}\,\mathrm{S2}(t) = I\,\omega\tilde{}\,\left(\frac{\partial}{\partial t}\,\mathrm{S2}(t)\right) - \frac{1}{2}\,\omega\tilde{}\,\nu\,\mathrm{S2}(t)$$
$$-\nu^2\,D^2\,\mathrm{S2}(t) - \frac{1}{4}\,\nu^2\,\mathrm{S2}(t)$$

Notice that we have converted to a single second-order equation with constant coefficients. We pass this equation to `dsolve`:

```
> sol:=dsolve({eq2p,S2(0)=0,D(S2)(0)=0},S2(t));
  Error, (in dsolve) unable to convert from D notation to diff
```

We should not have used D for a parameter. In Maple D is reserved for the differentiation of functions. Use the new name `Del` instead:

```
> eq2p:=subs(D=Del,eq2p);
```

$$\mathrm{eq2p} := \frac{\partial^2}{\partial t^2}\,\mathrm{S2}(t) = I\,\omega\tilde{}\,\left(\frac{\partial}{\partial t}\,\mathrm{S2}(t)\right) - \frac{1}{2}\,\omega\tilde{}\,\nu\,\mathrm{S2}(t)$$

$$- \nu^2 \,\mathrm{Del}^2\, \mathrm{S2}(t) - \frac{1}{4} \nu^2\, \mathrm{S2}(t)$$

```
> sol:=dsolve({eq2p,S2(0)=0,D(S2)(0)=I*nu*Del},S2(t)):
```

We are interested in the square of the amplitude, i.e., the occupation probability for S_2. First we convert the solution to exponentials. We omit the lengthy display, but results with explicit parameter choices are given below.

```
> sol1:=simplify(convert(sol,exp));
```

$$\mathrm{sol1} := \mathrm{S2}(t) = -I \Big[\cos\big(\frac{1}{2}\omega\tilde{}\ t\big)\, \mathrm{e}^{(1/2\sqrt{\%1}\,t)} - \cos\big(\frac{1}{2}\omega\tilde{}\ t\big)\, \mathrm{e}^{(-1/2\sqrt{\%1}\,t)} + I\, \mathrm{e}^{(1/2\sqrt{\%1}\,t)} \sin\big(\frac{1}{2}\omega\tilde{}\ t\big) - I\, \mathrm{e}^{(-1/2\sqrt{\%1}\,t)} \sin\big(\frac{1}{2}\omega\tilde{}\ t\big) \Big] \nu\, \mathrm{Del} \sqrt{\%1} / \big[\omega\tilde{}^{\,2} + 2\,\omega\tilde{}\ \nu + 4\,\nu^2\, \mathrm{Del}^2 + \nu^2 \big]$$

$$\%1 := -\omega\tilde{}^{\,2} - 2\,\omega\tilde{}\ \nu - 4\,\nu^2\, \mathrm{Del}^2 - \nu^2$$

```
> P2:=evalc(abs(rhs(sol1))^2):
```

Now we wish to explore how the magnetic moment of the particle (which is hidden in the variable $\nu = 2\mu B_0$) can be determined. The parameter `Del` represents the ratio of $B_1/(2B_0)$, i.e., a quantity that can be tuned in the experiment. One is interested in a time t_0 at which P_2 becomes a maximum.

Let us look at the choice $B_0 = B_1 = 1$ and insert the approximate proton magnetic moment of 2.8 nucleon magnetons. We work in units adapted to the nucleon problem in which

$$\frac{e\hbar}{2m_{\mathrm{p}}c} = 1 \quad .$$

```
> P2case1:=evalf(subs(Del=0.5,nu=2*2.8,omega=1.,P2));
```

$$\mathrm{P2case1} := .00017816(-24.236 \sin(9.1556\,t) + 24.236 \sin(.50000\,t))^2 + .00017816(24.236 \cos(9.1556\,t) - 24.236 \cos(.50000\,t))^2$$

```
> plot(Re(evalf(P2case1)),t=0..5);
```

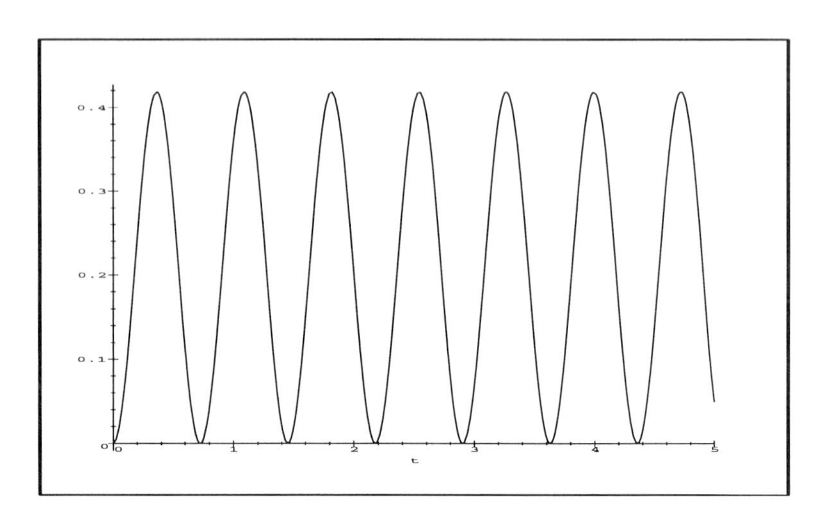

Now we try the same parameters except that we supply the neutron magnetic moment of approximately -1.91 nucleon magnetons.

```
> P2case2:=evalf(subs(Del=0.5,nu=-2*1.91,omega=1.,P2));
```

$$\text{P2case2} := .64726\cos(.50000\,t)^2\sin(2.3741\,t)^2 \\ + .64726\sin(2.3741\,t)^2\sin(.50000\,t)^2$$

```
> plot(Re(evalf(P2case2)),t=0..5);
```

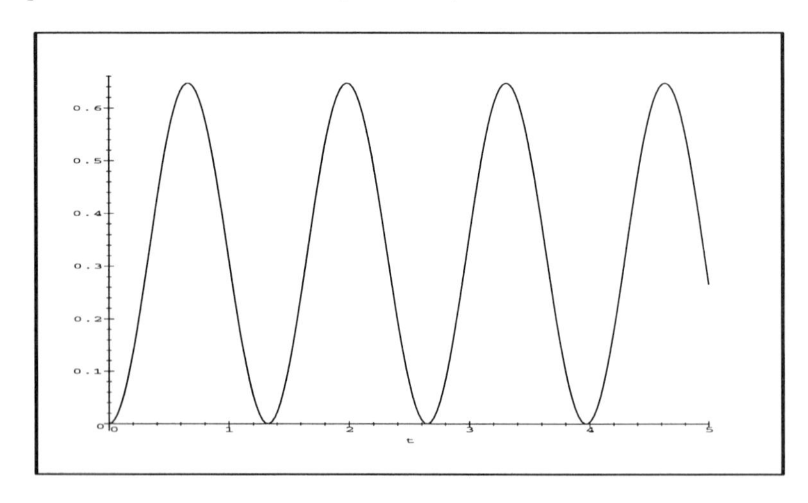

It is evident that the first maximum in the opposite spin polarization is obtained at a different time, i.e., for a different flight path for a constant particle beam velocity. In the Rabi experiment one makes use of the sensitivity of the spin flip operation caused by the time-dependent magnetic field to determine the value of the magnetic moment. If a complete spin flip is possible after some flight path, then an initially spin-polarized beam can be arranged to either pass or not pass through a focusing system depending on the parameters of the magnetic field.

To establish some dependencies of the time evolution of the spin projection on the field parameters we vary the ratio between the oscillating x and y components and the constant z component of the magnetic field:

```
> P2case3:=evalf(subs(Del=0.05,nu=-2*1.91,omega=1.,P2));
```

$$\text{P2case3} := .018019\cos(.50000\,t)^2\sin(1.4229\,t)^2 + .018019\sin(1.4229\,t)^2\sin(.50000\,t)^2$$

```
> plot(Re(evalf(P2case3)),t=0..5);
```

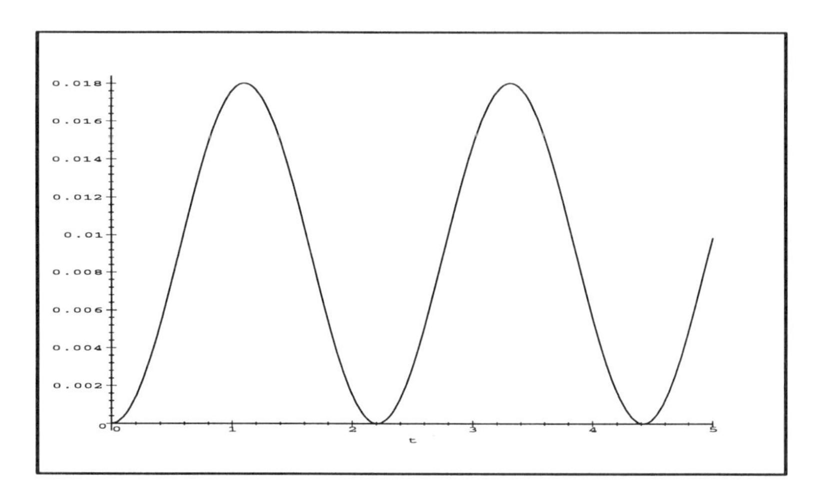

The spin-flip operation is not very effective this way, so let's change the ratio $B_1/(2B_0)$ the other way around:

```
> P2case4:=evalf(subs(Del=5.,nu=-2*1.91,omega=1.,P2));
```

$$\text{P2case4} := .99458\cos(.50000\,t)^2\sin(19.152\,t)^2 + .99458\sin(19.152\,t)^2\sin(.50000\,t)^2$$

The graph shows that it is indeed possible to obtain a complete spin flip with this parameter choice, but also that the time dependence has changed to a much higher frequency. Subsequently we vary the frequency of the magnetic AC field to show that the oscillation frequency of the spin projection can be reduced proportionally.

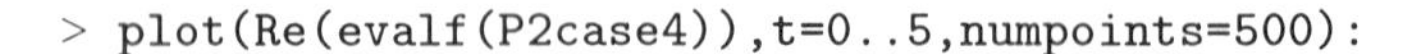

```
> plot(Re(evalf(P2case4)),t=0..5,numpoints=500):
```

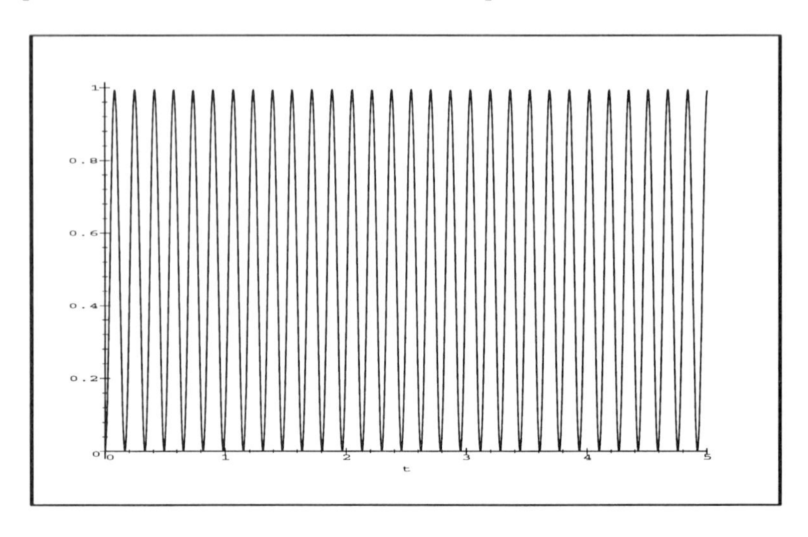


```
> P2case5:=evalf(subs(Del=2.,nu=-2*1.91,omega=0.5,P2)):
> plot(Re(evalf(P2case5)),t=0..5):
```

One can dial `Del` together with ω such that for a given value of μ (ν) the spin polarization changes after some particular time t_0 (flight path s_0) to the opposite value from the initial one. This is used in the Rabi experiment to determine very accurately the magnetic moments of the neutron and the proton. From this information it becomes evident that the proton and the neutron are not structureless spin-$\frac{1}{2}$ particles, but composite objects. Their structure can be understood within quark models that provide a consistent picture of the baryons as composed of three spin-$\frac{1}{2}$ objects called quarks [Na90, HM84].

The understanding of baryon (and meson) structure in terms of quarks is supported by the determination of properties such as the magnetic moments, charge radius/distribution, in addition to the baryon spectroscopy (energy levels in the GeV range) and decay constants. This is important as the building blocks themselves, the quarks, and the mediators of the strong interactions, the gluons, cannot be isolated, and can only be observed indirectly.

Exercise 5.2.1: The hyperon Λ has a mass of 1115.6 MeV as compared to the mass of the proton of 938.3 MeV and the neutron mass of 939.6 MeV. Its mean lifetime is reported to be 2.6×10^{-10} seconds and the magnetic moment is approximately -0.6 nucleonic magnetons $e\hbar/(2m_{\mathrm{p}}c)$ [PDG92]. Observe at what time spin flip is most effective in comparison with the proton and neutron results for some choice of magnetic field parameters. How does this time compare with the lifetime of the Λ?

5.3 Temporary Perturbation

This section serves to demonstrate the use of time-dependent perturbation theory (TDPT) for a two-level system ([Fl71, problem 74]). This approximation technique is required for problems in which energy is not conserved due to the fact that an explicitly time-dependent part of the Hamiltonian $W(t)$ is mixing the eigenstates of the stationary part H_0 that usually dominates the problem. Time-dependent, in contrast to energy-conserving, interactions arise usually in the context of semiclassical descriptions in which a part of the problem is described by classical physics instead of quantum mechanics. The cases of atom – laser interactions or atomic collisions may serve as examples, in which only the electronic excitation problem is treated with QM, while the radiation field or the nuclear motion are treated classically. The changes in the classically described external source are not accounted for in any detail, and it serves as an unlimited supply of energy to the electronic problem. This is usually justified by the amount of energy contained in the source.

We begin with the TD Schrödinger equation (TDSE) and convert it to the energy representation of the H_0 problem. The TDSE

$$\mathrm{i}\hbar\frac{\partial\psi(x,t)}{\partial t} - H_0\psi(x,t) + W(t)\psi(x,t)$$

is represented in terms of $H_0|m\rangle = E_m|m\rangle$ eigenstates with an appropriate phase factor for the time dependence given by

$$\psi_m(x,t) = \phi_m(x)\exp\left(-\frac{\mathrm{i}}{\hbar}E_m t\right) = \phi_m(x)\exp(-\mathrm{i}\omega_m t)\,.$$

As the eigenstates form a complete set any TD wavefunction can be expanded as

$$\psi(x,t) = \sum_{m=1}^{\infty} c_m(t)\psi_m(x,t)\,.$$

One inserts this ansatz and projects the equation onto $\psi_k(x,t)$ for $k = 1, 2, ..., \infty$. This converts the TDSE to an infinite system of ordinary differential equations (ODEs) for the expansion coefficients c_k:

$$\mathrm{i}\hbar\dot{c}_k(t) = W_{km}(t)\exp\left[\mathrm{i}(\omega_k - \omega_m)t\right],$$

where $W_{km}(t) = \langle\phi_k|W(t)|\phi_m\rangle$, and the transition frequencies are defined as $\omega_{km} = (E_k - E_m)/\hbar$.

Under special circumstances the interaction $W(t)$ efficiently couples only two substates out of the usually infinite set $\{\phi_m\}$. This can be caused by the following:

- (a) selection rules (direct couplings permitted only between states of different symmetries);

- (b) the Fourier transform of $W(t)$ from the time to the frequency domain $W'(\omega)$ implies that only one pair of states couples strongly, i.e., $W'(\omega_{12})$ is large, while all other $W'(\omega_{1n})$ are small.

The orthonormality conditions for the basis functions imply for a real interaction $W(t)$ that the norm of $\psi(x,t)$ must be conserved. For a two-level system this translates into

$$|c_1(t)|^2 + |c_2(t)|^2 = 1\,.$$

The magnitude-squares of the expansion coefficients for asymptotically vanishing interaction $W(t \to \infty) = 0$ give the constant final occupation probabilities

$$P_i(t \to \infty) = |c_i(t \to \infty)|^2\,.$$

The TD normalization constraint is necessary to ensure the conservation of probability. The system of ODEs is of such a structure that one can prove the conservation of norm for real $W(t)$, and see how norm (probability in the states considered) is lost if $W(t)$ has an imaginary part. The latter is useful for a so-called optical potential approach in which the interaction is used to describe loss of norm to channels that are not incorporated explicitly [LAD87].

For two channels we have four (three independent) matrix elements for the perturbation $W(t)$, namely $W_{11}(t)$, $W_{12}(t) = W_{21}(t)$, and $W_{22}(t)$. We could allow a more general perturbation with W_{11}, W_{22} real functions and $W_{12} = W_{21}^*$, i.e., a hermitean operator $W(t)$.

We can demonstrate the main features of the solution without being too specific about the wavefunctions and the perturbation, although we emphasize that often the spatial and temporal dependencies in the operator $W(t)$ factorize into $W_0(x)f(t)$. We consider the case of a constant perturbation ($f(t) = 1$) and demonstrate the exact evolution of the wavefunction within the two-state approximation.

In the following we set $\hbar = 1$. The original system

$$\mathrm{i}\dot{c}_1(t) = c_1(t)W_{11} + c_2(t)\exp\left[-\mathrm{i}(\omega_{21})t\right]W_{12}$$

$$\mathrm{i}\dot{c}_2(t) = c_1(t)\exp\left[-\mathrm{i}(\omega_{12})t\right]W_{12} + c_2(t)W_{22}$$

can be transformed into a system with constant coefficients.

```
> eq1:=I*diff(c1(t),t) = c1(t)*W11 + c2(t)*exp(-I*om21*t)*W12;
```

$$\mathrm{eq1} := I\left(\frac{\partial}{\partial t}\,\mathrm{c1}(t)\right) = \mathrm{c1}(t)\,\mathrm{W11} + \mathrm{c2}(t)\,e^{(-I\,om21\,t)}\,\mathrm{W12}$$

```
> eq2:=I*diff(c2(t),t) = c2(t)*W22 + c1(t)*exp(I*om21*t)*W12;
```

$$\mathrm{eq2} := I\left(\frac{\partial}{\partial t}\,\mathrm{c2}(t)\right) = \mathrm{c2}(t)\,\mathrm{W22} + \mathrm{c1}(t)\,e^{(I\,om21\,t)}\,\mathrm{W12}$$

We attempt a direct special solution:

```
> sol:=dsolve({eq1,eq2,c1(0)=1,c2(0)=0},{c1(t),c2(t)});
```

$$\text{sol} :=$$

as well as a general solution without specified initial conditions:

```
> sol:=dsolve({eq1,eq2},{c1(t),c2(t)});
```

$$\text{sol} :=$$

It is too bad that Maple couldn't find the solution directly. Thus, we follow the development as suggested by Flügge [F171], i.e., we use the redefined coefficients $b_1 = c_1$, $b_2 = c_2 \exp(-\mathrm{i}\omega_{21}t)$. Note that the difference between b_2 and c_2 amounts only to a phase factor, and the probability content remains the same.

```
> eq1p:=simplify(subs(c1(t)=b1(t),c2(t)=b2(t)*exp(I*om21*t),
> eq1));
```

$$\text{eq1p} := I\left(\frac{\partial}{\partial t}\,\text{b1}(t)\right) = \text{b1}(t)\,\text{W11} + \text{b2}(t)\,\text{W12}$$

```
> eq2p:=simplify(subs(c1(t)=b1(t),c2(t)=b2(t)*exp(I*om21*t),
> eq2));
```

$$\text{eq2p} := I\,\mathrm{e}^{(I\,om21\,t)}\left(\frac{\partial}{\partial t}\,\text{b2}(t)\right) - \mathrm{e}^{(I\,om21\,t)}\,\text{b2}(t)\,om21 =$$
$$\text{b2}(t)\,\mathrm{e}^{(I\,om21\,t)}\,\text{W22} + \text{b1}(t)\,\mathrm{e}^{(I\,om21\,t)}\,\text{W12}$$

```
> eq2p:=simplify(eq2p*exp(-I*om21*t));
```

$$\text{eq2p} := I\left(\frac{\partial}{\partial t}\,\text{b2}(t)\right) - \text{b2}(t)\,om21 = \text{b2}(t)\,\text{W22} + \text{b1}(t)\,\text{W12}$$

```
> sol:=dsolve({eq1p,eq2p,b1(0)=1,b2(0)=0},{b1(t),b2(t)}):
```

We assign the solution to `b1(t)` and `b2(t)` and calculate the occupation probabilities:

```
> assign(sol);    b1sq:=abs(b1(t))^2:    b2sq:=abs(b2(t))^2:
```

We graph an example of the time-dependent occupation probabilities $P_1(t)$ and $P_2(t)$ for a case where the perturbation is strong and constant in time. Without the analytic derivation done in Flügge's problem 74 (it is general in W_{ij}, ω_{ij} and in much neater form) we observe some basic features of the time evolution. The perturbing matrix elements W_{ij} are of the same order of magnitude as the transition energies, or frequencies ω_{ij}. This causes a significant repopulation between the two states.

The selection of transition matrix elements and frequencies is made through a substitution in the general expressions for the time-dependent occupation probabilities `b1sq` and `b2sq`:

```
> plot({Re(evalf(subs(W11=1/2,W22=1/2,W12=1/4,om21=1/2,
> b1sq))),Re(evalf(subs(W11=1/2,W22=1/2,W12=1/4,om21=1/2,
> b2sq)))},t=0..10);
```

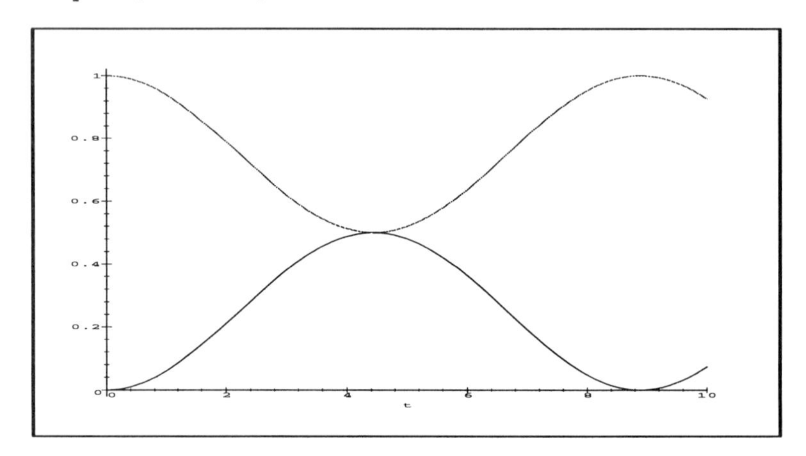

The perturbation $\{W_{11}, W_{12}, W_{22}\}$, which is constant in time, induces temporal oscillations in the occupation probabilities for states 1 and 2. For asymptotically constant probabilities to appear the perturbations either have to go to zero for large times, or alternatively for $W(t \to \infty) \to W_a$ one should look at occupation probabilities for eigenstates of the time-constant Hamiltonian after $t = t_a$, i.e., diagonalize $H_0 + W_a$, and calculate the occupation probabilities for its eigenfunctions $\chi_j(x)$.

We consider the simple case of a sudden change of Hamiltonians at $t = t_a = 0$. Assuming that for $t < 0$ the system was in the eigenstate of H_0, namely $\psi_1(x,t)$, the transition probabilities are obtained simply from the projections

$$P_j(t > 0) = |\langle\chi_j|\psi_1\rangle|^2 , \quad j = 1, 2, \ldots \quad .$$

We emphasize that normally the system winds up in a superposition of such eigenstates, and that even though the occupation probabilities are constant, the coordinate space probability density oscillates. For a two-state final system we have:

$$\rho(x,t) = |c_1(t)\chi_1(x)\exp(-\mathrm{i}\lambda_1 t) + c_2(t)\chi_2(x)\exp(-\mathrm{i}\lambda_2 t)|^2 \quad .$$

Oscillations occur even for constant $c_1(t) = c_1, c_2(t) = c_2$ due to the different eigenenergies λ_i that follow from

$$(H_0 + W_a)|\chi_j\rangle = \lambda_j|\chi_j\rangle \quad .$$

We can demonstrate this feature for the case of a superposition of two harmonic oscillator eigenstates. The probability (charge density) oscillations have a frequency that is related to the energy difference $(\lambda_2 - \lambda_1)/\hbar$. In fact, we provided the result in Sect. 1.3: oscillations of a QM particle appear even in a single-well harmonic oscillator potential, if the particle is not in a pure state, but in a superposition of eigenstates. The period of oscillation is inversely proportional to the spacing of the two energy levels involved. In the case of near-degenerate levels the transit time becomes very long and means that there is a potential barrier between two equilibrium positions.

Applications of this phenomenon are the tunneling oscillations that occur for nearly degenerate ground and first excited states such as:

- the ammonia molecule NH_3: oscillation of the nitrogen atom through the plane spanned by the 3 hydrogen atoms [Ga79, Mer70];
- resonant charge transfer in proton-hydrogen collisions [LE62, LD81] (or quasi-resonant oscillations of K shell electrons in almost symmetric systems, such as fluorine-neon [HC+82, TH+87]. This amounts to oscillations of an electron from one centre to the other and back.

Both problems have in common that the potential has a two-centre characteristic and it is not the symmetric ground state that is occupied. Instead a nearly-degenerate linear combination of ground and anti-symmetric first excited states that results in a single-hump wavefunction over one of the centers is formed. Due to the time evolution of the two phase factors with slightly different energies the probability density evolves and describes the tunneling phenomenon to the opposite centre. In the case of the ammonia molecule the somewhat excited linear combination is formed by thermal excitations. In the collision case with charge transfer the system is set up asymmetrically, as the symmetric two-centre state does not correspond to the setup of an ion beam colliding with a neutral gas target.

Finally, we consider the case of a weak perturbation for which a solution of the coupled ODEs may not be required, but a successive approximation scheme suffices. In that case the pair of equations derived for the two-level case can be solved formally by equating in zeroth order the coeffcients to the initial conditions

$$c_1^{(0)}(t) = 1 \quad , \quad c_2^{(0)}(t) = 0 \quad ,$$

substituting in the ODEs on the right-hand side and integrating from 0 to t:

$$c_1^{(1)}(t) - 1 = \frac{1}{\mathrm{i}} \int_0^t [c_1^{(0)}(s) W_{11}(s) + c_2^{(0)}(s) \mathrm{e}^{-\mathrm{i}\omega_{21} s} W_{12}(s)] \mathrm{d}s \quad ,$$

$$c_2^{(1)}(t) - 0 = \frac{1}{\mathrm{i}} \int_0^t [c_1^{(0)}(s) \mathrm{e}^{-\mathrm{i}\omega_{12} s} W_{12}(s) + c_2^{(0)}(s) W_{22}(s)] \mathrm{d}s \quad .$$

The answer to first order can be expected to be in reasonable agreement with the exact solution to the pair of ODEs for short times t and weak interactions $W_{mn}(t)$ when the level populations have not had time to deviate from the initial conditions. It is evident now that one can improve the solution iteratively using the following scheme:

$$c_j^{(\nu)}(t) = \delta_{j,1} + \frac{1}{\mathrm{i}} \int_0^t \big[\sum_{k=1}^{n} c_k^{(\nu-1)}(s) \mathrm{e}^{-\mathrm{i}\omega_{kj}s} W_{kj}(s)\big] \mathrm{d}s \quad .$$

This scheme results in a recursion that can be solved formally by an ν-fold integral.

I provide a Maple procedure to perform a single step in the recursion, but leave it as an exercise for the reader to make use of it. One can verify how TDPT can be used to build up a Taylor approximation to the numerically obtained exact solution shown in the previous figure. This can serve as an example where TDPT is not very practical as very high order is required to obtain the solution at large times accurately. More successful demonstrations of TDPT can be performed on multi-level quantum systems that are excited by weak time-dependent periodic fields that represent a coherent radiation source.

The transition frequencies for a three-level system with energy gaps $\omega_{21} = 1/2$ and $\omega_{32} = 1/4$ are coded as follows:

```
> om[1,1]:=0: om[2,2]:=0: om[3,3]:=0: om[2,1]:=1/2:
> om[1,2]:=-om[2,1]: om[3,1]:=3/4: om[1,3]=-om[3,1]:
> om[3,2]:=1/4: om[2,3]:=-om[3,2]:
```

The procedure that performs a single iteration step assumes that the subscripted string variables `Wi` and `om` contain the interaction and transition frequencies, respectively, and they are made available by a **`global`** declaration. The arguments for the procedure are a string containing the previous set of coefficients $\{c_m^{(\nu-1)}(t)\}$ stored in a string of length **n** passed as the second argument, and the time parameter t on which they depend is passed as the third argument.

```
> Pert:=proc() global Wi,om;

> cold:=args[1]; n:=args[2]; t:=args[3];

> for k from 1 to n do:   Kdelta:=0;

> if k=1 then Kdelta:=1 fi;

> arg:=sum(cold[i]*exp(-I*om[i,k]*t)*Wi[i,k],i=1..n);

> cnew[k]:=Kdelta+subs(s=t,evalc(int(arg,t=0..s)/I));

> od;     evalm(cnew);     end;
```

Now we define a time-dependent interaction that represents a laser field with frequency $\omega = 1/3$ and non-zero coupling strengths between levels 1 and 2, and levels 2 and 3, respectively. All other couplings vanish as a result of selection rules for the dipole operator.

```
> Wi[1,1]:=0: Wi[2,2]:=0: Wi[3,3]:=0: Wi[1,2]:=sin(t/3)/20:
> Wi[2,1]:=Wi[1,2]: Wi[2,3]:=sin(t/3)/20: Wi[3,2]:=Wi[2,3]:
> Wi[1,3]:=0: Wi[3,1]:=Wi[1,3]:
```

The system is set up at $t = 0$ to be entirely in level 1:

```
> c1[1]:=1: c1[2]:=0: c1[3]:=0:
```

To first order the answer is calculated by the following call:

```
> c2:=Pert(c1,3,t):
```

It can be graphed by the next command:

```
> plot({(abs(c2[1]))^2,(abs(c2[2]))^2,(abs(c2[3]))^2},
> t=0..100,P=0..1,title=`1st order`):
```

In some instances the internal Maple integrator generates a mess in the bookkeeping of terms that are generated and is incapable of doing more than just two orders even though the integrals that need to be performed are entirely trivial. For the purpose of succeeding in such a problem that should be a showcase for the application of CASs I have included in the worksheet file a procedure that performs an iteration without an explicit call to `int`. Instead the program analyzes the structure of the expression to be integrated and assembles the integral through its own bookkeeping effort. This should serve as a warning to the reader that it is easy to condemn a CAS as being of little use on a particular problem, while using it as a black box with its canned procedures (such as `int`). It is more constructive to try to understand why the CAS 'hangs up' and of finding a workaround within the CAS itself.

Exercise 5.3.1: Calculate the time evolution of a two-level system that corresponds to the two lowest harmonic oscillator eigenstates with basic frequency $\omega_0 = 1$, occupied initially in the ground state ($c_1(0) = 1, c_2(0) = 0$) under the perturbation $W_{11} = 1, W_{12} = W_{21} = 1/4, W_{22} = 1/2$ acting for $0 < t < 1$. What are the final occupation probabilities at $t = 1$? Calculate the time evolution of the position expectation value for $t > 1$.

Exercise 5.3.2: Consider the following sudden transition problem: a positively charged helium atom in the (hydrogen-like) ground state He$^+$(1s) undergoes a change whereby a neutron becomes a proton (fictitious β decay without considering the emitted electron). As a result the Coulomb Hamiltonian changes the charge value from $Z = 2$ to $Z = 3$. What are the probabilities to find the atom in the Li^{++}(1s), Li^{++}(2s), and Li^{++}(3s) states? Calculate the time evolution of the radial position expectation value using the three lowest eigenstates in the expansion.

Exercise 5.3.3: Use the included procedure `Pert` to generate symbolic expressions for the TDPT results of orders 1 to 10 for the problem solved exactly at the beginning of the section. Observe how TDPT violates probability conservation for large times, and how this is corrected with increasing order.

5.4 Collisional Excitation

In this section we discuss the problem of excitation of a bound atomic electron by charged particle impact within the framework of time-dependent perturbation theory (TDPT). A number of approximations to the original Schrödinger equation for the entire many-particle problem leads to a semiclassical impact parameter description in which use is made of the impinging particle's classical trajectory $\mathbf{R}(t)$. This introduces an explicit time dependence into the Schrödinger equation and serves as the energy source for excitations. For many purposes it is sufficient to consider a rectilinear nuclear trajectory given by $x = 0$, $y = b$, and $z(t) = v_0 t$, where b is the impact parameter and v_0 the constant impact velocity. Time is considered for the interval $-t_0 < t < t_0$, such that the closest approach occurs at $t = 0$.

The atomic Hamiltonian for the active electron located at $\mathbf{r}$, i.e., $H_0 = T + V(r)$, is supplemented by the Coulomb interaction with the projectile of charge Z_P

$$V_P(\mathbf{r}, t) = -\frac{Z_P e^2}{|\mathbf{r} - \mathbf{R}(t)|} .$$

The energy is not conserved in the electronic excitation problem as the projectile trajectory $\mathbf{R}(t)$ is determined a priori. Thus, we are dealing with a case where the system is not described as a closed QM problem in which energy would be conserved, i.e., the stationary Schrödinger equation would be sufficient for its description. The classical trajectory is introduced in order to avoid the description of the the projectile's and target's nuclear motion at the level of QM. It is a general feature of semiclassical descriptions of problems of this kind that the QM problem becomes explicitly time-dependent, i.e., energy is not conserved in the part described by QM.

To simplify matters it is possible to consider a multipole expansion of the projectile's potential. Due to the cylindrical symmetry (strictly with respect to the internuclear axis after a separation in center-of-mass and relative nuclear motion) the electron transition probabilities will be weighted with the impact parameter b. Thus, the probabilities receive a large contribution from intermediate to large b values, for which it is sufficient to consider the lowest-order multipoles in the interaction. According to our findings in the previous section, TDPT should be adequate at least for small Z_P (small coupling) and for a high impact velocity (short-time residual interaction).

The specific limiting case of a dipole interaction that allows transitions between $l = 0$ and $l = 1$ states is considered here. The impact parameter is set to unity ($b = 1$ a.u.) and the dependence on the collision velocity (or energy of the colliding particle) of the temporal excitation function is examined. The dipole operator has two components: a longitudinal one along the direction of the scattering particle's velocity vector (z axis), and a transverse component aligned with the impact parameter vector (y axis). We consider only the longitudinal component, for which the selection rule $\Delta l = 1$ and $\Delta m =$

0 holds. A complete treatment with the analytic calculation of transition probabilities can be found in the literature, e.g., [Br70].

The longitudinal dipole component of the perturbing potential is derived from a projection of $V_P(\mathbf{r},t)$ onto the spherical harmonic $Y_{1,0}(\Omega)$. It vanishes at the closest approach and takes on a maximum value at some time before and after this point on the projectile's trajectory. For the purpose of calculating the electronic excitation we assume that the target atom (strictly speaking its center of mass, not just the nucleus) remains at rest. The longitudinal dipole interaction is responsible for a displacement or polarization of the electronic wavefunction towards the positively charged projectile, i.e., it induces an electric dipole moment along the z axis. This process can be understood as a virtual excitation of the original typically spherically symmetric electron state, e.g., in hydrogen a transition 1s $\rightarrow$ 2p$_0$ and back. For a projectile moving on a rectilinear trajectory with velocity v_0 the dipole component of the interaction is proportional to

```
> fcoll:=(v0,t)->(v0*t)/(1+(v0*t)^2)^(3/2);
```

$$\text{fcoll} := (\,v0, t\,) \rightarrow \frac{v0\,t}{(\,1 + v0^2\,t^2\,)^{3/2}}$$

Within first-order time-dependent perturbation theory the transition amplitude that connects the H_0 eigenstates $|m\rangle \rightarrow |n\rangle$ is given by the Fourier transform of the temporal excitation function $W_{mn}(t)$ into the frequency domain, evaluated at the transition frequency $\omega_{mn} = (E_m - E_n)/\hbar$. For the excitation function provided above the Fourier transform results in a Bessel function [Br70], but Maple is unable to perform the integral.

We can therefore ask: is there a simple temporal profile of similar shape for which the Fourier transform can be done easily? We would like to keep the dependence on the impact velocity v_0. First we investigate the temporal excitation function and observe how it changes with v_0

```
> with(plots):
```

```
> animate(fcoll(v,t),t=-25..25,v=0.1..1.6,color=red):
```

We display the excitation function for the velocities $v = 0.2, 0.5$, and 1.0.

We observe that the amplitude of the temporal excitation function remains the same for the three velocities. It depends only on the projectile charge Z_P and the impact parameter b. For slow collisions the perturbation acts over a long time period. One might think that in this case the strogest transitions are induced in the target atom. While this general observation appears to be correct, it does not take into account the resonance phenomenon discussed below. The Fourier transform to the frequency (or energy) domain shows that the strength of electronic excitations depends also on a matching between external projectile velocity and internal electronic motion.

```
> plot(subs(v=0.2,fcoll(v,t)),t=-25..25,color=red);
```

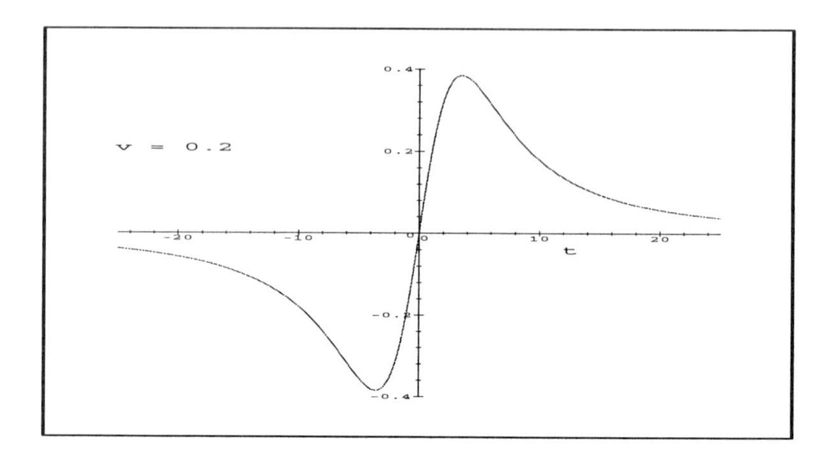
v = 0.2
0.4
0.2
-0.2
-0.4
-20
-10
0
10
20
t

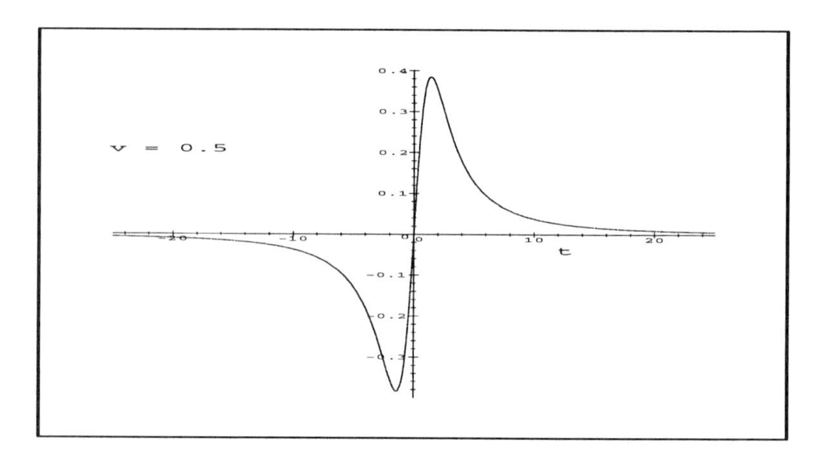
v = 0.5
0.4
0.3
0.2
0.1
-0.1
-0.2
-0.3
-20
-10
0
10
20
t

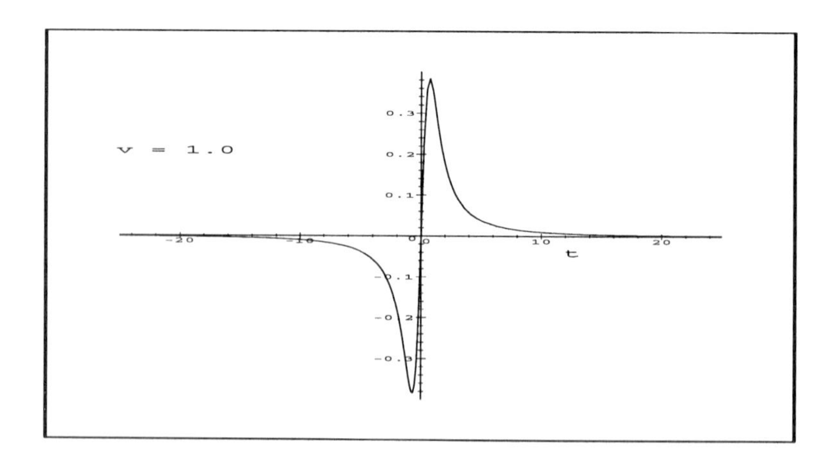
v = 1.0
0.3
0.2
0.1
-0.1
-0.2
-0.3
-20
-10
0
10
20
t

An alternative excitation function that is less realistic due to its short range, but which we prefer in this example, since its Fourier transform can be calculated, is given by:

```
> fcoll:=(v*t)*exp(-v^2*t^2);
```

$$\text{fcoll} := v\,t\,e^{(-v^2\,t^2)}$$

```
> animate(fcoll,t=-25..25,v=0.1..1.6):
```

For comparison with the original temporal excitation profile we show the result for two of the velocities considered before, namely, $v = 0.2$, and $v = 0.5$ a.u.:

```
> plot(subs(v=0.2,fcoll),t=-25..25,color=red);
```

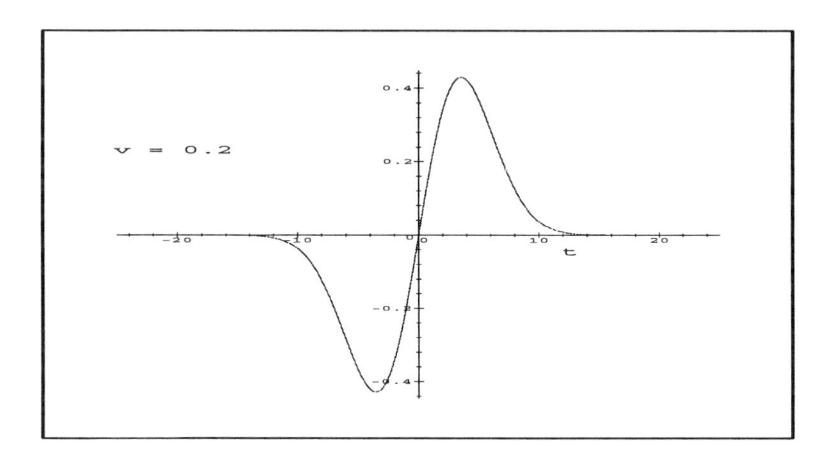

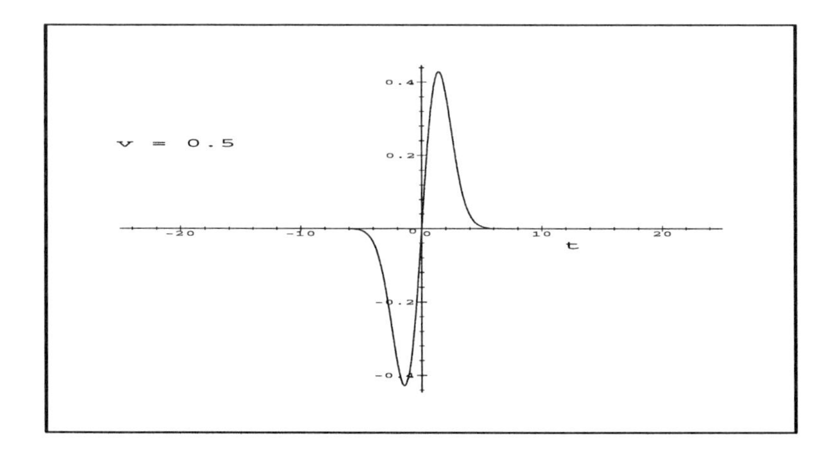

We define the Fourier transform as a procedure:

```
> FT:=(f,t)->int(f*exp(I*omega*t),t=-infinity..infinity)/(2*Pi);
```

$$FT := (f, t) \to \frac{1}{2} \frac{\text{int}(f\, e^{(I\,\omega\, t)}, t = -\infty..\infty)}{\pi}$$

We now pick all parameters except for the velocity (in atomic units):

```
> assume(v>0); assume(omega>0);
> FT(fcoll,t);        fomega:=abs(")^2;
```

$$\frac{1}{4} \frac{I\, e^{\left(-1/4 \frac{\omega\tilde{}^2}{v\tilde{}^2}\right)} \omega\tilde{}}{v\tilde{}^2 \sqrt{\pi}} \qquad \text{fomega} := \frac{1}{16} \frac{\left(e^{\left(-1/4 \frac{\omega\tilde{}^2}{v\tilde{}^2}\right)}\right)^2 \omega\tilde{}^2}{v\tilde{}^4\, \pi}$$

In a Maple session one can observe the change in the excitation energy profile as a function of impact velocity using the **animate** feature. In the text we provide graphs for $v = 0.2, 0.5$, and 1.0.

```
> animate(fomega,omega=0..2,v=0.1..1.6,color=black):
> plot(subs(v=0.2,fomega),omega=0..2,color=red);
```

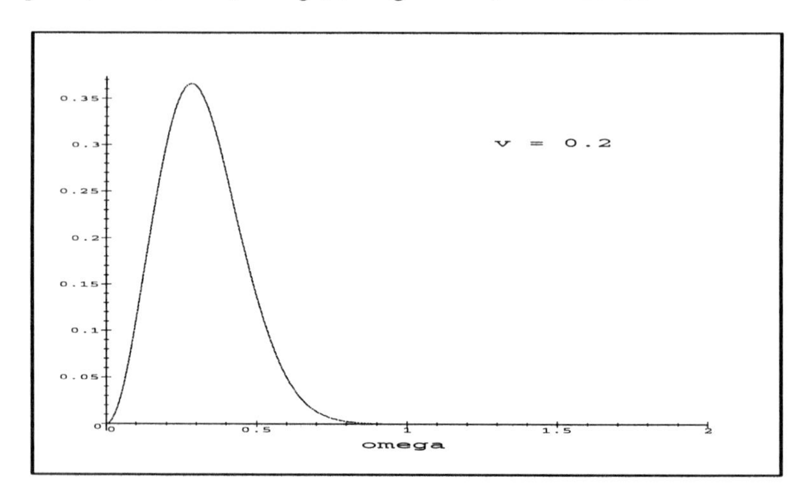


```
> plot(subs(v=0.5,fomega),omega=0..2,color=blue);
```

Note the change of scale on the abscissa of the third graph, i.e., the case of $v = 1.0$:

```
> plot(subs(v=1.0,fomega),omega=0..4,color=blue);
```

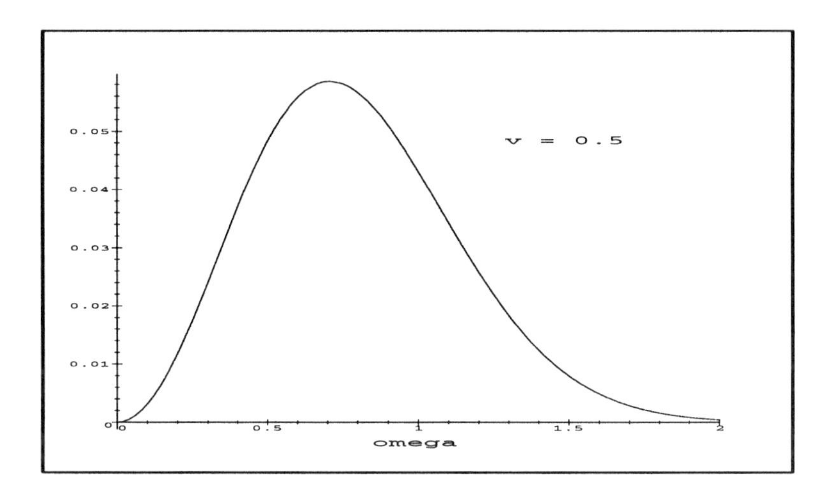

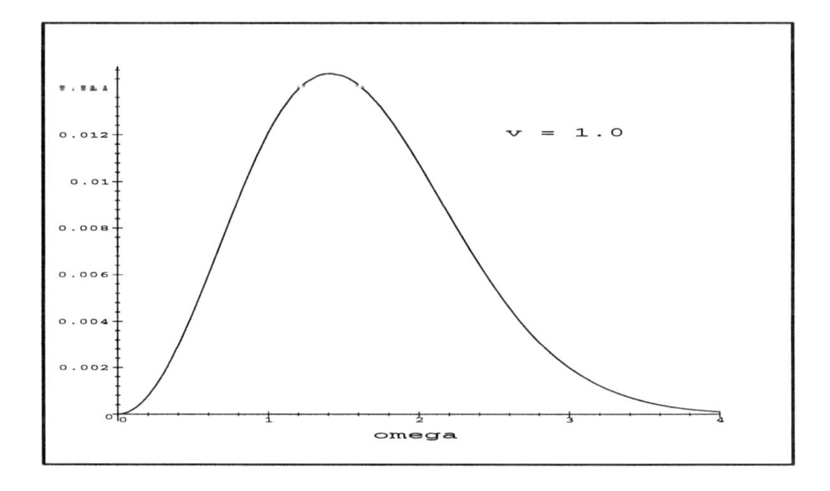

As a function of impact velocity the Fourier transform of the temporal excitation profile favours different excitation frequencies. As a result, for any given transition $|m\rangle \longrightarrow |n\rangle$ there is an optimum velocity at which the collision maximizes the transition probability P_{mn}.

The total cross section for the transition follows from a summation over impact parameters (a replacement for the polar angle) as well as over azimuthal angles (which gives a factor of 2π due to the independence of the projectile potential on the angles). The weight b in the summation over impact parameters results from the Jacobian in cylindrical coordinates:

$$\sigma_{mn}(v) = 2\pi \int_0^\infty P_{mn}(b, v) b \mathrm{d}b \,.$$

The very general implication from this simple calculation is that excitation cross sections as a function of impact energy (or velocity) show a maximum at some energy. At this energy the collision velocity matches an average velocity of the internal bound particle. This resonant behaviour represents the analog of a classical oscillator with damping.

We make the following observation: at very high collision speeds the interaction is short in time and is dominated by high Fourier components. While it can generate high-lying excitations (with a small probability), it is tuned out of resonance with typical transition energies, say between the ground state of the atom and closely lying excited levels. At very low speeds only low Fourier components are present in the spectrum of the temporal pulse and the relevant transitions cannot be excited efficiently. Somewhere in between is a point where the transition probability is maximized. The individual strengths for the transition to specific states depend on the radial matrix element of the dipole operator.

The reasoning is so general that it is not just confined to atomic collision processes. Most total cross sections are indeed of such a shape as a function of energy (for finite-range interactions):

$$\sigma(E \to 0) \to 0\,, \quad \sigma(E \to \infty) \to 0\,, \quad \sigma(E_0) = \text{maximum}\,.$$

Exercise 5.4.1: Consider a simple collision system such as protons colliding with atomic hydrogen using the simplified, i.e., Gaussian, temporal excitation function. Calculate for fixed velocity the dependence of the transition probability on the impact parameter for $1\text{s} \to 2\text{p}_0$ as well as for $1\text{s} \to 2\text{p}_{\pm 1}$ transitions.

6. Scattering in 3D

6.1 Partial Waves

This section deals with the expansion of a free plane wave moving in the z direction into spherical harmonics. In stationary scattering theory one considers the free Schrödinger equation in the form

$$(\nabla^2 + k^2)u(r, \theta, \phi) = 0\,,$$

with $k^2 = 2mE/\hbar^2$ determined by the known collision energy E. For the problem of scattering from a potential the Schrödinger equation is to be solved with boundary conditions representing an incident free beam and an outgoing spherical wave modulated by the scattering amplitude $f(\theta, \phi)$ that carries the scattering information. For central scattering potentials $V(r)$ this complex-valued amplitude does not depend on the azimuthal angle ϕ due to angular momentum conservation. The differential scattering cross section is related to the scattering amplitude by

$$\frac{\mathrm{d}\sigma}{\mathrm{d}\theta} = |f(\theta)|^2\,,$$

and represents the generalization of reflection and transmission coefficients from the 1D to the 3D situation. The partial wave expansion discussed for free plane waves in this section provides the foundation for the separation-of-variables method to attack the scattering-state Schrödinger equation, i.e., to rewrite the partial differential equation into an infinite set of ordinary differential equations.

We seek an expansion of the solution in the form

$$u(r, \theta, \phi) = \sum_{L,m} F_l(kr) Y_{l,m}(\theta, \phi)\,.$$

The plane wave describing a beam of free particles along the z direction is given by

$$\exp(\mathrm{i}kz) = \exp(\mathrm{i}kr\cos\theta)\,,$$

and does not depend on ϕ. Thus, we can simplify the general expansion given above to

$$u(r,\theta)=\sum_{l=0}^{\infty}C_l f_l(r) P_l(\cos\theta)\,.$$

The expansion coefficients C_l and functions $f_l(r)$ that depend on the magnitude of the wavevector $k = |\mathbf{k}|$ are obtained from projection onto Legendre polynomials. First, we make the polynomials available:

```
> assume(k>0);    with(orthopoly):    P(1,x);
```

$$x$$

We can define a plane wave either in terms of the polar angle

```
> psi:=(r,theta)->exp(I*k*r*cos(theta));
```

$$\psi := (\,r,\theta\,) \to \mathrm{e}^{(\,I\,k\,r\cos(\,\theta\,)\,)}$$

or more conveniently for the later development in terms of $x = \cos\theta$:

```
> phi:=(r,x)->exp(I*k*r*x);
```

$$\phi := (\,r,x\,) \to \mathrm{e}^{(\,I\,k\,r\,x\,)}$$

We project the plane wave onto the Legendre polynomials to obtain the expansion coefficients. The projection is accomplished using an appropriately defined inner product.

```
> int(P(0,x)*phi(r,x),x=-1..1);
```

$$-\frac{I\,\mathrm{e}^{(\,I\,k\tilde{}\,r\,)}}{k\tilde{}\,r}+\frac{I\,\mathrm{e}^{(\,-I\,k\tilde{}\,r\,)}}{k\tilde{}\,r}$$

```
> simplify(");
```

$$2\,\frac{\sin(\,k\tilde{}\,r\,)}{k\tilde{}\,r}$$

This is the lowest term in the expansion, i.e., $C_0 f_0(r)$. We recognize $f_0 = j_0(kr)$ as the zero-order spherical Bessel function and proceed with the next order $l = 1$:

```
> int(P(1,x)*phi(r,x),x=-1..1);
```

$$-\frac{\mathrm{e}^{(\,I\,k\tilde{}\,r\,)}\,(\,I\,k\tilde{}\,r-1\,)}{k\tilde{}^{2}\,r^{2}}-\frac{\mathrm{e}^{(\,-I\,k\tilde{}\,r\,)}\,(\,I\,k\tilde{}\,r+1\,)}{k\tilde{}^{2}\,r^{2}}$$

```
> simplify(");
```

$$-\frac{I\,(\,\mathrm{e}^{(\,I\,k\tilde{}\,r\,)}\,r\,k\tilde{}+I\,\mathrm{e}^{(\,I\,k\tilde{}\,r\,)}+\mathrm{e}^{(\,-I\,k\tilde{}\,r\,)}\,k\tilde{}\,r-I\,\mathrm{e}^{(\,-I\,k\tilde{}\,r\,)}\,)}{k\tilde{}^{2}\,r^{2}}$$

```
> evalc(");
```

$$-\frac{I\,(2\cos(\,k\tilde{}\,r\,)\,r\,k\tilde{} - 2\sin(\,k\tilde{}\,r\,))}{k\tilde{}^{2}\,r^{2}}$$

We recognize that the projections are proportional to the spherical Bessel functions, which are elementary functions that arise as special cases of the general Bessel function with half-integer order:

```
> BesselJ(0+1/2,xi);
```

$$\frac{\sqrt{2}\sin(\,\xi\,)}{\sqrt{\pi}\,\sqrt{\xi}}$$

```
> BesselJ(1+1/2,xi);
```

$$-\frac{\sqrt{2}\,(\,-\sin(\,\xi\,) + \cos(\,\xi\,)\,\xi\,)}{\xi^{3/2}\,\sqrt{\pi}}$$

Up to a factor of $\sqrt{\xi} = \sqrt{kr}$ our projection gives as a radial function an 'expansion coefficient' in terms of the Bessel functions of half-integer order. These can be built up by repeated differentiation of j_0 according to [Ar85]

$$j_l(x) = (-1)^l x^l \left(\frac{\mathrm{d}}{x\mathrm{d}x}\right)^l \left(\frac{\sin x}{x}\right) .$$

A procedure for the spherical Bessel functions can be written as follows:

```
> jsB:=proc();
> l:=args[2];
> if whattype(l)=integer then
> x:=args[1];
> sin(x)/x;
> if l=0 then RETURN(");
> else
> res:=";
> for i from 1 to l do
> res:=expand(diff(res,x)/x);
> od;
> expand(res*(-1)^l*x^l);
> fi;
> else
```

```
> RETURN('l is not an integer: ',l);

> fi;     end;
```

We try the procedure for $l = 0, 1, 2$ and graph $j_2(x)$:

```
> jsB(x,0);    jsB(x,1);     jsB(x,2);
```

$$\frac{\sin(x)}{x} \qquad -\frac{\cos(x)}{x}+\frac{\sin(x)}{x^2} \qquad -3\frac{\cos(x)}{x^2}-\frac{\sin(x)}{x}+3\frac{\sin(x)}{x^3}$$

```
> plot(jsB(x,2),x=0..10,j2=-0.4..0.4,axes=boxed);
```

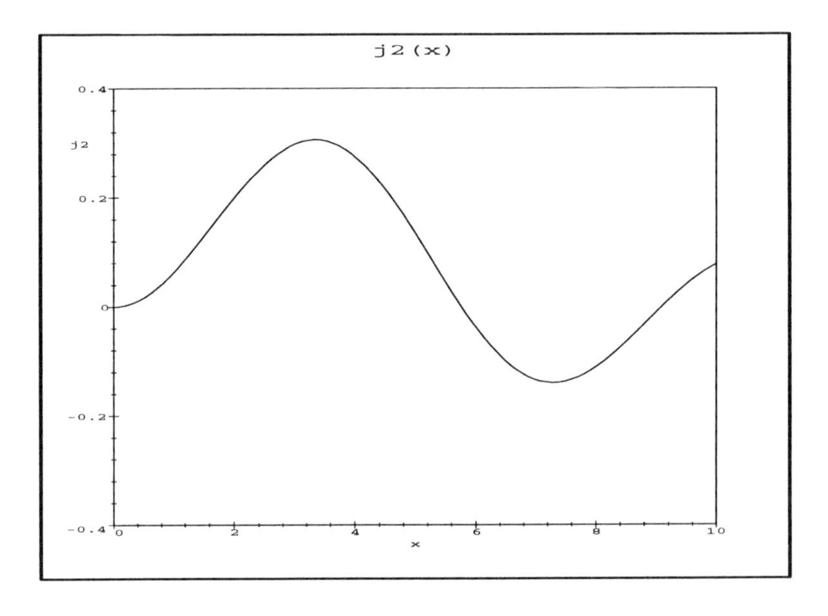

Now we try to plot the function at twice its previous argument:

```
> plot(jsB(2*x,2),x=0..10,j2=-0.4..0.4);
  Error, (in jsB)
  wrong number (or type) of parameters in function diff
```

Maple complains due to the differentiation with respect to the argument x that occurs inside the procedure. This can be fixed either by a re-design of the procedure, or by the application of a substitution trick:

```
> plot(subs(xi=2*x,jsB(xi,2)),x=0..10,j2=-0.4..0.4,title=
> 'j2(2*x)',axes=boxed);
```

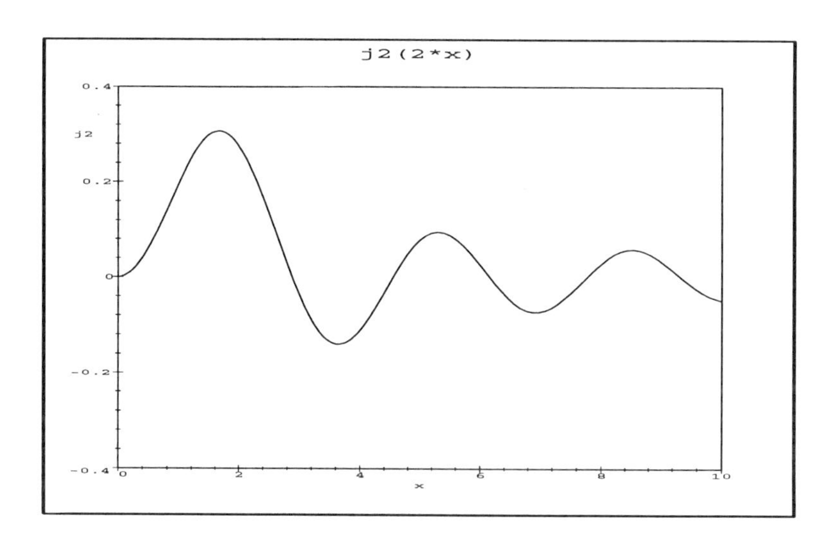

We can explore now the convergence properties of the following expansion for various values of k

$$\exp(ikz) = \exp(ikr\cos\theta) = \sum_{l=0}^{\infty} (2l+1) i^l j_l(kr) P_l(\cos\theta)\,.$$

We continue to use the abbreviation $x = \cos\theta$ which, of course, has nothing to do with the x coordinate. We make an attempt at a procedure to calculate the partial wave expansion truncated at order $l_{\max}$:

```
> TruncPW:=proc(k,r,x,l_max);

> if type(l_max,integer) <> true then RETURN

> else

> sum(subs(xi=k*r,jsB(xi,l))*I^l*(2*l+1)*P(l,x),l=0..l_max);

> fi;    end:
```

Our attempt to evaluate with this procedure fails, however:

```
> TruncPW(k,r,x,3);
  Error, (in TruncPW)
  wrong number (or type) of parameters in function subs
```

The problem seems to be that the `sum` command passes the variable l and the Bessel procedure chokes on it, because it sees no guarantee that l is an integer. Hopefully a release beyond R3 will have this fixed. We rewrite the procedure using an explicit loop construct.

```
> TruncPW:=proc(k,r,x,l_max);
```

```
> if type(l_max,integer) <> true then RETURN
> else
> suml:=0;
> for l from 0 to l_max do
> suml:=suml+subs(xi=k*r,jsB(xi,l))*I^l*(2*l+1)*P(l,x);
> od;
> fi;
> suml;          end:
> TruncPW(k,r,x,3);          assume(r>0);
```

$$\frac{\sin(k^{\sim} r)}{k^{\sim} r} + 3\,I\left(-\frac{\cos(k^{\sim} r)}{k^{\sim} r} + \frac{\sin(k^{\sim} r)}{k^{\sim 2} r^2}\right) x$$
$$-5\left(-3\frac{\cos(k^{\sim} r)}{k^{\sim 2} r^2} - \frac{\sin(k^{\sim} r)}{k^{\sim} r} + 3\frac{\sin(k^{\sim} r)}{k^{\sim 3} r^3}\right)\left(\frac{3}{2}x^2 - \frac{1}{2}\right)$$
$$-7\,I\left(-15\frac{\cos(k^{\sim} r)}{k^{\sim 3} r^3} - 6\frac{\sin(k^{\sim} r)}{k^{\sim 2} r^2} + \frac{\cos(k^{\sim} r)}{k^{\sim} r} + 15\frac{\sin(k^{\sim} r)}{k^{\sim 4} r^4}\right)$$
$$\left(\frac{5}{2}x^3 - \frac{3}{2}x\right)$$

We graph the real part of the plane wave for $k = 1$ and $x = 1/2$ as a function of r:

```
> plot(subs(k=1,Re(phi(r,0.5))),r=0..10,axes=boxed);
```

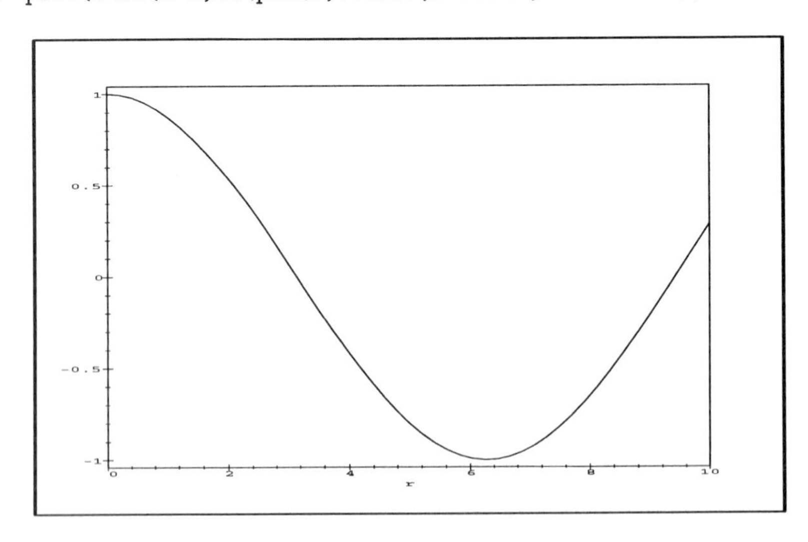

Obviously the answer is simple, i.e.,

```
> Re(phi(r,1/2));
```

$$\cos\left(\frac{1}{2}\,k\tilde{}\;r\tilde{}\right)$$

Now for comparison we calculate the partial wave expansion truncated at $l_{\max} = 2$, $l_{\max} = 4$, and $l_{\max} = 6$ and graph the results:

```
> plot({Re(TruncPW(1,r,0.5,2)),Re(TruncPW(1,r,0.5,4)),
> Re(TruncPW(1,r,0.5,6))},r=0..10,-1..1,axes=boxed);
```

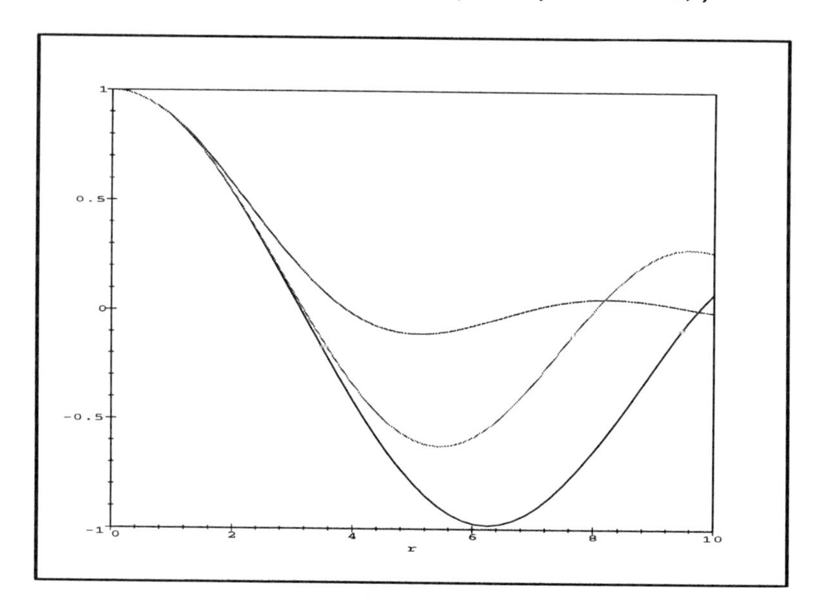

Clearly the expansion converges for a larger r range as $l_{\max}$ is increased. For large $l_{\max}$ problems of a numerical nature appear at small distances r due to the cancellations between the divergent terms at zero argument in the spherical Bessel functions. The problem can be remedied with an increase in the value of the `Digits` parameter. Programmers should be aware of this problem and refer to the literature as to how to calculate the spherical Bessel functions of higher order efficiently [AS65].

We can also show how close the answer is for a small momentum $k = 0.1$, moderate r and fixed $l_{\max} = 8$ by graphing the magnitude. A deviation from unity sets in at large r.

```
> plot({abs(TruncPW(0.1,r,0.5,8)),subs(k=0.1,abs(phi(r,0.5)))}
> ,r=0..200,Rphi=0..1);
```

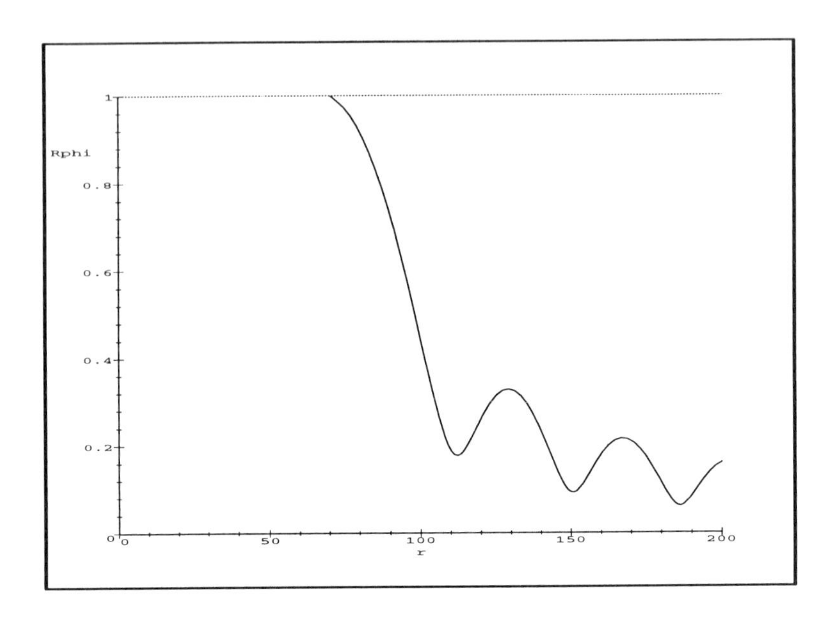

Next we show the $l = 0$ result only. It does not depend on x:

```
> with(plots):
> animate(Re(TruncPW(1,r,x,0)),r=0..10,x=-1..1);
```

The subsequent command generates the animation that shows what the plane wave visualized in this way should look like.

```
> animate(Re(subs(k=1,phi(r,x))),r=0..10,x=-1..1);
```

We can observe what the higher l values can do for us: $l = 1$ contributes to the imaginary part only.

```
> animate(Re(TruncPW(1,r,x,4)),r=0..10,x=-1..1);
```

We can display the exact and approximate result as a function of both r and x. We choose $k = 1$ and request a surface plot of the real part of the plane wave. I leave it as an exercise for the reader to explore the results for different wavenumbers. It is of interest to compare the convergence behaviour of the partial wave expansion at different energies.

```
> plot3d(Re(subs(k=1,phi(r,x))),r=0..10,x=-1..1,axes=boxed);
```

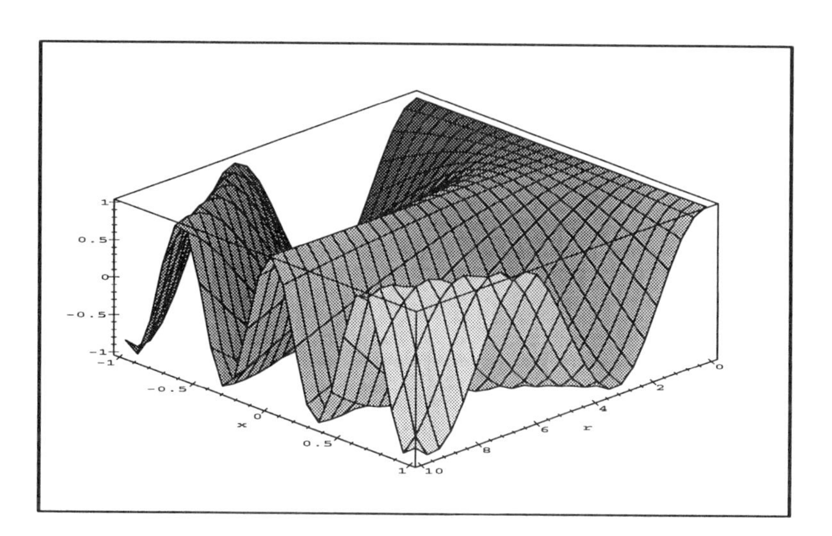

A comparison of the surface plots, above for the exact answer and below for the truncated expansion, provides the information about in which parts of the $r - x$ plane convergence can be reached quickly with the partial wave expansion. We generate the approximate result with $l_{\max} = 4$:

```
> plot3d(Re(TruncPW(1,r,x,4)),r=0..10,x=-1..1,axes=boxed);
```

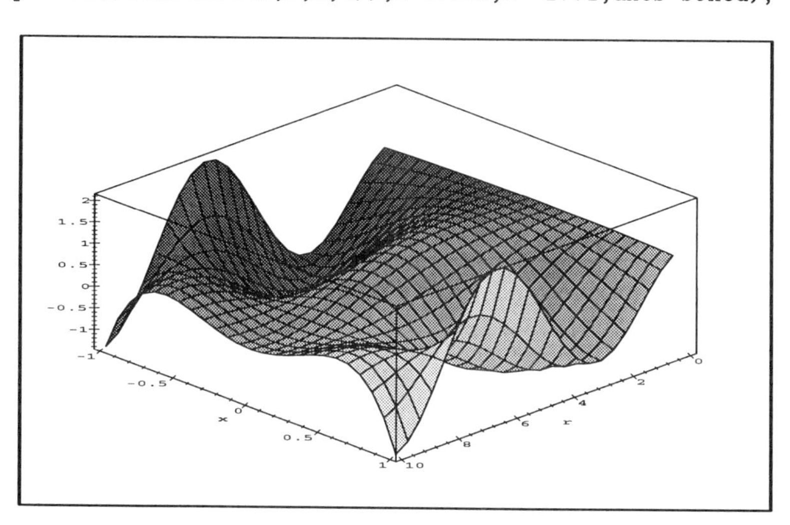

Now we increase to $l_{\max} = 8$, and observe that the result is more satisfying even away from $r = 0$ and for the full range of $x = \cos\theta = -1..1$:

```
> plot3d(Re(TruncPW(1,r,x,8)),r=0..10,x=-1..1,axes=boxed);
```

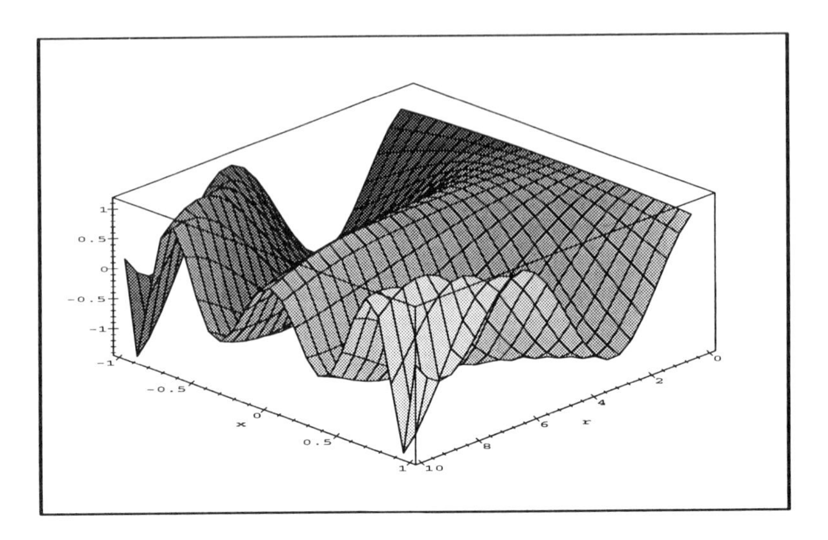

The partial wave expansion is successful if

- (a) the range of the interaction in the scattering problem a is short and the expansion of the plane wave in spherical harmonics is required only for small r;
- (b) the collision energy is not high and the scattered beam is not sharply focused in some direction. In particular, the zero-wavevector limit $k \to 0$ is dominated by the $l = 0$ partial wave.

Both criteria can be combined into a dimensionless parameter; if $ka < 1$ then the number of partial waves required to describe the scattering process to reasonable accuracy remains moderate.

Exercise 6.1.1: Perform a convergence analysis of the partial wave expansion of the plane wave in coordinate representation as a function of the particle's momentum for fixed direction angle θ.

Exercise 6.1.2: Write a procedure for the spherical Bessel functions on the basis of the recursion relation

$$j_{l+1}(\rho) + j_{l-1}(\rho) = \frac{2l+1}{\rho} j_l(\rho),$$

and compare the numerical reliability with the given procedure for small ρ and high l.

6.2 Potential Scattering

In this section we deal with simple potential scattering in three dimensions. We consider the case of a simple barrier or a well and solve the radial equation that arises in partial wave analysis (cf. Sect. 6.1)

$$\frac{\mathrm{d}^2 f_l}{\mathrm{d}r^2} + \left[K^2 - \frac{l(l+1)}{r^2}\right] f_l(r) = 0\,,$$

where $K^2 = k^2 \pm 2mV_0/\hbar^2$ with the plus applicable for the case of a barrier, and the minus sign for an attractive potential well.

The solutions to the radial equations for the partial waves $l = 0, 1, 2, ...$ contain in the asymptotic large-r domain the information about the scattering amplitude $F(\theta)$ in the form of so-called phase shifts $\eta_l(k)$ that parametrize the expansion in Legendre polynomials

$$F(\theta) = \sum_{l=0}^{\infty} P_l(\cos\theta)\,.$$

The phase shift represents the change of the free solution discussed in the previous section in the large-r domain (where the potential vanishes) due to the presence of a scattering potential at short distances. For s waves we have

$$f_0(r) \approx \sin(kr + \eta_0) = \cos(kr) + \tan\eta_0 \sin(kr)\,.$$

We solve the radial Schrödinger equation by finding solutions inside and outside the potential regions and by matching them at the potential boundary $r = R$. First we look at the free (asymptotic) solutions, i.e., $K = k$. As usual we set $\hbar = m = 1$.

```
> assume(l,integer);    E0:=k^2/2;
```

$$E0 := \frac{1}{2}k^2$$

```
> SEfree:=diff(f_l(r),r$2) + (k^2-l*(l+1)/r^2)*f_l(r)=0;
```

$$\mathrm{SEfree} := \left(\frac{\partial^2}{\partial r^2}\mathrm{f\_l}(r)\right) + \left(k^2 - \frac{l\tilde{}\,(l\tilde{}+1)}{r^2}\right)\mathrm{f\_l}(r) = 0$$

```
> solfree:=dsolve(SEfree,u_l(r));
```

$$\mathrm{solfree} := \mathrm{f\_l}(r) = \_C1\,\sqrt{r}\,\mathrm{BesselJ}\left(\frac{1}{2} + l\tilde{}, k\,r\right) + \_C2\,\sqrt{r}\,\mathrm{BesselY}\left(\frac{1}{2} + l\tilde{}, k\,r\right)$$

Let's look at the regular and irregular pieces:

```
> uJ:=(k,r,l)->sqrt(r)*BesselJ(1/2+l,k*r):
> vJ:=(k,r,l)->sqrt(r)*BesselY(1/2+l,k*r):
> plot(uJ(1,r,0),r=0..10);
```

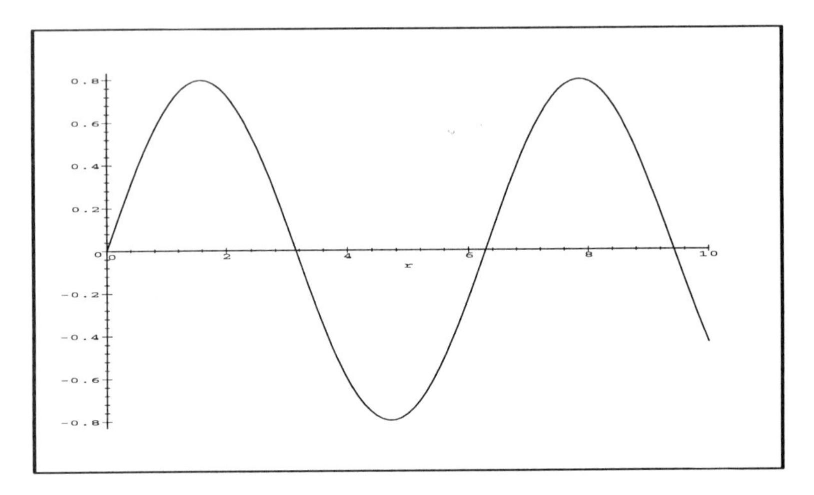


```
> plot(vJ(1,r,0),r=0..10);
```

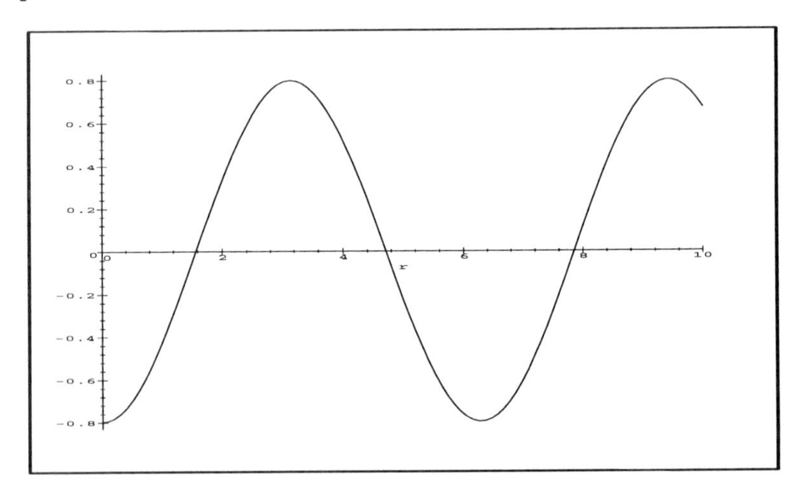

There is apparently nothing wrong; one should keep in mind, however, that this radial function is divided by r to provide the wavefunction. Therefore, only the sine function is acceptable at $r = 0$. Now we set up the matching conditions without paying attention to normalization:

```
> K2:=k^2-2*V0;     K:=sqrt(K2);
```

$$K2 := k^2 - 2\ V0 \qquad K := \sqrt{k^2 - 2\ V0}$$

```
> uJp:=(k,r,l)->diff(uJ(k,r,l),r):
> vJp:=(k,r,l)->diff(vJ(k,r,l),r):
```

We express the solution inside the barrier/well just in terms of the regular Bessel (-Ricatti) function $u_l(x)$, while in the asymptotic region we need a combination of Neumann (-Ricatti) $v_l(x)$ as well as Bessel (-Ricatti) functions $u_l(x)$:

$$f_l(r) = A_l u_l(kr) + B_l v_l(kr)\,.$$

We keep in mind that the normalization of the incident wave is arbitrary. One can either set $A_l = 1$, or express the wavefunction in terms of the ratio $R_l = B_l/A_l$, which is related to the phase shift according to $R_l = \tan\eta_l$.

```
> matchcond:=uJp(K,R0,0)/uJ(K,R0,0)=(uJp(k,R0,0)+BA_0
> *vJp(k,R0,0))/(uJ(k,R0,0)+BA_0*vJ(k,R0,0)):
> solmatch:=solve(matchcond,BA_0);
```

$$\begin{aligned}\text{solmatch} := -\big(&-2\,(k^2 - 2\ V0\,)\,R0^2 \cos(\sqrt{k^2 - 2\ V0}\ R0)\sin(k\ R0)\\ &+ 2\sqrt{k^2 - 2\ V0}\ R0^2\, k\cos(\,k\ R0\,)\sin(\sqrt{k^2 - 2\ V0}\ R0)\big)/\\ &\big[2\,(k^2 - 2\ V0\,)\,R0^2 \cos(\sqrt{k^2 - 2\ V0}\ R0)\cos(\,k\ R0\,)\\ &+ 2\sqrt{k^2 - 2\ V0}\ R0^2\, k\sin(\,k\ R0\,)\sin(\sqrt{k^2 - 2\ V0}\ R0)\big]\end{aligned}$$

```
> eta_0:=-arctan(solmatch):
```

We discuss separately the results for barrier potentials given by $V_0 > 0$ and for wells for which $V_0 < 0$ (alternatively we can keep V_0 positive and choose the appropriate sign in $K(k, V_0)$). In the graphs presented below we make the observation that for attractive potentials (wells) the phase shift is positive, while for repulsive potentials (barriers) it is negative for all scattering energies.

Other general observations allow the conclusion that the phase shift starts at zero energy (or wave number) at zero or a multiple of π, and one can prove that it goes to zero as the energy increases to infinity. The latter is consistent with the notion that at very high impact energy no scattering takes place. We leave it as an exercise to derive the behaviour of the phase shift for small and large k using a series expansion of the exact, analytic answer.

For those cases where the phase shift starts at zero for $k = 0$ (all repulsive potentials and attractive potentials that do not support bound states) the slope of the phase shift $\eta(k)$ at $k = 0$ has the interpretation of a scattering length (inverse wave number). The latter is negative for repulsive potentials and positive for attractive ones. To make it consistent with the interpretation that an attractive potential 'draws' the free wave towards $r = 0$, while a repulsive potential 'pushes' it outward, one changes the sign in the definition. The magnitude of the scattering length provides a measure of the strength of the interaction.

```
> barriereta_0:=subs(V0=2,R0=1,eta_0):
> plot(barriereta_0,k=0..25,title=`V0=2, R0=1`);
```

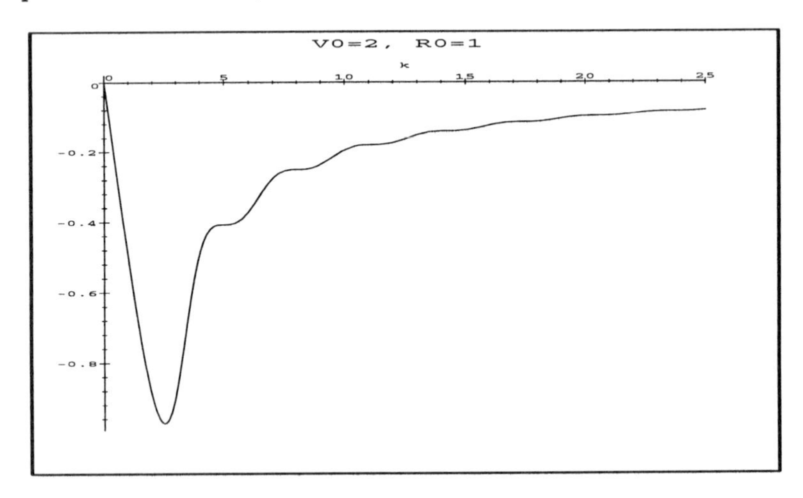


```
> welleta_0:=subs(V0=-0.5,R0=1,eta_0):
> plot(welleta_0,k=0..25,title=`V0=-0.5, R0=1`);
```

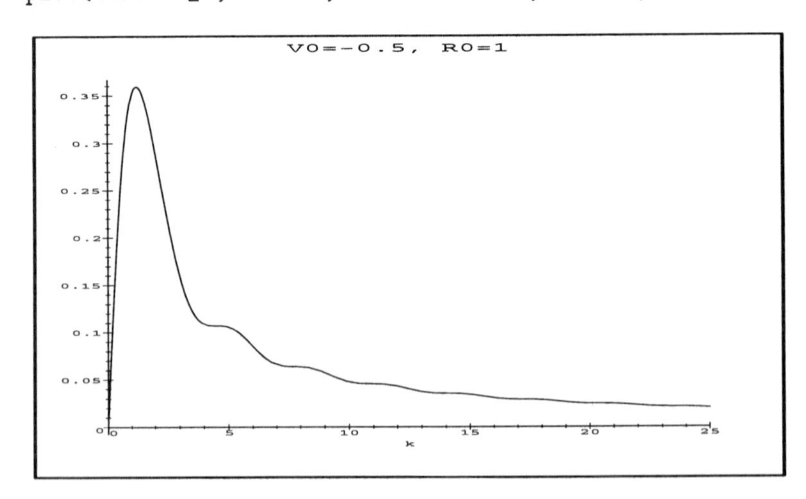

Now we deepen the well to accomodate some bound states:

```
> welleta_0:=subs(V0=-5,R0=1,eta_0):
```

```
> plot(welleta_0,k=0..25,title='V0=-5, R0=1');
```

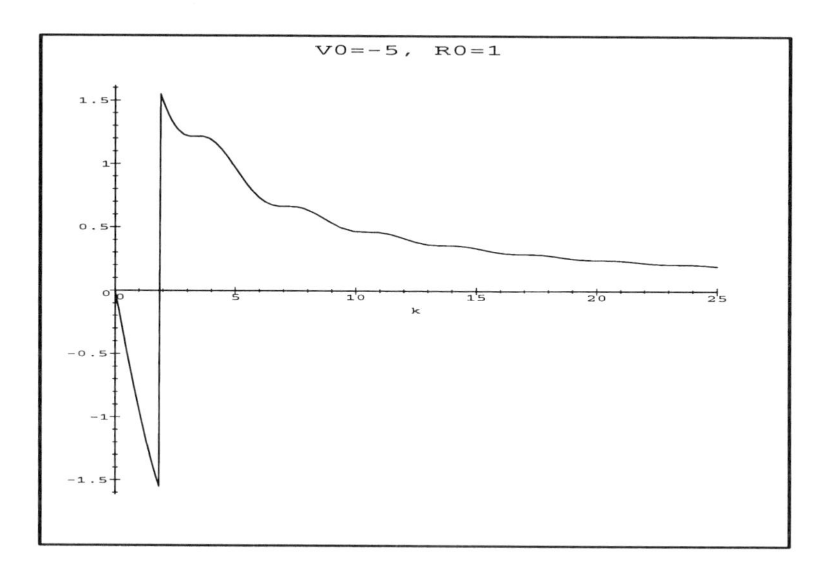

Exercise 6.2.1: Find the parameter which determines the number of bound states in a spherical potential well. What are the critical values to accomodate one, two, and three bound states? Why is the situation different from the one-dimensional potential well?

We make the potential well wider to accomodate even more bound states

```
> welleta_0:=subs(V0=-5,R0=2,eta_0);
```

$$\mathrm{welleta\_0} := \arctan\Big(\big(-8\,(k^2+10)\cos(2\sqrt{k^2+10})\sin(2\,k) + 8\sqrt{k^2+10}\,k\cos(2\,k)\sin(2\sqrt{k^2+10})\big) \Big/ \Big[8\,(k^2+10)\cos(2\sqrt{k^2+10})\cos(2\,k) + 8\sqrt{k^2+10}\,k\sin(2\,k)\sin(2\sqrt{k^2+10})\Big]\Big)$$

and again graph the phase shift as a function of wavenumber:

```
> plot(welleta_0,k=0..25,title='V0=-5, R0=2');
```

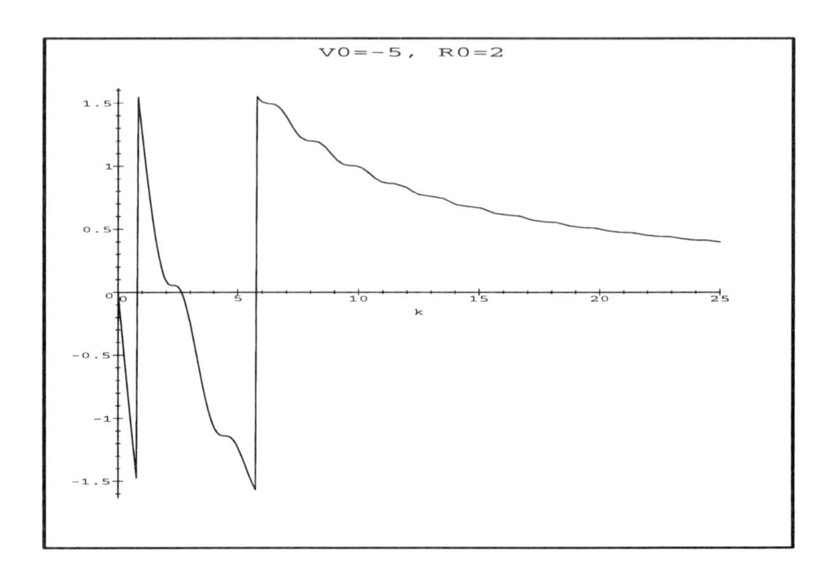

We observe that the phase shift for attractive potentials that can hold bound states begins at $k = 0$ at a multiple of π such that it goes to zero as the energy increases to infinity. The formal statement of this property, which can be used to infer the number of bound states supported by a potential from the behaviour of the phase shift with wave number, is referred to as Levinson's theorem.

We also wish to have a look at the wavefunctions. Assuming that $A_0 = 1$ we can deduce:

if $r < R_0$ then $\psi_0(r) = N_I u_J(K, r, 0)$, while

if $r > R_0$ then $\psi_0(r) = u_J(k, r, 0) + B_0 v_J(k, r, 0)$.

We assign the solution of the matching condition found above to `B0` and make a choice for the extent of the spherical potential well.

```
> B0:=solmatch:      R0:=1:
```

The implementation of an if-then-else construct to define a single function defined piecewise in various sections is not satisfactory up to release 3 of Maple. Therefore, we do not define a single matched function, but work with functions in both regions and convince ourselves from the graphs that they are matched properly. We still have to determine the constant N_I. In the subsequent figures we show the wavefunctions defined in the regions I and II for the entire r range. This serves as a demonstration of how these functions match at $r = R_0$ and shows the different wavelengths (and wave numbers) that appear in the potential-free and in the interaction region. Another quantity of interest is the ratio of the amplitudes for the wavefunctions in the two regions that arises as a consequence of matching.

```
> psiI:=(k,r)->uJ(K,r,0):
```

```
> psiII:=(k,r)->uJ(k,r,0)+B0*vJ(k,r,0):
> N_I:=k->psiII(k,R0)/psiI(k,R0):
> P1:=plot(subs(V0=0.5,k=2,N_I(k)*psiI(2,r)),r=0..10):
> P2:=plot(subs(V0=0.5,k=2,psiII(2,r)),r=0..10,color=red):
> with(plots):     display({P1,P2});
```

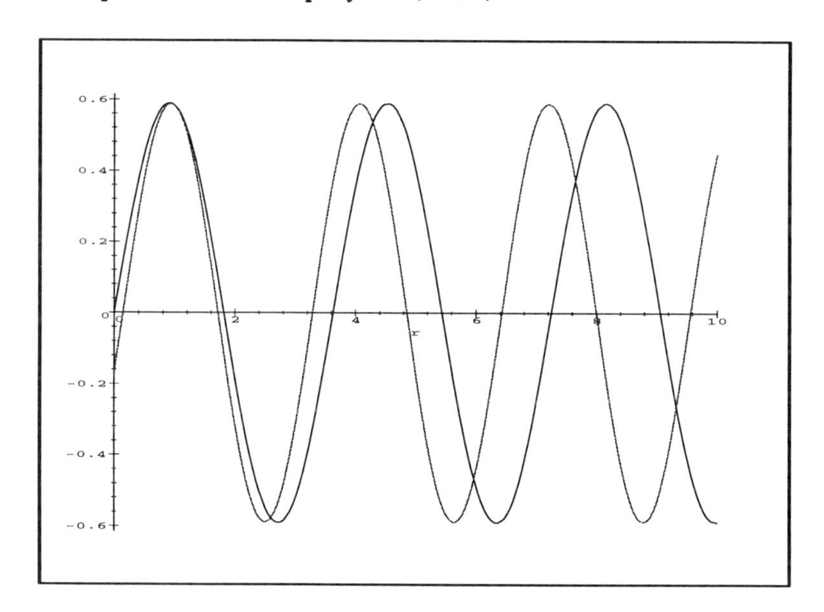

Note the different wave numbers for the solution valid inside the barrier (black, up to $r = R_0 = 1$) and the one for outside. Observe how they match in value and derivative at $r = 1$. Compared to a free solution with $k = 2$ the scattering from the potential barrier pushed the red curve outwards, i.e., it has a small negative value at $r = 0$ instead of zero.

We can also investigate the wavefunction for the case of an attractive square well potential to observe how the inside and outside parts of the wave match at the boundary point R_0.

```
> P1:=plot(subs(V0=-2,k=2,N_I(k)*psiI(2,r)),r=0..10):
> P2:=plot(subs(V0=-2,k=2,psiII(2,r)),r=0..10,color=red):
> display({P1,P2});
```

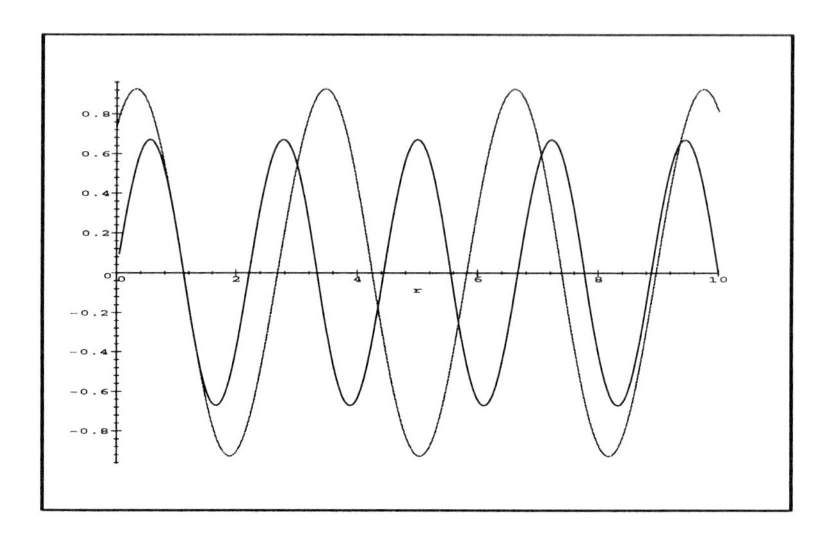

The well is quite attractive and has drawn the asymptotically free wavefunction into the well by matching onto the solution with higher wave number (again for $r < R_0 = 1$).

```
> P1:=plot(subs(V0=-2,k=1,N_I(k)*psiI(1,r)),r=0..10):
> P2:=plot(subs(V0=-2,k=1,psiII(1,r)),r=0..10,color=red):
> display({P1,P2});
```

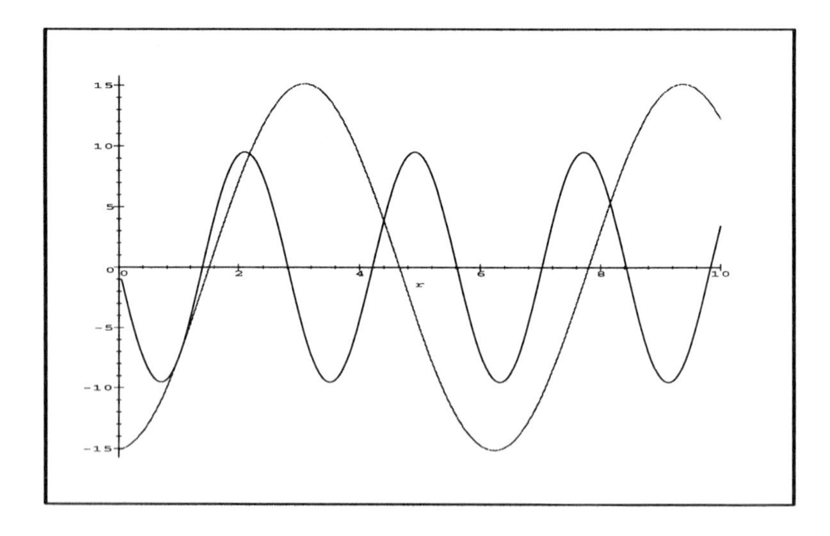

As we lowered the scattering energy the difference in wave numbers changed considerably. Below we look at a case of a deeper well.

```
> P1:=plot(subs(V0=-5,k=1,N_I(k)*psiI(1,r)),r=0..10):
> P2:=plot(subs(V0=-5,k=1,psiII(1,r)),r=0..10,color=red):
> display({P1,P2},title=`V0=-5, R0=1, k=1`);
```

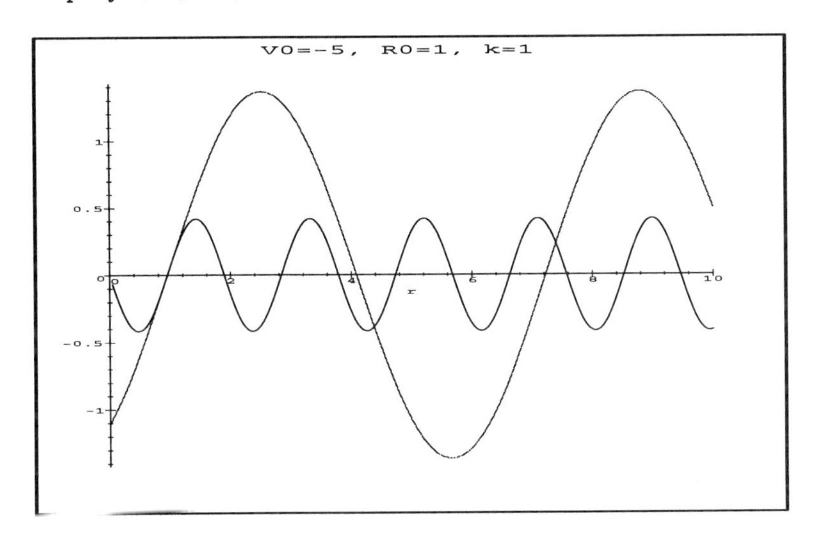


```
> P1:=plot(subs(V0=-5,k=2,N_I(k)*psiI(2,r)),r=0..10):
> P2:=plot(subs(V0=-5,k=2,psiII(2,r)),r=0..10,color=red):
> display({P1,P2},title=`V0=-5, R0=1, k=2`);
```

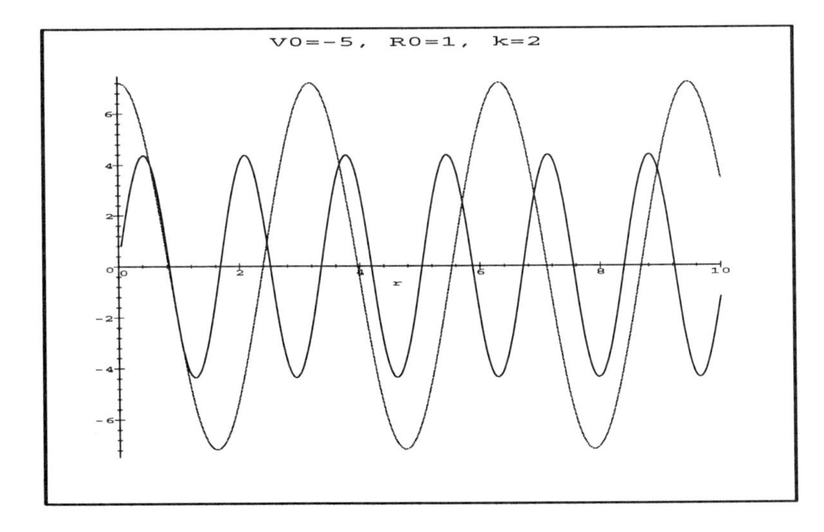

We display the amplitude of the inner part of the wavefunction as a function of wave number:

```
> plot(subs(V0=-5,N_I(k)),k=0..10,N_I=-10..10);
```

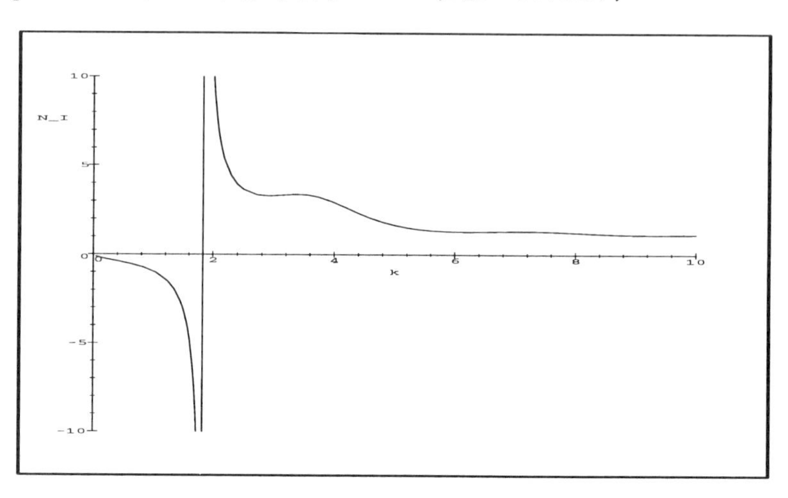


```
> P1:=plot(subs(V0=-5,k=1.9,N_I(k)*psiI(1.9,r)),r=0..10):
> P2:=plot(subs(V0=-5,k=1.9,psiII(1.9,r)),r=0..10,color=red):
> display({P1,P2},title='V0=-5, R0=1, k=1.9');
```

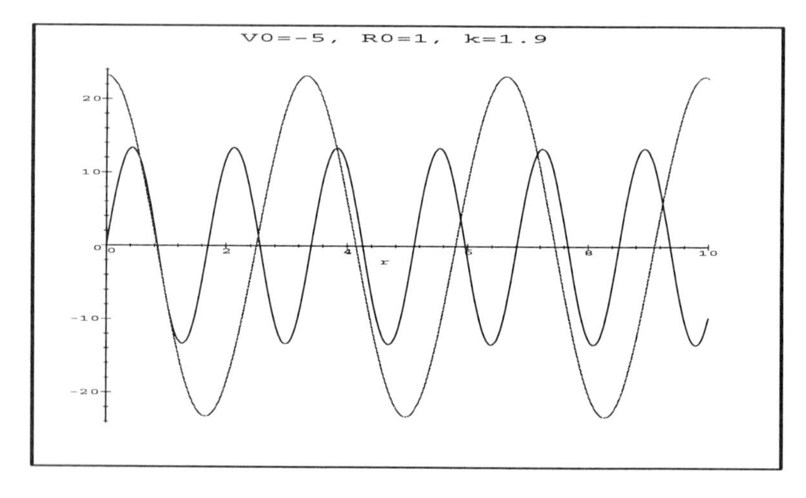

We note that this energy corresponds to a passage of the phase shift through $\pi/2$. This phenomenon is called a resonance. The amplitude of the wavefunction becomes very large for constant amplitude (with a value of unity) of the incoming wave $u_0(kr)$ in the asymptotic region (cf. Sect. 3.4). On resonance ($\eta = \pi/2$) the partial wave cross section takes on a maximum possible value, called the unitarity limit

$$\sigma_l = \frac{4\pi}{k^2} \sin^2 \eta_l \to \frac{4\pi}{k^2} \,.$$

6.3 Born Approximation

In this section we calculate elastic electron scattering from a model atom in the first Born approximation (FBA). This approximation is useful at high collision energies for which the partial wave expansion delat with in Sect. 6.1 and 6.2 is slowly converging, and thus impractical. The complex-valued scattering T matrix which is proportional to the scattering amplitude $f(\theta, \phi)$ can be shown to satisfy a Lippmann-Schwinger integral equation [RT67, GW64, Br70, MM71, Ne82]

$$T = V + VG_{+}T\,.$$

Here V represents the Fourier transform of the scattering potential to momentum space, and G_{+} is the free Schrödinger-Green's function with outgoing wave boundary conditions imposed. At high collision energies the potential operator weakens and the integral equation can be solved iteratively using the von Neumann method [Ar85], resulting in the Born series for the scattering matrix. In the lowest order one just sets $T_{\mathrm{FBA}} = V$.

We consider Rutherford scattering from the nucleus as well as the electronic (atomic) form factor. The form factor involves a Fourier–sine transform of the electronic charge density. We assume a simple form for the density function, namely a single exponential function with a screening parameter. Thus, for atoms more complex than hydrogen or helium we ignore the shell structure and assume that only the outermost shell has a significant impact on the projectile's deflection. The calculation ignores the complexity of the atom, i.e., it ignores polarization of the charge cloud, excitations, ionization, etc.

We introduce the difference vector between the incident and outgoing momentum vectors $\mathbf{q} = \mathbf{k_f} - \mathbf{k_i}$, and define its magnitude:

```
> q:=(k,theta)->4*k^2*sin(theta/2)^2:
```

We consider the scattering energy $E = k^2/2$ to be fixed, and use atomic units ($m_\mathrm{e} = \hbar = e = 1$). The magnitude of $\mathbf{q}$ varies as a function of the polar scattering angle. Rutherford scattering obeys the following differential cross section that arises both in a classical and in the quantum mechanical calculation

$$\frac{\mathrm{d}\sigma}{\mathrm{d}\theta_\mathrm{r}} = \Big(\frac{Z_1 Z_2}{2\mu v^2 \sin^2(\theta_\mathrm{r}/2)}\Big)^2\,.$$

Here Z_1 and Z_2 are the charges of the two colliding particles, μ is their reduced mass (which becomes equal to the projectile particle's mass if the target atom is assumed to be infintely heavy), θ_r is the polar scattering angle in relative coordinates.

```
> sigR:=(k,theta)->4*Z^2/q(k,theta)^2:

> plot(subs(Z=1,sigR(1,theta)),theta=0..Pi,q_R=0..100);
```

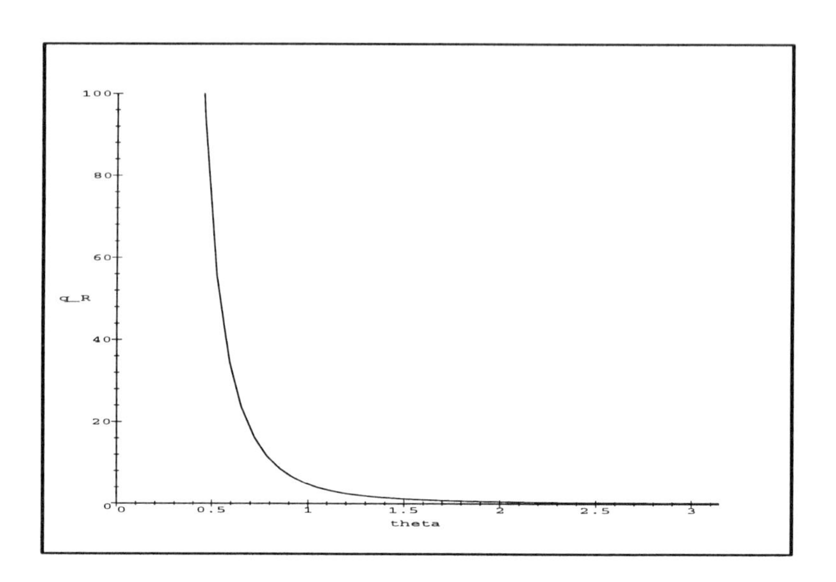

The cross section diverges badly in the forward direction due to the infinite range of the interaction. The problem is that even for very distant collisions (large impact parameter) the impinging particles experience a deflection. This has also repercussions in the total cross section as a function of energy in a divergence at $E = 0$. The form factor is to be calculated for given $q(k, \theta)$. We pick a spherical charge density, which is exact for hydrogen atoms,

$$\rho(r) = N_0 \exp(-2r) .$$

```
> assume(q0>0);
> Form:=q0->int(sin(q0*r)/(q0*r)*exp(-2*r)*r^2,r=0..infinity);
```

$$Form := q0 \to \int_0^\infty \frac{\sin(\, q0\, r\,)\, r\, e^{(-2\,r)}}{q0}\, dr$$

```
> Form(q0):   simplify(");
```

$$4\, \frac{1}{(\, 4 + q0^{\sim 2}\,)^2}$$

Therefore, screened Rutherford scattering yields the differential cross section expressed in terms of the screening constant a for $\rho = N_0 \exp(-r/a)$:

```
> sigSR:=(k,theta)->Z^2/(4*k^4)*(1-1/(1+4*k^2*a^2*sin(theta/2)
> ^2)^2)^2/sin(theta/2)^4;
```

$$sigSR := (k,\theta) \rightarrow \frac{1}{4}\,\frac{Z^2\left(1-\dfrac{1}{\left(1+4\,k^2\,a^2\sin\left(\dfrac{1}{2}\,\theta\right)^2\right)^2}\right)^2}{k^4\sin\left(\dfrac{1}{2}\,\theta\right)^4}$$

```
> plot(subs(a=1/2,Z=1,sigSR(1,theta)),theta=0..Pi,q_SR=0..1);
```

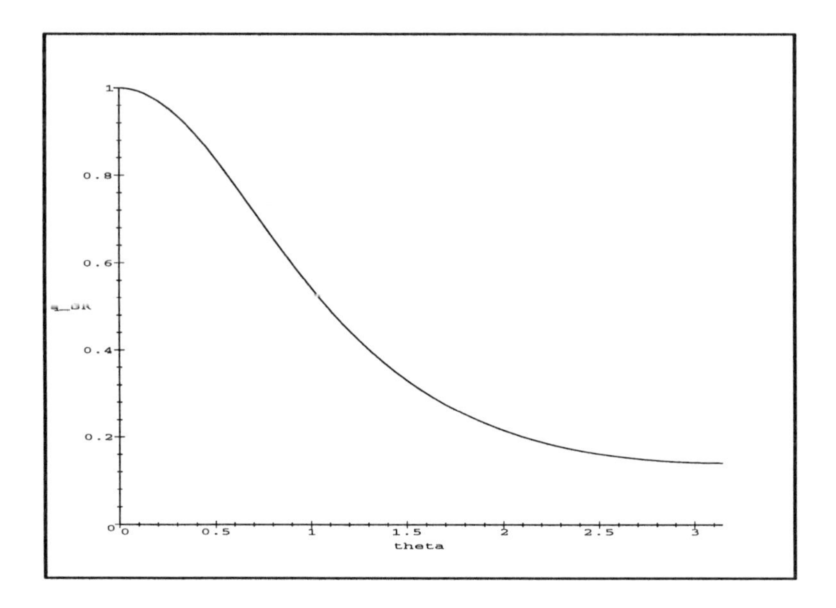

This is a typical shape for a differential cross section in elastic light particle – spherical atom scattering. We can demonstrate how the screening alters the cross section. Note that we smear the same amount of charge ($N_{\mathrm{el}} = Z$) over different distances away from the nucleus. A small screening parameter a means that the electronic charge is close to the nucleus, a larger value of a implies a more loosely bound electron density.

The dependence of the differential cross section on the electronic screening allowed experimental confirmation not only of energy levels, but of probability distributions for electrons in atoms.

```
> with(plots):
```

```
> animate(subs(Z=1,sigSR(1,theta)),theta=0..Pi,a=0.5..2);
```

To compare the results on a single graph we show the cross section on a logarithmic scale:

```
> plot({subs(Z=1,a=1/2,log(sigSR(1,theta))),subs(Z=1,a=1,
> log(sigSR(1,theta))),subs(Z=1,a=2,log(sigSR(1,theta)))},
> theta=0..Pi,axes=boxed,title=`log(sigma) for a=1/2, 1, 2`);
```

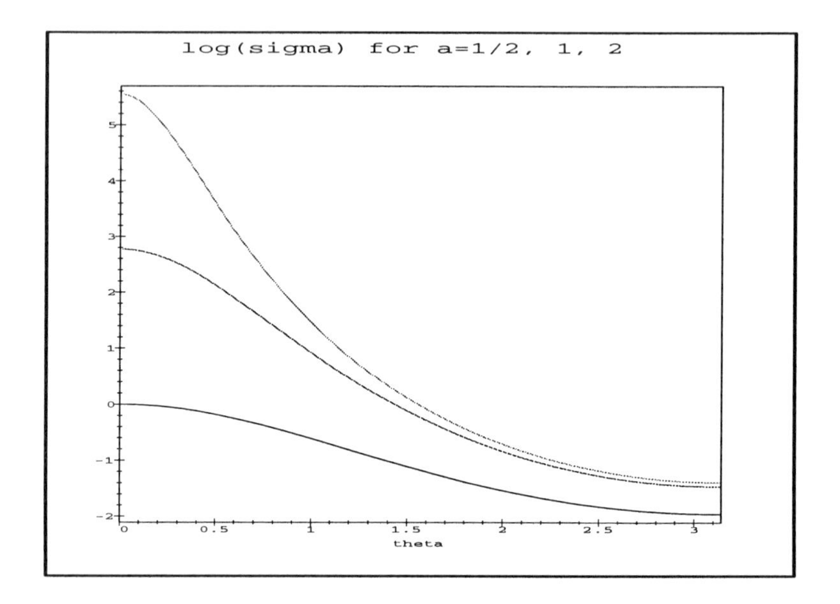

Exercise 6.3.1: The measured elastic differential cross section for ions scattering from neon atoms has been found to agree with a screened Rutherford result for the choice of screening parameter $a = 0.53$ atomic units [HC+82]. Assume that the relative error in the cross-section measurement for projectile scattering angles in the range of 0.2 to 2 mrad is of the order of 20 %. How accurately can the screening parameter a be determined from such a measurement?

6.4 Variational Method

In this section we explore the Kohn variational method for potential scattering, as well as the numerical solution of a single-channel radial Schrödinger equation. The problem considered here is elastic s-wave scattering for electrons or positrons from an exponential potential [MM71, p. 123] We are interested in a functional that permits the approximate calculation of scattering solutions to the radial Schrödinger equation that arises in partial wave analysis, namely

$$G''(r) + (k^2 - U(r))G(r) = 0 ,$$

with the boundary conditions

$$G(r=0) = 0 , \quad \text{and for} \quad r \to \infty \quad G \approx A \sin(kr + \eta) .$$

Here η is the s wave phase shift, and the potential satisfies the condition

$$\lim_{r \to \infty} r^2 U(r) = 0 .$$

It is possible to consider the normalization A either as a constant ($A = 1$) or as a function of η. This leads to different variational principles (Hulthén's, Kohn's methods, etc.). We explore Kohn's method in which $A = \sec(\eta)$.

The wavefunction can also be written in the asymptotic domain as

$$G(r) = \sin(kr) + a \cos(kr) ,$$

with $a = \tan \eta$ providing the scattering information. In contrast to the stationary problem, where the variational principle is based on the expectation value of the Hamiltonian itself, we consider a functional based on the expectation value of the difference between the scattering energy and the Hamiltonian $E - H$:

$$I = \int_0^\infty G(r) \left[\frac{\mathrm{d}^2}{\mathrm{d}r^2} + k^2 - U(r) \right] G(r) \mathrm{d}r .$$

This functional vanishes for the exact solution $G(r)$ and remains finite for approximate trial functions $G_\mathrm{t}(r)$ that satisfy the correct scattering boundary conditions. As before, the idea is to have a trial function G_t that depends on some parameters $\{c_j\}$ in addition to a, to look for stationary points in $I_\mathrm{t} = I[G_\mathrm{t}(r)]$ and to extract the phase shift from a. The variation with respect to the parameters $\{c_j\}$ and a result in the equations

$$\frac{\partial I_\mathrm{t}}{\partial c_j} = 0 , \quad \text{and} \quad \frac{\partial I_\mathrm{t}}{\partial a} = -k .$$

In Kohn's version I_t is not zero (for non-exact trial functions G_t) and an improved phase shift can be calculated from the stationarity of the expression $I + k \tan \eta$, which yields

$$\tan \eta \approx \tan \eta_{\text{var}} = a + \frac{I_\text{t}}{k}\,.$$

We consider the potential for a charged particle (electron without exchange) scattering from a static exponential potential which represents a part of the potential that appears in the scattering from frozen hydrogen atoms in their ground-state (Hartree-potential).

```
> U:=r->-exp(-2*r):
```

The operator for the variational principle takes the form (assuming that the passed function **arg** depends on r, and that the wave number is given as k):

```
> Lop:=arg->diff(arg,r$2)+(k^2-U(r))*arg;
```

$$Lop := \text{arg} \rightarrow \text{diff}(\,\text{arg}, r\,\$2\,) + (\,k^2 - \text{U}(\,r\,)\,)\,\text{arg}$$

The functional is given as

```
> I_t:=arg->int(arg*Lop(arg),r=0..infinity);
```

$$I\_t := \text{arg} \rightarrow \int_0^\infty \text{arg}\,\text{Lop}(\,\text{arg}\,)\,dr$$

Now we specify a trial state $G_\text{t}(r)$ for the problem at hand. The state includes one free parameter b in addition to the one that carries the scattering information. We make use of a Maple expression and not a function. Notice how the additional parameter b is not affecting the wavefunction in the asymptotic region.

```
> G_t:=sin(k*r)+(a+b*exp(-r))*(1-exp(-r))*cos(k*r);
```

$$G\_t := \sin(\,k\,r\,) + (\,a + b\,\mathrm{e}^{(-r)}\,)\,(\,1 - \mathrm{e}^{(-r)}\,)\cos(\,k\,r\,)$$

```
> assume(k>0);     Ifunc:=I_t(G_t);
```

$$\begin{aligned}
-\frac{1}{120}(&-243000\,k^{\sim 2} + 145800\,b^2 + 259200\,a\,b - 120120\,b\,k^{\sim 3} + 1769820\,a\,k^{\sim 3}\\
&- 94500\,k^{\sim}\,b - 27300\,b\,k^{\sim 5} + 702000\,k^{\sim}\,a + 1223700\,a\,k^{\sim 5} + 9480\,k^{\sim 8}\,a^2\\
&+ 405000\,a^2 - 1680\,k^{\sim 7}\,b + 348360\,a\,k^{\sim 7} - 234630\,k^{\sim 4} - 77070\,k^{\sim 6}\\
&- 10320\,k^{\sim 8} - 480\,k^{\sim 10} + 1920\,a\,k^{\sim 11} + 43200\,a\,k^{\sim 9} + 256\,k^{\sim 10}\,b\,a\\
&+ 7680\,k^{\sim 8}\,b\,a + 87168\,k^{\sim 6}\,b\,a + 430420\,b\,k^{\sim 4}\,a + 761076\,k^{\sim 2}\,b\,a\\
&+ 400\,k^{\sim 10}\,a^2 + 84225\,k^{\sim 6}\,a^2 + 352570\,a^2\,k^{\sim 4} + 682425\,a^2\,k^{\sim 2}\\
&+ 144\,b^2\,k^{\sim 10} + 4680\,b^2\,k^{\sim 8} + 51057\,b^2\,k^{\sim 6} + 194670\,b^2\,k^{\sim 4}\\
&+ 218349\,b^2\,k^{\sim 2})\,/\\
&[(\,4\,k^{\sim 5} - 40\,I\,k^{\sim 4} - 155\,k^{\sim 3} + 290\,I\,k^{\sim 2} + 261\,k^{\sim} - 90\,I\,)\\
&(\,4\,k^{\sim 5} + 40\,I\,k^{\sim 4} - 155\,k^{\sim 3} - 290\,I\,k^{\sim 2} + 261\,k^{\sim} + 90\,I\,)]
\end{aligned}$$

The appearance of a complex-valued result is puzzling, since all of our input is real-valued. Let's try to simplify:

```
> evalc(Ifunc):      Ifunc:=simplify(");
```

$$\mathrm{Ifunc} := -\frac{1}{120}(-243000\,k\tilde{}^{2} + 145800\,b^2 + 259200\,a\,b - 120120\,b\,k\tilde{}^{3} + 1769820\,a\,k\tilde{}^{3} - 94500\,k\tilde{}\,b - 27300\,b\,k\tilde{}^{5} + 702000\,k\tilde{}\,a + 1223700\,a\,k\tilde{}^{5} + 9480\,k\tilde{}^{8}\,a^2 + 405000\,a^2 - 1680\,k\tilde{}^{7}\,b + 348360\,a\,k\tilde{}^{7} - 234630\,k\tilde{}^{4} - 77070\,k\tilde{}^{6} - 10320\,k\tilde{}^{8} - 480\,k\tilde{}^{10} + 1920\,a\,k\tilde{}^{11} + 43200\,a\,k\tilde{}^{9} + 256\,k\tilde{}^{10}\,b\,a + 7680\,k\tilde{}^{8}\,b\,a + 87168\,k\tilde{}^{6}\,b\,a + 430420\,b\,k\tilde{}^{4}\,a + 761076\,k\tilde{}^{2}\,b\,a + 400\,k\tilde{}^{10}\,a^2 + 84225\,k\tilde{}^{6}\,a^2 + 352570\,a^2\,k\tilde{}^{4} + 682425\,a^2\,k\tilde{}^{2} + 144\,b^2\,k\tilde{}^{10} + 4680\,b^2\,k\tilde{}^{8} + 51057\,b^2\,k\tilde{}^{6} + 194670\,b^2\,k\tilde{}^{4} + 218349\,b^2\,k\tilde{}^{2}) / [16\,k\tilde{}^{10} + 360\,k\tilde{}^{8} + 2913\,k\tilde{}^{6} + 10390\,k\tilde{}^{4} + 15921\,k\tilde{}^{2} + 8100]$$

We are ready to calculate the derivatives with respect to the variational parameters in order to find the stationary points of the functional I_{t}.

```
> eq1lhs:=diff(Ifunc,b);
```

$$\mathrm{eq1lhs} := -\frac{1}{120}(291600\,b + 259200\,a - 120120\,k\tilde{}^{3} - 94500\,k\tilde{} - 27300\,k\tilde{}^{5} - 1680\,k\tilde{}^{7} + 256\,k\tilde{}^{10}\,a + 7680\,k\tilde{}^{8}\,a + 87168\,k\tilde{}^{6}\,a + 430420\,k\tilde{}^{4}\,a + 761076\,k\tilde{}^{2}\,a + 288\,b\,k\tilde{}^{10} + 9360\,b\,k\tilde{}^{8} + 102114\,b\,k\tilde{}^{6} + 389340\,b\,k\tilde{}^{4} + 436698\,b\,k\tilde{}^{2}) / [16\,k\tilde{}^{10} + 360\,k\tilde{}^{8} + 2913\,k\tilde{}^{6} + 10390\,k\tilde{}^{4} + 15921\,k\tilde{}^{2} + 8100]$$

```
> eq2lhs:=diff(Ifunc,a);
```

$$\mathrm{eq2lhs} := -\frac{1}{120}(259200\,b + 1769820\,k\tilde{}^{3} + 702000\,k\tilde{} + 1223700\,k\tilde{}^{5} + 18960\,k\tilde{}^{8}\,a + 810000\,a + 348360\,k\tilde{}^{7} + 1920\,k\tilde{}^{11} + 43200\,k\tilde{}^{9} + 256\,b\,k\tilde{}^{10} + 7680\,b\,k\tilde{}^{8} + 87168\,b\,k\tilde{}^{6} + 430420\,b\,k\tilde{}^{4} + 761076\,b\,k\tilde{}^{2} + 800\,k\tilde{}^{10}\,a + 168450\,k\tilde{}^{6}\,a + 705140\,k\tilde{}^{4}\,a + 1364850\,k\tilde{}^{2}\,a) / [16\,k\tilde{}^{10} + 360\,k\tilde{}^{8} + 2913\,k\tilde{}^{6} + 10390\,k\tilde{}^{4} + 15921\,k\tilde{}^{2} + 8100]$$

Now we solve for the optimum values of the variational parameters a and b.

```
> soln:=solve({eq1lhs=0,eq2lhs=-k},{a,b});
```

$$\mathrm{soln} := \{b = 450k\tilde{}(1557649\,k\tilde{}^{4} + 405000 + 897981\,k\tilde{}^{8} + 175324\,k\tilde{}^{10} + 16240\,k\tilde{}^{12} + 2126545\,k\tilde{}^{6} + 576\,k\tilde{}^{14} - 1013175\,k\tilde{}^{2}) / (4694760000$$

$$
\begin{aligned}
&+ 41216\,k^{\sim 18} + 1883136\,k^{\sim 16} + 34066272\,k^{\sim 14} + 310416112\,k^{\sim 12} \\
&+ 7695712809\,k^{\sim 4} + 9399855600\,k^{\sim 2} + 4221860943\,k^{\sim 8} \\
&+ 6667246679\,k^{\sim 6} + 1537843233\,k^{\sim 10}),\ a = -30(704\,k^{\sim 14} + 10384\,k^{\sim 12} \\
&- 30268\,k^{\sim 10} - 1154653\,k^{\sim 8} - 7832936\,k^{\sim 6} - 26472781\,k^{\sim 4} - 46160550\,k^{\sim 2} \\
&- 50220000)k^{\sim} / (4694760000 + 41216\,k^{\sim 18} + 1883136\,k^{\sim 16} \\
&+ 34066272\,k^{\sim 14} + 310416112\,k^{\sim 12} + 7695712809\,k^{\sim 4} + 9399855600\,k^{\sim 2} \\
&+ 4221860943\,k^{\sim 8} + 6667246679\,k^{\sim 6} + 1537843233\,k^{\sim 10})\}
\end{aligned}
$$

The found solution for a and b has to be assigned such that Maple expressions that depend on a and b are calculated making use of the solution.

```
> assign(soln);    plot(a,k=0..2);
```

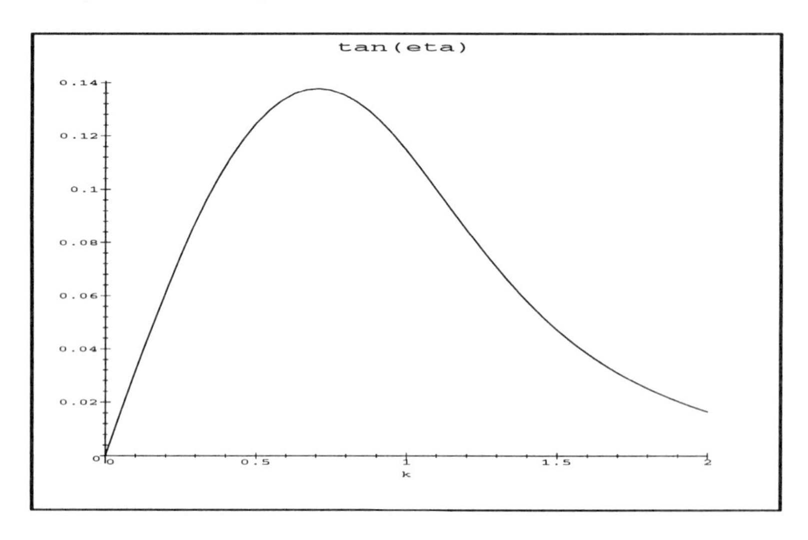

For small phase shifts η the difference between $\tan\eta$ and η is not appreciable as can be seen from the graph below.

We note that the phase shift is positive, displays linear behaviour at $k = 0$ and tends towards zero at large wave number k, or scattering energy. I leave it as an exercise for the reader to extract the analytic form of the small-k and large-k dependencies from the symbolic expression using a Taylor and series expansion, respectively.

I remind the reader of the general discussion of the behaviour of phase shifts for attractive and repulsive potentials, respectively, provided in Sect. 6.2 in the context of scattering from a spherical potential well or barrier. We observe a confirmation of the conjectured general statement that attractive potentials induce positive phase shifts as we have chosen a sign of the potential that corresponds to attraction.

```
> plot(arctan(a),k=0..2);
```

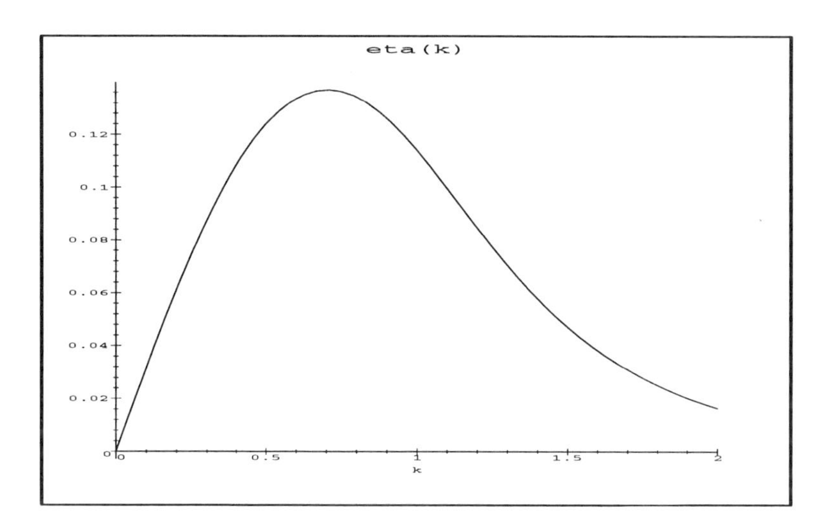

We calculate the correction to $\tan\eta$ that arises in Kohn's variational method and compare with the original estimate. Note that the assigning operation fixed the values of a and b to the solution and the functional `Ifunc` is calculated with the correct, optimum values for a and b.

```
> P2:=plot(a+Ifunc/k,k=0..2,color=red):
> P1:=plot(a,k=0..2,color=black):
> with(plots):    display({P1,P2});
```

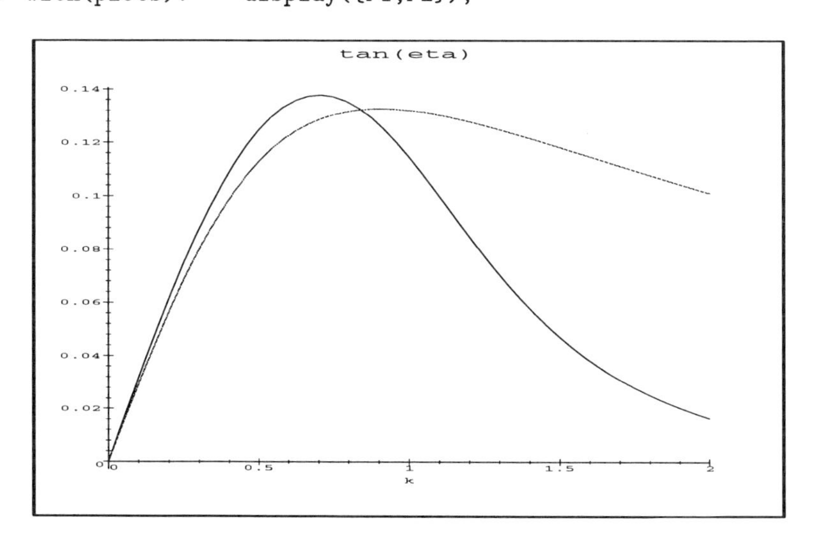

There is an apparently appreciable correction of the simple guess for the phase shift at higher scattering energies.

We demonstrate for a few collision energies the result of solving the Schrödinger equation numerically. This can be done by stepping out two linearly independent solutions and superimposing them in such a way as to obtain the correct asymptotic behaviour. We define the Schrödinger equation and a procedure that generates its numerical solution.

```
> SEq:=Lop(psi(r));
```

$$\mathrm{SEq} := \left(\frac{\partial^2}{\partial r^2}\psi(r)\right) + (k^{\sim 2} + \mathrm{e}^{(-2r)})\,\psi(r)$$

```
> Sol1:=dsolve({subs(k=1,SEq),psi(0)=0,D(psi)(0)=1},psi(r),
> type=numeric):
```

```
> Sol1(10);
```

$$\left[r = 10, \psi(r) = -.57495, \frac{\partial}{\partial r}\psi(r) = -.67377\right]$$

We are interested in an extraction of the value for the wavefunction:

```
> rhs(Sol1(10)[2]);
```

$$-.5749510318504475$$

Now we generate a sequence of points for fast plotting:

```
> for i from 1 to 100 do
> rm[i]:=i/10;    Wf[i]:=rhs(Sol1(rm[i])[2]);    od:
> plot([[rm[j],Wf[j]]$j=1..100]);
```

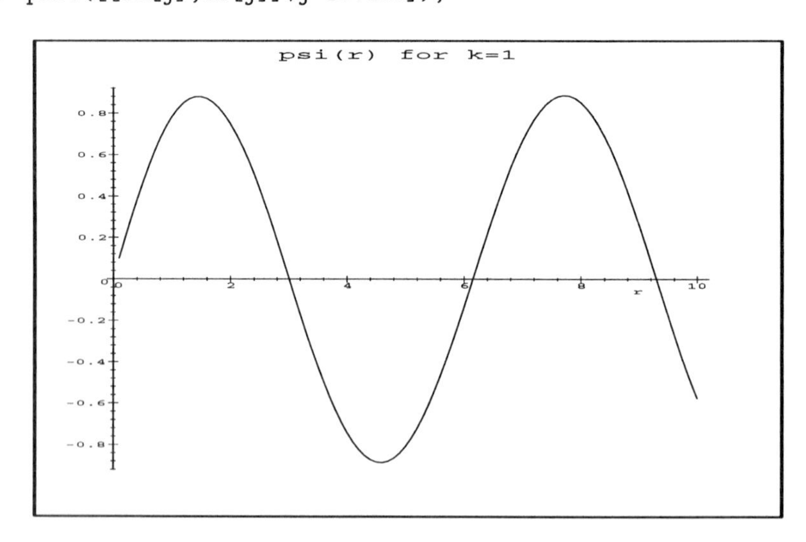

```
> P1:=plot([[rm[j],Wf[j]]$j=1..100]):
> P2:=plot(sin(r),r=0..10,color=red):
> display({P1,P2});
```

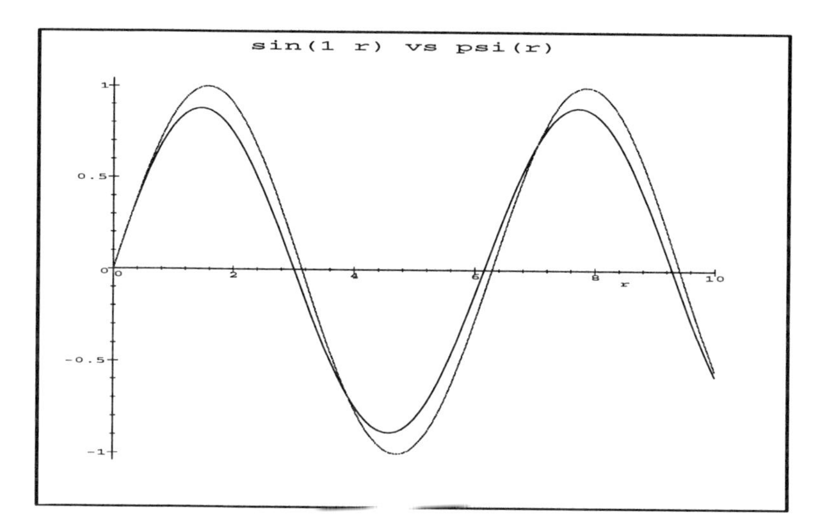

The scattering solution is not simply a sine function. To extract the phase shift at some large value $r = R$ we assume that $\psi(R) = \sin(kR) + a\cos(kR)$ with $a = \tan\eta$. We have, however, arbitrary normalization due to the arbitrary choice of the slope for the wavefunction at the origin. We can solve this problem by comparing the solutions for two different slopes at the origin.

Now we consider the extraction of the phase shift at $r = R$. Another way to write the solution in the asymptotic domain is $\psi(r) = A\sin(kr+\eta)$. Thus, if we have two independent calculations ψ_1 and ψ_2, we find that $\psi_1(R)/\psi_2(R) = A_1/A_2$. We can form the ratio

$$\frac{\psi_1(R)}{\psi_2'(R)} = \frac{A_1}{kA_2}\tan(kR+\eta),$$

from which η can be determined as k and R are known beforehand and A_1/A_2 has just been determined. Let us implement this trick and note that R has to be large enough such that the wavefunction has acquired its asymptotic form. For the short-range potential considered here this is not a problem.

We define the second solution by picking a different value for the slope at $r = 0$:

```
> Sol2:=dsolve({subs(k=1,SEq),psi(0)=0,D(psi)(0)=2},psi(r),
> type=numeric):
> val2:=Sol2(10);
```

$$\text{val2} := \left[r = 10, \psi(r) = -1.14990, \frac{\partial}{\partial r}\psi(r) = -1.34753\right]$$

```
> val1:=Sol1(10);
```

$$\text{val1} := \left[r = 10, \psi(r) = -.57495, \frac{\partial}{\partial r}\psi(r) = -.67377 \right]$$

```
> A1overA2:=rhs(val1[2])/rhs(val2(10)[2]);
```

$$\text{A1overA2} := .5000000116$$

```
> k1:=1:
```

We are ready to evaluate $\tan\eta$ by solving the appropriate equation at $R = 10$ atomic units:

```
> tan(arctan(rhs(val1[2])/rhs(val2[3])/A1overA2*k1)-k1*10);
```

$$.1319649746$$

This value is close to the improved variational result presented above. We can calculate a sequence of phase shift values:

```
> for m from 1 to 20 do:

> k1:=m/10;     kval[m]:=k1;

> Sol1:=dsolve({subs(k=k1,SEq),psi(0)=0,D(psi)(0)=1},psi(r),
> type=numeric);

> Sol2:=dsolve({subs(k=k1,SEq),psi(0)=0,D(psi)(0)=2},psi(r),
> type=numeric);

> val2:=Sol2(10);

> val1:=Sol1(10);

> A1overA2:=rhs(val1[2])/rhs(val2(10)[2]);

> taneta[m]:=tan(arctan(rhs(val1[2])/rhs(val2[3])/A1overA2*k1)
> -k1*10);

> od:
```

We display these numbers along with the variationally obtained answer for arbitrary k:

```
> P1:=plot([[kval[j],taneta[j]]$j=1..20],color=blue,
> style=point,symbol=diamond):

> P2:=plot(a+Ifunc/k,k=0..2,color=red):

> display({P1,P2});
```

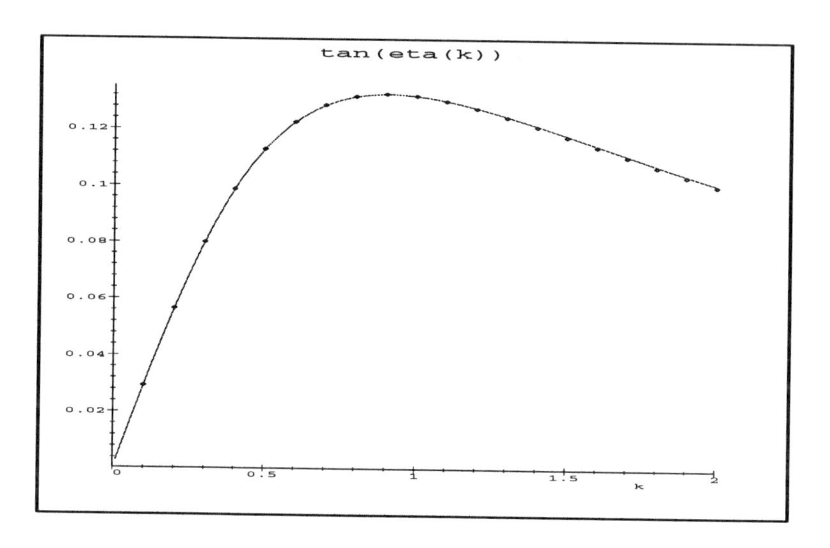

The results are identical to graphing accuracy. The difference between the numerical and variational result for this problem is indeed small as shown below in a display of the error (which has been rescaled by a factor of 10^3 to simplify the labels on the y axis). For long-range interactions one has to be careful with the extraction point R in the numerical procedure, particularly for small k values.

```
> plot([[kval[j],10^3*(taneta[j]-subs(k=kval[j],a+Ifunc/k))]
> $j=1..20],style=point);
```

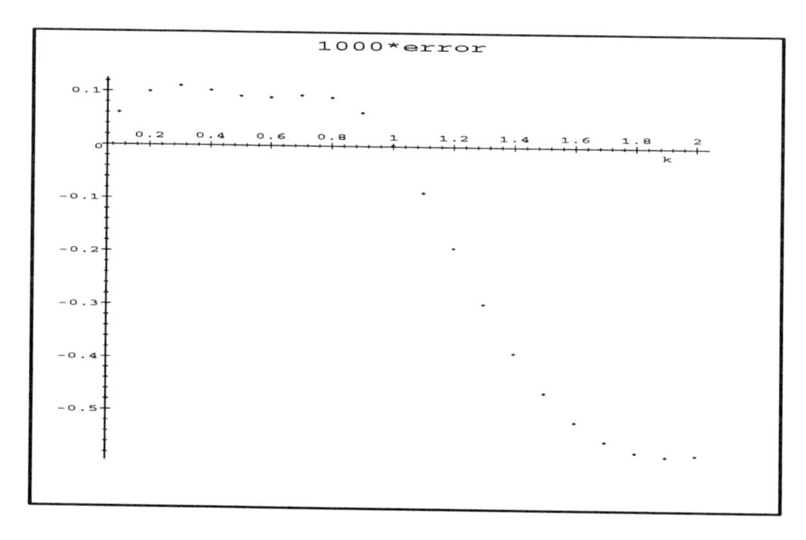

We note that the variational method is still popular in coupled-channel problems that arise when scattering is considered not from a simple potential, but from a QM system that undergoes changes (virtual excitations, polarization) as a result of interactions with the projectile. Kohn's variational method is practical if a larger number of linear parameters ($b \rightarrow \{b_1..b_n\}$) is used. One can also make use of functionals that concentrate on the short-range correlation terms [DH83].

We remark also on a technical subject in Maple and on a practical philosophy in scattering calculations. A straightforward variational calculation of phase shifts as presented here is possible if the functional can be calculated in closed form. Already the problem of static electron – hydrogen scattering without electron exchange (i.e., potential scattering from $U(r) = -(1 + 1/r)\exp(-2r)$) presented in [MM71] with the trial function employed here presents some difficulties in Maple. One can appreciate, however, the bookkeeping effort involved in a numerical environment, where all the required integrals have to be assembled. The numerical techniques for the numerical solution of the Schrödinger equation are more straightforward to implement and have become the method of choice in coupled-channel scattering calculations. Thus, variational methods no longer enjoy the same popularity as in previous times. Nevertheless, they allow us to develop more directly an intuition for the scattering wavefunction with its asymptotic part containing the incident and scattered wave and the short-distance correlation terms that describe the adjustment of the scattering state to the interaction potential.

Exercise 6.4.1: Repeat the numerical phase shift calculation for the case of p waves by adding the corresponding centrifugal potential to the Hamiltonian. Note that the wavefunction does not start linearly at $r = 0$ in this case.

Exercise 6.4.2: Perform variational and numerical s-wave phase shift calculations for the exponential potential with both signs (attraction and repulsion). Increase the strength of the potential and observe the changes in the phase shift. Vary the range of the interaction, by changing the exponential parameter.

7. Many-Particle Problems

7.1 Helium Atom

This section presents an approximate calculation of the energy levels for the helium atom as shown in [Fl71, problem 90]. We make use of a pure Coulomb Hamiltonian, i.e., no spin-dependent forces are included. The wavefunction separates into a spin part and a spatial part. We cannot ignore the spin part, however, as the requirement of an anti-symmetric (AS) two-electron wavefunction (Pauli principle) has to be satisfied. The wavefunction is given as a product of the coordinate space part that can be chosen to be symmetric (S), ψ_S, or AS, ψ_{AS}, under particle interchange times a two-particle spinor that again can be either S or AS, labeled χ_S and χ_{AS}, respectively. Thus, we can have an overall AS combination $\psi_S \times \chi_{AS}$ or $\psi_{AS} \times \chi_S$.

We consider the case of zero total (two-electron) orbital angular momentum, $L = 0$. There are two possibilities to couple the spins:

- anti-parallel: $\hat{S} = \hat{s}_1 \oplus \hat{s}_2 = |s_1 - s_2| = 0$, with $m_S = 0$ (one state); and
- parallel: $\hat{S} = \hat{s}_1 \oplus \hat{s}_2 = s_1 + s_2 = 1$, with $m_S = 1, 0, -1$ (three states).

There are three ways to make symmetric spin wavefunctions and one way to make an anti-symmetric one. Thus, the spin singlet ($S = 0$) has the AS spin wavefunction, while the spin triplet ($S = 1$) uses the symmetric ones.

In coordinate space we wish to make use of the simplest possible wave functions, i.e., trivial superpositions of products. This ansatz would be exact if (!) there were no two-particle operators in the Hamiltonian. However, the electron–electron repulsion $e^2/|\mathbf{r}_1 - \mathbf{r}_2|$ is a two-particle operator, and, therefore, the antisymmetrized products represent an approximation.

The Hamiltonian contains:

- the kinetic energies for both electrons,

$$T_1 + T_2 = -\frac{\hbar^2}{2m}\left(\nabla_1^2 + \nabla_2^2\right),$$

- the interactions with a nucleus of charge $Z = 2$:

$$V_{\text{nuc}} = -\frac{2e^2}{r_1} - \frac{2e^2}{r_2},$$

– and, the mutual electron – electron repulsion:

$$V_{\mathrm{e-e}} = \frac{e^2}{|\mathbf{r}_1 - \mathbf{r}_2|}.$$

We ignore the recoil motion of the nucleus.

The Hamiltonian can be regrouped as $H = H_1 + H_2$, where H_i are the He^+ atom Hamiltonians for electrons 1 and 2, respectively. This hydrogen-like problem can be solved exactly, and we can make use of perturbation theory to obtain approximate answers for the helium atom. However, this regrouping may not be very smart as the electron–electron interaction is responsible for screening. It may be advantageous to introduce an artificial splitting of the Hamiltonian, which prepares the ground for a variational solution.

The individual single-electron Hamiltonians H_i contain the respective kinetic energies T_i and the interactions with a nucleus of charge $Z_{\mathrm{eff}} = (2 - s)$, i.e., $V_i(r_i) = -(2-s)e^2/r_i$. Due to this separation – with s still undetermined – the residual interaction becomes:

$$W_{12} = \frac{e^2}{|\mathbf{r}_1 - \mathbf{r}_2|} - \frac{se^2}{r_1} - \frac{se^2}{r_2}.$$

The $H_i(s)$ single-electron problems have the exact ground-state solution in atomic units:

```
> E0:=s->-(2-s)^2/2;
```

$$E0 := s \to -\frac{1}{2}(2-s)^2$$

```
> psi0:=(r,g)->g^(3/2)/sqrt(Pi)*exp(-g*r);
```

$$\psi 0 := (r, g) \to \frac{g^{3/2}\, \mathrm{e}^{(-g\, r)}}{\mathrm{sqrt}(\pi)}$$

These eigenfunctions are normalized. The two-particle ground state for the spin singlet is symmetric. If we take equal orbitals for both electrons, the simple product is symmetric:

```
> Psi0:=(r1,r2,g)->psi0(r1,g)*psi0(r2,g);
```

$$\Psi 0 := (r1, r2, g) \to \psi 0(r1, g)\, \psi 0(r2, g)$$

Insertion into the Schrödinger equation yields

$$(2E_0 - E + W_{12})\Psi_0(r_1, r_2) = 0,$$

where E_0 is a function of s and not just the He^+ ground-state energy. The equation above is not satisfied exactly, as the product ansatz is not exact due to the fact that W_{12} is a two-particle operator. Instead we use perturbation theory, the result of which can also be derived from the demand that the projection of the equation onto the state Ψ_0 ought to vanish. While we can't satisfy the Schrödinger equation $(\hat{H} - E)|\Psi_0\rangle = 0$ exactly, we demand the projection $\langle\Psi_0|(\hat{H} - E)|\Psi_0\rangle$ to vanish, which determines the best-possible wavefunction within the space of employed trial functions Ψ_0.

This yields the perturbative result

$$E = 2E_0 + \langle\Psi_0|W_{12}|\Psi_0\rangle\,.$$

The latter matrix element has to be calculated, which can be done handily in elliptic coordinates. In the present implementation these coordinates are chosen such that the nucleus and the electron that is currently not integrated over serve as focal points (draw a diagram and find $r_{12} = |\mathbf{r}_1 - \mathbf{r}_2|$).

Using $r_1 = 2c$, we have $r_2 = c(\xi-\eta)$, $r_{12} = c(\xi+\eta)$ with the ranges $1 \leq \xi < \infty$, and $-1 \leq \eta \leq 1$, and the volume element $\mathrm{d}^3r = c^3(\xi^2-\eta^2)\mathrm{d}\xi\mathrm{d}\eta\mathrm{d}\phi$, and ϕ denoting the azimuthal angle. First we perform just the inner integration over r_2 and perform the ϕ integral trivially (giving 2π), and then the η integral:

```
> assume(g>0);

> W12int:=2*Pi*c^2*int((Psi0(r1,c*(xi-eta),g))^2*(xi^2-eta^2)
> *(1/(xi+eta)-s/2-s/(xi-eta)),eta=-1..1):
```

$$\mathrm{W12int} := 2\pi\, c^2 \Big(-\frac{1}{8}(\mathrm{e}^{(-g\tilde{}\, c\,(\xi-1))})^2 \Big(-2\,g\tilde{}\, c + 4\,g\tilde{}^{\,2} c^2 - s - 4\,g\tilde{}^{\,2} c^2 \xi + 4\,\xi\, s\, g\tilde{}^{\,2} c^2 + 2\,s\, g\tilde{}^{\,2} c^2 \xi^2 + 2\,s\, g\tilde{}^{\,2} c^2\Big)$$
$$g\tilde{}^{\,3}\,(\mathrm{e}^{(-g\tilde{}\, r1)})^2 \Big/ (c^3\pi^2) + \frac{1}{8}(\mathrm{e}^{(-g\tilde{}\, c\,(\xi+1))})^2 \Big(-2\,g\tilde{}\, c - 4\,g\tilde{}^{\,2} c^2 \xi + 4\,\xi\, s\, g\tilde{}^{\,2} c^2 - 4\,g\tilde{}^{\,2} c^2 - s + 2\,s\, g\tilde{}^{\,2} c^2 \xi^2 - 6\,s\, g\tilde{}^{\,2} c^2 - 4\,s\, g\tilde{}\, c\Big)\, g\tilde{}^{\,3}$$
$$(\mathrm{e}^{(-g\tilde{}\, r1)})^2 \Big/ (c^3\pi^2)\Big)$$

```
> W12int1:=int(W12int,xi):     assume(c>0);

> W12int2:=simplify(limit(W12int1,xi=infinity)
> -subs(xi=1,W12int1));
```

$$\mathrm{W12int2} := -\frac{1}{2} g\tilde{}^{\,3}\, \mathrm{e}^{(-2\,g\tilde{}\, r1)}$$
$$(-1 + s + 2\,s\, g\tilde{}\, c\tilde{} + \mathrm{e}^{(-4\,g\tilde{}\, c\tilde{})} + 2\,g\tilde{}\, c\tilde{}\, \mathrm{e}^{(-4\,g\tilde{}\, c\tilde{})})\, /$$
$$(c\tilde{}\,\pi)$$

Now we substitute $c = r_1/2$, and perform the r_1 integration in spherical polar coordinates with the $\mathrm{d}\Omega$ part done trivially (yielding a 4π factor):

```
> W12int3:=4*Pi*int(subs(c=r1/2,W12int2)*r1^2,r1=0..infinity);
```

$$\text{W12int3} := -\frac{1}{8}(-5+16\,s)\,g\tilde{}$$

Next we assemble the total energy and find out, how the result depends on s. Note that the wavefunction parameter g is a known function of s, *viz.*, $g = 2 - s$:

```
> Etot:=2*E0(s)+subs(g=2-s,W12int3);
```

$$\text{Etot} := -(2-s)^2 - \frac{1}{8}(-5+16\,s)(2-s)$$

```
> Etot:=simplify(Etot);
```

$$\text{Etot} := -\frac{11}{4} - \frac{5}{8}s + s^2$$

```
> plot(Etot,s=0..1,E=-3..-2);
```

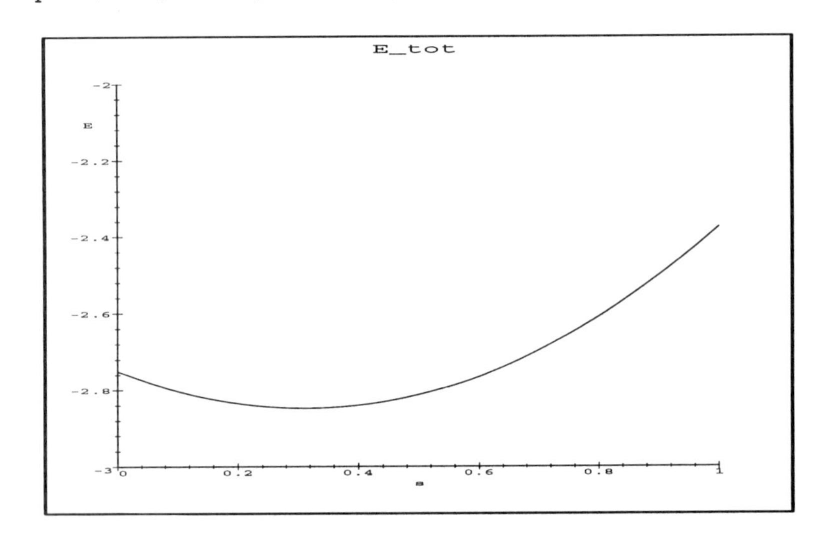


```
> s0:=solve(diff(Etot,s),s);     E0:=subs(s=s0,Etot);
```

$$s0 := \frac{5}{16} \qquad E0 := \frac{-729}{256}$$

```
> evalf(E0);        evalf(E0*27.12 eV);
```

$$-2.847656250 \qquad -77.22843750 \quad \text{eV}$$

Thus, we find the approximation to the total ground-state energy that corresponds to a $1s^2$ configuration of parahelium lying above the experimental energy of $E_{\text{He}} \approx -78.6$ eV, in accord with the Rayleigh-Ritz theorem.

The remaining energy difference is called a correlation effect. It arises as a consequence of the fact that the electron–electron interaction is properly diagonalized by a wavefunction that contains a more complicated dependence on r_1 and r_2 than the simple product form.

In fact, Hylleraas [Hy30] found more accurate energy values by going beyond the ansatz of an identical dependence of the two orbitals on r_i. Such a form has to be symmetrized in coordinate space and yields a picture of He atoms in which one electron is closely bound (basically a $He^+(1s)$ state), while the other plays the role of a valence electron. In an extension of the present Hartree (-Fock) scheme one has to work towards such a picture by mixing single-particle configurations.

We note that experimentally it is difficult to measure total energies. A more accessible quantity in the structure problem is the ionization potential of an atom, i.e., in helium the energy required to remove one of the two electrons. It is measured to be 24.46 eV for ground-state parahelium. Theoretically this is related to the total energy difference of neutral He and the energy of He^+. The latter is known exactly based on the H-like atom expression $E_n = -Z^2/(2n^2)$, which for our case results in a value of -2 atomic units. This yields the approximate ionization potential for He of 0.84766 a.u. = 23.0 eV, i.e., about 6 % less than the experimental valuo.

We note that the value of $s = 5/16$ turns the perturbation to zero:

```
> subs(s=5/16,subs(g=2-s,W12int3));
                                0
```

Therefore, the original product ansatz with this choice of break-up of the Hamiltonian represents a particularly good approximation to the full problem. To improve upon it one would have to go to the next order in perturbation theory.

We will, however, look at the problem also from the point of view of perturbation theory for non-optimal choices of s. We diagonalize the problem and discuss some excited $L = 0$ states of helium. For that purpose we consider $s = 0$, i.e., pure He^+ Hamiltonians for both unperturbed electrons, but allow them to be in any $l = 0$ eigenstate $|m\rangle$, not just in the ground state.

We can have a product state $\Psi_1 = |n\rangle|m\rangle$ that is energetically degenerate with $\Psi_2 = |m\rangle|n\rangle$ with the energy following from the H_0 problem as $e_m + e_n$, where the e_i are the single-particle energies.

The perturbation Hamiltonian consists now only of $W_\mathrm{P} = e^2/|\mathbf{r}_1 - \mathbf{r}_2|$ (due to $s = 0$) and there are three possible matrix elements:

$$W_{\mathrm{P}11} = \langle\Psi_1|W_\mathrm{P}|\Psi_1\rangle, \quad W_{\mathrm{P}12} = \langle\Psi_1|W_\mathrm{P}|\Psi_2\rangle, \quad W_{\mathrm{P}22} = \langle\Psi_2|W_\mathrm{P}|\Psi_2\rangle\,.$$

To break the degeneracy we consider linear combinations of $|\Psi_1\rangle$ and $|\Psi_2\rangle$ and diagonalize. This leads to a homogeneous system of two linear equations for the expansion coefficients c_1 and c_2.

$$(e_m + e_n + W_{\mathrm{P}11} - E)c_1 + W_{\mathrm{P}12}c_2 = 0\,,$$

and

$$W_{\mathrm{P}21}c_1 + (e_m + e_n + W_{\mathrm{P}22} - E)c_2 = 0\,.$$

With the substitution $\lambda = E - (e_m + e_n)$ the characteristic equation becomes (we use $W_{\mathrm{P}21} = W_{\mathrm{P}12}$):

```
> with(linalg):
> HPmat:=matrix(2,2,[[WP11-lambda,WP12],[WP12,WP22-lambda]]);
```

$$\mathrm{HPmat} := \begin{bmatrix} \mathit{WP11} - \lambda & \mathit{WP12} \\ \mathit{WP12} & \mathit{WP22} - \lambda \end{bmatrix}$$

```
> ChEq:=det(HPmat);
```

$$\mathrm{ChEq} := \mathit{WP11}\ \mathit{WP22} - \mathit{WP11}\ \lambda - \lambda\ \mathit{WP22} + \lambda^2 - \mathit{WP12}^2$$

```
> Evals:=solve(ChEq,lambda);
```

$$\mathrm{Evals} := \frac{1}{2}\,\mathit{WP11} + \frac{1}{2}\,\mathit{WP22} + \frac{1}{2}\sqrt{\mathit{WP11}^2 - 2\,\mathit{WP11}\ \mathit{WP22} + \mathit{WP22}^2 + 4\,\mathit{WP12}^2},\ \frac{1}{2}\,\mathit{WP11} + \frac{1}{2}\,\mathit{WP22} - \frac{1}{2}\sqrt{\mathit{WP11}^2 - 2\,\mathit{WP11}\ \mathit{WP22} + \mathit{WP22}^2 + 4\,\mathit{WP12}^2}$$

For the degenerate case we can make the substitution $W_{\mathrm{P}22} = W_{\mathrm{P}11}$:

```
> Evals:=solve(subs(WP22=WP11,ChEq),lambda);
```

$$\mathit{Evals} := \mathit{WP11} + \mathit{WP12},\ \mathit{WP11} - \mathit{WP12}$$

This shows how the perturbation splits the degeneracy. One can also calculate the eigenvectors:

```
> eigenvects(subs(WP22=WP11,op(HPmat)));
```

$$[-\lambda + \mathit{WP12} + \mathit{WP11}, 1, \{[1\ 1]\}], [-\lambda - \mathit{WP12} + \mathit{WP11}, 1, \{[-1\ 1]\}]$$

From the eigenvectors we conclude that the eigenvalue $E_{\mathrm{s}} = e_m + e_n + W_{\mathrm{P}11} + W_{\mathrm{P}12}$ corresponds to the symmetric wavefunction in coordinate space (parahelium, $S = 0$), while $E_{\mathrm{t}} = e_m + e_n + W_{\mathrm{P}11} - W_{\mathrm{P}12}$ belongs to the spin triplet, i.e., orthohelium ($S = 1$).

In the ground state, however, i.e., $m = n = 1$, there can be no AS wavefunction in coordinate space, as $\Psi = 0$ in this case. Thus, a degeneracy (without perturbation W_{P12} that implies a level splitting) occurs earliest at the level of $m = 1$, $n = 2$, i.e., an (1s,2s) configuration. The symmetric combination describes an excited state of parahelium, while the AS combination represents the ground state of orthohelium, which is almost 20 eV above the parahelium ground state. The orthohelium ground state is not allowed to decay to the parahelium ground state by a strong selection rule (a spin flip – electron exchange – is required) and has a half-life of about two months.

The two brands of helium have different macroscopic properties, parahelium is diamagnetic, while orthohelium is paramagnetic. This is not surprising as magnetism is a manifestation of the spin degrees of freedom of atoms. In our case one atom is spin-saturated and does not interact strongly with a magnetic field, while the other has a spin.

We try to estimate the splitting of the degeneracy between the (1s,2s) configurations. We need a hydrogen-like 2s state:

```
> psi1:=(r,g)->(1-g/2*r)*exp(-g/2*r):
```

This function is not yet normalized, but orthogonal already:

```
> int(psi1(r,g)*psi0(r,g)*r^2,r=0..R);  limit(",R=infinity);
```

$$\frac{1}{3}\,\frac{g^{\sim 3/2}\,R^3\,(\,e^{(-1/2\,g^{\sim}\,R)}\,)^3}{\sqrt{\pi}} \qquad 0$$

Now we normalize: first we check the normalization of ψ_0 (it contains $Y_{00}(\Omega)$):

```
> int(psi0(r,2)^2*r^2,r=0..infinity);
```

$$\frac{1}{4}\,\frac{1}{\pi}$$

We introduce a normalization constant c_1 for the 2s state and normalize the state consistently with the 1s state:

```
> c1^2*int(r^2*psi1(r,g)^2,r=0..infinity)=1/(4*Pi);
```

$$2\,\frac{c1^2}{g^{\sim 3}} = \frac{1}{4}\,\frac{1}{\pi}$$

```
> solc1:=solve(",c1);
```

$$\text{solc1} := \frac{1}{4}\,g^{\sim 2}\,\sqrt{2}\,\sqrt{\frac{1}{g^{\sim}\,\pi}},\; -\frac{1}{4}\,g^{\sim 2}\,\sqrt{2}\,\sqrt{\frac{1}{g^{\sim}\,\pi}}$$

The 2s state **psi1** is redefined using the above answer and graphed together with the 1s state.

```
> psi1:=(r,h)->subs(g=h,solc1[1]*(1-g/2*r)*exp(-g/2*r)):
> plot({r*psi0(r,2),r*psi1(r,2)},r=0..5,psi=-0.3..0.3);
```

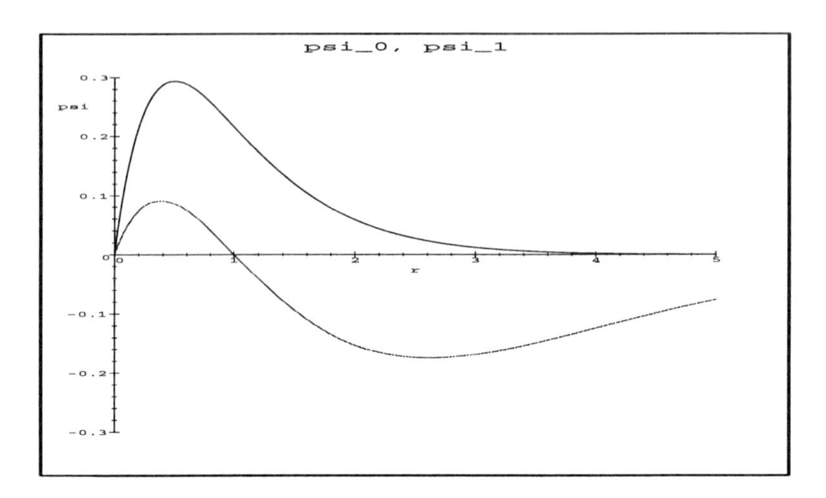

The probability densities up to the 4π factor have the following shape:

```
> plot({(r*psi0(r,2))^2,(r*psi1(r,2))^2},r=0..5);
```

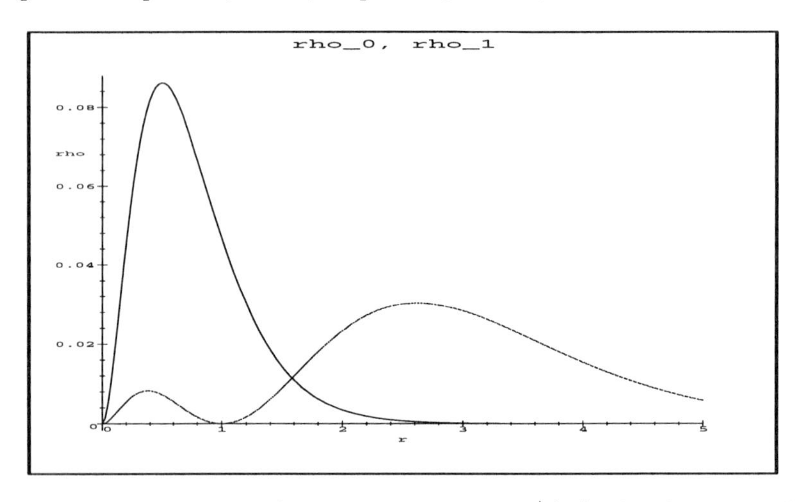

Note that the Bohr radius for our unperturbed He^{+}(1s) orbital is $r_B = 1/Z = 1/2$ in this calculation.

We have to repeat the calculation of the matrix elements for the perturbation. The two-electron wavefunctions for charge $Z = 2$ are given as

```
> PsiP1:=(r1,r2)->psi0(r1,2)*psi1(r2,2):
```

```
> PsiP2:=(r1,r2)->psi0(r2,2)*psi1(r1,2):
```

The trick is done again in elliptic coordinates using the nucleus and the not currently integrated electron's position as focal points. For the splitting we need to calculate $W_{\rm P12}$ only. First the inner integral: $r_1 = 2c$, $r_2 = c(\xi-\eta)$, etc. The η integration can be performed, but we suppress the output:

```
> WP12inn0:=2*Pi*c^2*int(PsiP1(r1,c*(xi-eta))*PsiP2(r1,
> c*(xi-eta))*(xi^2-eta^2)/(xi+eta),eta=-1..1):
```

Next we perform the ξ integral.

```
> WP12inn1:=int(WP12inn0,xi=1..infinity);
```

$$\mathrm{WP12inn1} := -\frac{32}{27}\,\frac{e^{(-2\,r1)}\,(-1+r1)\,e^{(-r1)}\,(e^{(-c^{\sim})})^3\,(6\,c^{\sim}+1)}{\pi\,(e^{c^{\sim}})^3}$$

The outer r_1-integration is doable in spherical polar coordinates, but we need to substitute $r_1 = 2c$ back:

```
> WP12:=4*Pi*int(r1^2*subs(c=r1/2,WP12inn1),r1=0..infinity);
```

$$\mathrm{WP12} := \frac{32}{729}$$

In electron volts we have for the energy difference $E_{\rm 1s2s} - E_{\rm 1s1s}$, which equals $(W_{\rm P11} + W_{\rm P12}) - (W_{\rm P11} - W_{\rm P12}) = 2W_{\rm P12}$:

```
> evalf(2*WP12*27.12 eV);
```

$$2.3809 \quad \mathrm{eV}$$

This should be 0.7 eV only! We conclude that the calculations for excited states are more difficult than for ground states. We finish with the remark that one can also consider double excitations, e.g., He(2s^2). However, one has to consider carefully whether such states are not embedded in the continuum of He$^+$. One can understand the occurence of autoionization processes as one electron deexciting to He$^+$(1s), while the other gains this energy to go to the continuum, i.e.,

$$\mathrm{He}^0(2\mathrm{s},2\mathrm{s}) \quad \rightarrow \quad \mathrm{He}^+(1\mathrm{s}) + 1\mathrm{e}^- \quad ,$$

with a characteristic kinetic energy for the ejected electron. Autoionization processes play an important role in atomic collisions with charged particles, particularly in collisions of highly charged ions with neutral atoms. Atoms are likely to wind up in multiply excited states which decay after the collision.

Exercise 7.1.1: Repeat the calculation for the energy splitting $E_{\rm 1s2s} - E_{\rm 1s1s}$ with an improved helium wavefunction based on the variational parameter choice $s = 5/16$ in the single-electron orbitals.

Exercise 7.1.2: Perform a ground-state calculation for the Li$^+$ ion. Calculate the ionization energy in the given approximation. Obtain the ionization energy and total ground-state energy as a function of the nuclear charge Z for a few discrete Z values in the He(1s^2) isoelectronic sequence (cf. [CR74]).

7.2 Hydrogen Molecule

In this section we solve the problem of the hydrogen molecule and show the electronic binding energies as a function of internuclear separation for the singlet and triplet symmetries. It is useful to start with the previous section on the helium atom before studying the present one in order to understand the approximation to the two-electron problem.

We investigate the permutation symmetry for the wavefunction under the operation of exchange of nuclei A and B, which are indistinguishable in a homonuclear diatomic molecule, for a product ansatz for the wavefunction

$$\Phi(\mathbf{r}_1, \mathbf{r}_2) = \phi_{\mathrm{A}}(\mathbf{r}_1)\phi_{\mathrm{B}}(\mathbf{r}_2)\,.$$

It yields the same energy as the state $\phi_{\mathrm{A}}(\mathbf{r}_2)\phi_{\mathrm{B}}(\mathbf{r}_1)$, if one makes use of a zero-order Hamiltonian that contains only the single-particle Hamiltonians for atoms A and B respectively.

Thus, we have to consider the linear combinations as possible eigensolutions with the same eigenvalue in zeroth order. We pick the symmetric and anti-symmetric spatial wavefunction combinations, as they form an overall AS state when combined with an AS spin function in the first case (spin-singlet, anti-parallel, called para), and with a symmetric spin function (three projections for spin=1) in the second case (spin-triplet, parallel, called ortho).

The residual interactions between electon 1 and nucleus B, electron 2 and nucleus A, and between both electrons, when added to the zero-order Hamiltonian break the degeneracy and two very different energy levels emerge as a function of internuclear separation R. Our aim is to calculate this singlet–triplet splitting from first principles. We use atomic units: $\hbar = m_{\mathrm{e}} = e = 1$, distances are measured in Bohr radii. The hydrogen atom ground-state eigenfunction is given as

```
> psi:=r->exp(-r)/sqrt(Pi):
```

We begin with the calculation of the overlap between the two possible two-particle products, $\Phi_1 = \phi_{\mathrm{A}}(\mathbf{r}_1)\phi_{\mathrm{B}}(\mathbf{r}_2)$ and $\Phi_2 = \phi_{\mathrm{B}}(\mathbf{r}_1)\phi_{\mathrm{A}}(\mathbf{r}_2)$, namely

$$S^2 = \int\int \Phi_1(\mathbf{r}_1, \mathbf{r}_2)\Phi_2(\mathbf{r}_1, \mathbf{r}_2)\mathrm{d}^3 r_1 \mathrm{d}^3 r_2\,,$$

which is a function of the internuclear separation R. We can make use of the axial symmetry with respect to the z axis. We use special coordinates to deal with the two-centre character of the integral. The overlap integral is a square, i.e.,

$$S = \int \phi_{\mathrm{A}}(\mathbf{r}_1)\phi_{\mathrm{B}}(\mathbf{r}_1)\mathrm{d}^3 r_1\,,$$

represents a 3D integral.

The coordinates use the azimuthal angle ϕ around the z axis, and an orthogonal pair whose lines of constant value describe hyperbolae and ellipses with the nuclei (protons) in the focal points. The internuclear separation R appears, therefore as a parameter. For further details see [Fl71, problem 91].

The conversion to Cartesians from ξ, η, ϕ using the half-separation $\mathtt{Rh} = R/2$ is given as

$$x = \mathtt{Rh}\sqrt{(1-\eta^2)(\xi^2-1)}\cos\phi\,,$$

$$y = \mathtt{Rh}\sqrt{(1-\eta^2)(\xi^2-1)}\sin\phi\,,$$

and

$$z = \mathtt{Rh}\xi\eta\,.$$

Of interest are the radius vectors in the (xy) plane to a point separated from nucleus A and B by the distances r_A and r_B respectively:

$$r_\mathrm{A} = \mathtt{Rh}(\xi+\eta)\,,\; r_\mathrm{B} = \mathtt{Rh}(\xi-\eta)\,,\; r_\mathrm{A}+r_\mathrm{B} = R\xi\,.$$

The ranges are: $1 < \xi < \infty$ (ξ controls the size of ellipses); and $-1 \leq \eta \leq 1$ (η determines the hyperbolae). The volume element is given by $\mathrm{d}^3r = \mathtt{Rh}^3(\xi^2 - \eta^2)\mathrm{d}\xi\mathrm{d}\eta\mathrm{d}\phi$. Thus, the overlap matrix element S involves an integral over $\psi(r_\mathrm{A})\psi(r_\mathrm{B}) = \psi(r_\mathrm{A}+r_\mathrm{B})$.

```
> assume(Rh>0);
> S:=2*Pi*Rh^3*int(int(psi(Rh*(xi+eta))*psi(Rh*(xi-eta))*
> (xi^2-eta^2),eta=-1..1),xi=1..infinity);
```

$$S := \frac{1}{3}\,\frac{4\,Rh^{\sim 2}+3+6\,Rh^{\sim}}{(\mathrm{e}^{Rh^{\sim}})^2}$$

```
> plot(S,Rh=0..10);
```

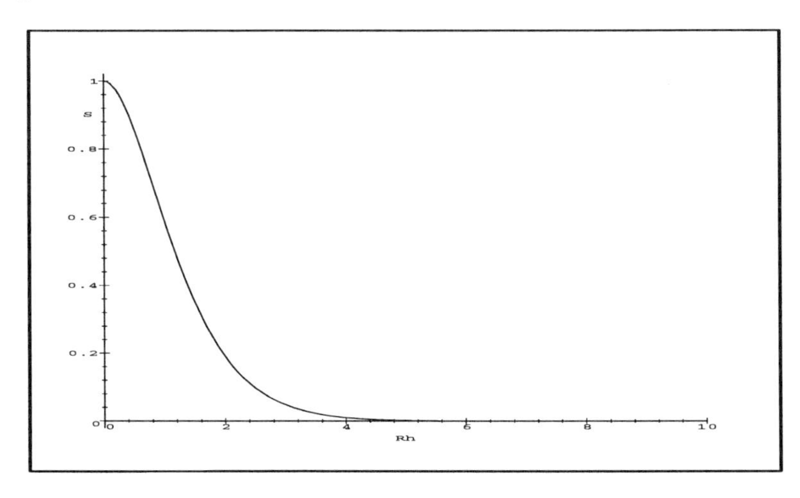

The graph with $\mathtt{Rh}$ in a.u., i.e., Bohr radii shows up to which distances $R/2$ the $\psi_{1\mathrm{s}}$ orbitals centered on nucleus A and B respectively still have an appreciable overlap. The result can be generalized easily to a wavefunction with a free scale parameter.

We now have to calculate the direct interaction matrix element for the residual potential. The interaction of the electron density with the nuclei belonging to the other atom, respectively,

$$\int\int [\psi_A(r_1)\psi_B(r_2)]^2 (\frac{1}{r_{1B}} + \frac{1}{r_{2A}}) d^3r_1 d^3r_2$$

has two equal contributions that are identical. This becomes obvious when one relabels the integration variable. We try to calculate:

```
> Kp:=2*2*Pi*Rh^3*int(int((psi(Rh*(xi+eta)))^2/(Rh*(xi-eta))*
> (xi^2-eta^2),eta=-1..1),xi=1..infinity);
  Error, (in int/cook/IIntd1) cannot evaluate boolean
```

OK, maybe we have to help with the limit in the integrand. We cancel the common $(\xi - \eta)$ factor in numerator and denominator of the integrand and try again.

```
> Kp:=2*2*Pi*Rh^3*int(int((psi(Rh*(xi+eta)))^2/(Rh)*
> (xi+eta),eta=-1..1),xi=1..infinity);
  Error, (in int/cook/IIntd1) cannot evaluate boolean
```

Let's try by changing the order of integration.

```
> Kp:=4*Pi*Rh^2*int(int((psi(Rh*(xi+eta)))^2*(xi+eta),
> xi=1..infinity),eta=-1..1);
  Error, (in int/cook/IIntd1) cannot evaluate boolean
```

We verify that, in fact, the wavefunction ensures convergence of the ξ integral.

```
> (psi(Rh*(xi+eta)))^2;
```

$$\frac{(e^{(-Rh\tilde{}\,(\xi+\eta))})^2}{\pi}$$

Since it is the ξ integral that seems to create the problems, we perform the η integral first.

```
> int((psi(Rh*(xi+eta)))^2*(xi+eta),xi=1..infinity);
  Error, (in int/cook/IIntd1) cannot evaluate boolean
```

```
> int((psi(Rh*(xi+eta)))^2*(xi+eta),eta=-1..1);
```

$$-\frac{1}{4}\frac{(e^{(-Rh\tilde{}\,(\xi+1))})^2(2\,Rh\tilde{}\,\xi + 2\,Rh\tilde{} + 1)}{Rh\tilde{}^{\,2}\,\pi} + \frac{1}{4}\frac{(e^{(-Rh\tilde{}\,(\xi-1))})^2(2\,Rh\tilde{}\,\xi - 2\,Rh\tilde{} + 1)}{Rh\tilde{}^{\,2}\,\pi}$$

```
> int(",xi=1..infinity);
  Error, (in int/cook/IIntd1) cannot evaluate boolean
```

Is it the restriction that ξ is limited in range to be greater than one isn't recognized? We make use of **assume** even though the integration range is restricting ξ, but we suspect that something might happen at $\xi = 1$. We note that assumptions on variables that are integrated over are not ignored by Maple, i.e., the integration variable is not treated as an independent dummy index.

```
> assume(xi>1);
```

```
> Kp:=4*Pi*Rh^2*int(int((psi(Rh*(xi+eta)))^2*(xi+eta),
> eta=-1..1),xi=1..infinity);
  Error, (in int/cook/IIntd1) cannot evaluate boolean
```

Are we stumbling over a simple integral indeed? We store the η integral and perform the ξ integral as an indefinite integral.

```
> val:=int((psi(Rh*(xi+eta)))^2*(xi+eta),eta=-1..1);
```

$$\mathrm{val} := -\frac{1}{4}\frac{(\mathrm{e}^{(-Rh\tilde{}\,(\xi\tilde{}+1))})^2\,(2\,Rh\tilde{}\,\xi\tilde{}+2\,Rh\tilde{}+1)}{Rh\tilde{}^{\,2}\,\pi} + \frac{1}{4}\frac{(\mathrm{e}^{(-Rh\tilde{}\,(\xi\tilde{}-1))})^2\,(2\,Rh\tilde{}\,\xi\tilde{}-2\,Rh\tilde{}+1)}{Rh\tilde{}^{\,2}\,\pi}$$

```
> int(val,xi);
```

$$\frac{1}{4}\frac{(\mathrm{e}^{(-Rh\tilde{}\,(\xi\tilde{}+1))})^2\,Rh\tilde{}\,(\xi\tilde{}+1)+(\mathrm{e}^{(-Rh\tilde{}\,(\xi\tilde{}+1))})^2}{Rh\tilde{}^{\,3}\,\pi} - \frac{1}{4}\frac{(\mathrm{e}^{(-Rh\tilde{}\,(\xi\tilde{}-1))})^2\,Rh\tilde{}\,(\xi\tilde{}-1)+(\mathrm{e}^{(-Rh\tilde{}\,(\xi\tilde{}-1))})^2}{Rh\tilde{}^{\,3}\,\pi}$$

Now we substitute the boundaries and obtain the desired answer.

```
> Kp:=4*Pi*Rh^2*(limit(",xi=infinity)-limit(",xi=1));
```

$$Kp := -\frac{2\,(\mathrm{e}^{(-2\,Rh\tilde{})})^2\,Rh\tilde{}-1+(\mathrm{e}^{(-2\,Rh\tilde{})})^2}{Rh\tilde{}}$$

```
> expand(");
```

$$-2\,\frac{1}{(\mathrm{e}^{Rh\tilde{}})^4}+\frac{1}{Rh\tilde{}}-\frac{1}{Rh\tilde{}\,(\mathrm{e}^{Rh\tilde{}})^4}$$

Maple shouldn't have stumbled over this one! We graph the energy associated with the attraction of the electron density from atom A with nucleus B and vice versa.

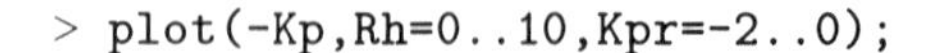

```
> plot(-Kp,Rh=0..10,Kpr=-2..0);
```

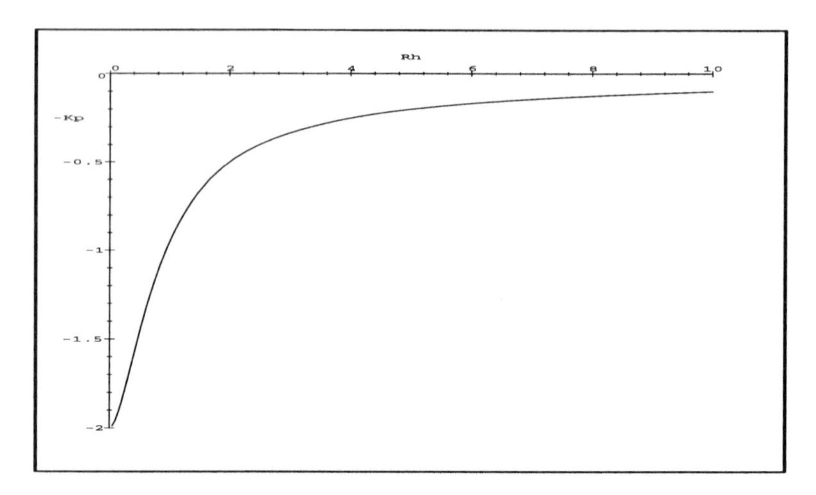

The gain in energy due to the attraction of the electron clouds by the opposite nuclei is long ranged. Next we calculate the electron–electron repulsion. This is trickier as both particles' coordinates are now involved.

First we perform the integration with respect to the coordinates of particle 2 in elliptic coordinates that use electron 1 and nucleus B (on which particle 2 is centered) as focal points. In these coordinates the interelectronic separation becomes

$$r_{12} = \frac{r_{1\mathrm{B}}}{2}(\xi + \eta)\,.$$

We have learned our lesson about the ξ integration and proceed as before with indefinite integration and substitution of boundaries.

```
> resint1:=2*Pi*r1Bh^3*int(int((psi(r1Bh*(xi-eta)))^2*
> (xi^2-eta^2)/(r1Bh*(xi+eta)),eta=-1..1),xi):
```

We suppress also the output from the substitution of limits into the indefinite ξ integral, and display only the simplified result below:

```
> resint:=limit(resint1,xi=infinity)-subs(xi=1,resint1):
```

```
> assume(r1Bh>0);        resint:=simplify(resint);
```

$$\mathrm{resint} := -\frac{1}{2}\,\frac{-1 + 2\,\mathrm{e}^{(-4\,r1Bh\tilde{})}\,r1Bh\tilde{} + \mathrm{e}^{(-4\,r1Bh\tilde{})}}{r1Bh\tilde{}}$$

Now we have to do the remaining integral in elliptic coordinates for which the nuclei are focal points. That implies the substitution $r_{1\mathrm{B}} = \frac{R}{2}(\xi - \eta)$

```
> 2*Pi*Rh^3*int(int(subs(r1Bh=Rh*(xi-eta)/2,resint)*(xi^2
> -eta^2)*(psi(Rh*(xi+eta)))^2,eta=-1..1),xi=1..infinity):
```

```
> J:=simplify(");
```

$$J := -\frac{1}{24}(16\,Rh^{\sim 3}\,e^{(2\,Rh^{\sim})} + 12\,e^{(2\,Rh^{\sim})}$$
$$+36\,Rh^{\sim 2}\,e^{(2\,Rh^{\sim})} - 12\,e^{(6\,Rh^{\sim})} + 33\,e^{(2\,Rh^{\sim})}\,Rh^{\sim})$$
$$e^{(-6\,Rh^{\sim})}\Big/Rh^{\sim}$$

Fortunately, the (undisplayed) imaginary parts (an artefact of the Maple computation) cancelled. We expand the expression and graph it below.

```
> expand(J);
```

$$-\frac{2}{3}\frac{Rh^{\sim 2}}{(e^{Rh^{\sim}})^4} - \frac{1}{2}\frac{1}{Rh^{\sim}\,(e^{Rh^{\sim}})^4} - \frac{3}{2}\frac{Rh^{\sim}}{(e^{Rh^{\sim}})^4} + \frac{1}{2}\frac{1}{Rh^{\sim}} - \frac{11}{8}\frac{1}{(e^{Rh^{\sim}})^4}$$

```
> plot(J,Rh=0..10,Rep=0..1);
```

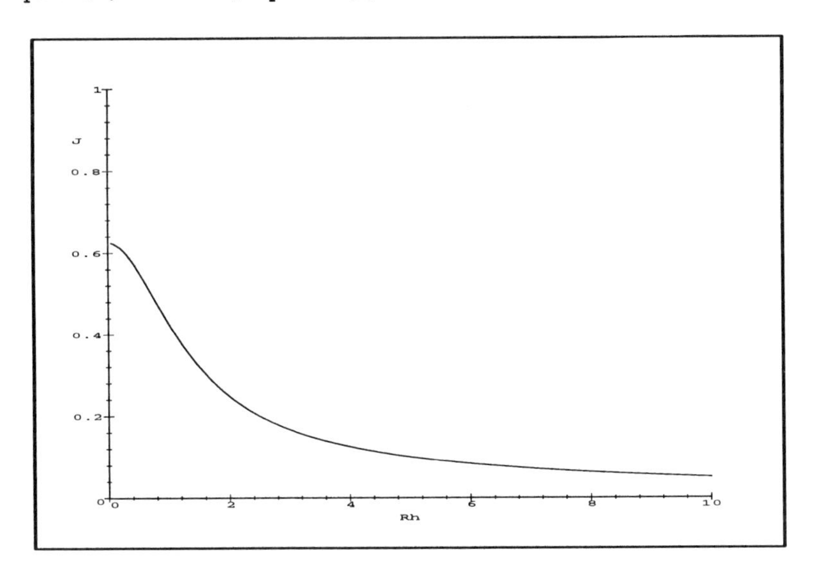

The latter graph shows the long-ranged Coulomb repulsion between the two electrons. Taken together the direct contribution gives the result:

```
> K:=simplify(-Kp+J):
> expand(K);
```

$$\frac{5}{8}\frac{1}{(e^{Rh^{\sim}})^4} - \frac{1}{2}\frac{1}{Rh^{\sim}} + \frac{1}{2}\frac{1}{Rh^{\sim}\,(e^{Rh^{\sim}})^4} - \frac{2}{3}\frac{Rh^{\sim 2}}{(e^{Rh^{\sim}})^4} - \frac{3}{2}\frac{Rh^{\sim}}{(e^{Rh^{\sim}})^4}$$

```
> plot(K,Rh=0..10,KD=-1.4..0);
```

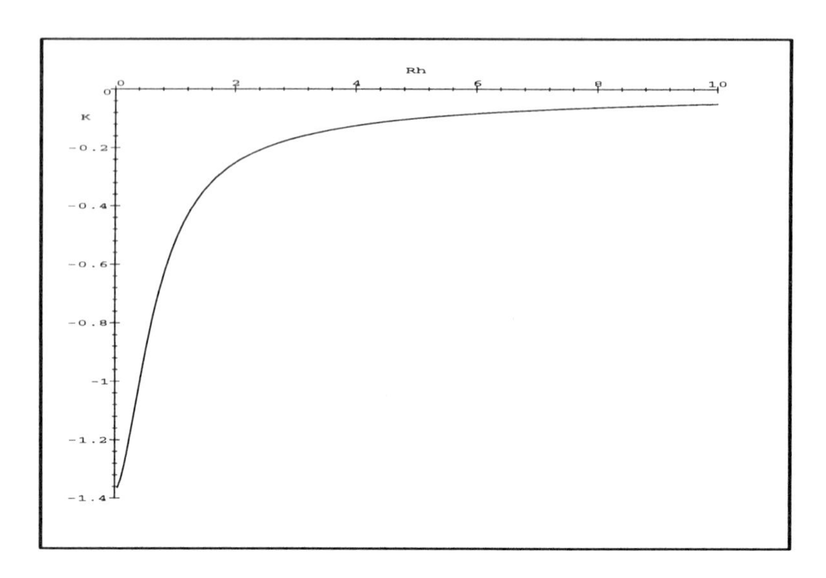

The exchange contribution is harder to calculate and is no longer representable solely in terms of elementary functions. Thus, we only list the final result from [Fl71]. It consists of two pieces, Ap $= A'$ is easy to calculate using the same tricks as done before for Kp $= K'$.

```
> Ap:=2*exp(-4*Rh)*(1+4*Rh+16/3*Rh^2+8/3*Rh^3);
```

$$Ap := 2\,e^{(-4\,Rh\tilde{})}\left(1 + 4\,Rh\tilde{} + \frac{16}{3}\,Rh\tilde{}^{2} + \frac{8}{3}\,Rh\tilde{}^{3}\right)$$

while the second piece contains S and a similar expression Sp $= S'$.

```
> Sp:=exp(2*Rh)*(1-2*Rh+4/3*Rh^2);
```

$$Sp := e^{(2\,Rh\tilde{})}\left(1 - 2\,Rh\tilde{} + \frac{4}{3}\,Rh\tilde{}^{2}\right)$$

We need the Euler-Mascheroni constant as well as the exponential integral $\mathrm{Ei}(x)$. We list Euler's constant γ, a predefined variable in Maple:

```
> c_E:=gamma;        evalf(c_E);
```

$$c\_E := \gamma \qquad .5772156649$$

```
> Jx:=1/5*(exp(-4*Rh)*(25/8-23/2*Rh-12*Rh^2-8/3*Rh^3)+3/Rh
> *(S^2*ln(2*Rh)+c_E*S^2+Sp^2*Ei(-8*Rh)-2*S*Sp*Ei(-4*Rh))):
```

```
> A:=Jx-Ap:            plot(A,Rh=0..10);
```

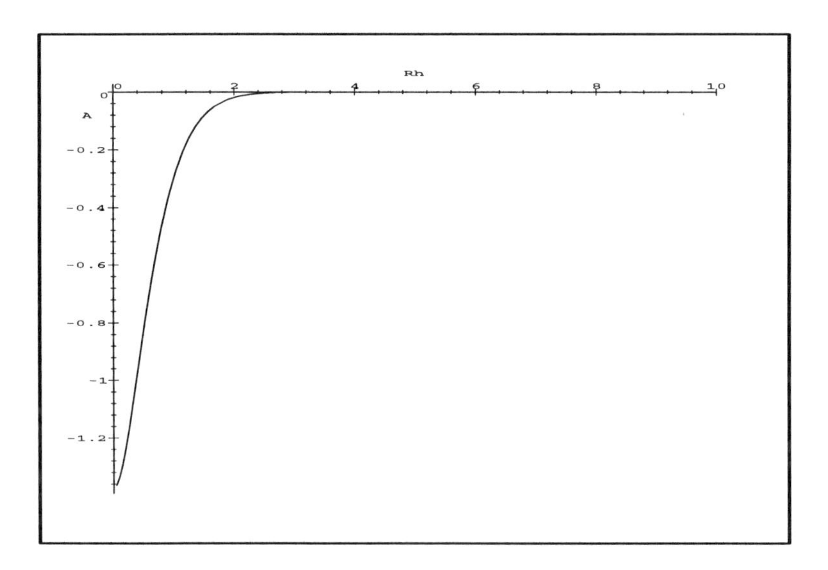

We see that the exchange contribution to the interaction energy has the following properties.

- *(i)* It is short-ranged. This seems reasonable: as the overlap between the two electronic orbitals decreases, we don't expect anti-symmetrization to be of any importance.
- *(ii)* It shows a surprisingly strong short-distanced behaviour. The exchange energy values appear to be high for a supposedly small effect that is related to the spin structure in the two-electron wavefunction. Naively one might think that spin effects are negligible, as the motion is non-relativistic and the Hamiltonian has no spin dependence. Nevertheless, symmetries in the wavefunctions can be very powerful.

Now we combine all contributions separately for the singlet and triplet cases. We choose the energy of two ground-state hydrogen atoms at infinite separation as the zero-reference point. This corresponds to $E = 0$ asymptotically for the dissociation limit.

```
> Etriplet:=1/(2*Rh) + (K-A)/(1+S^2):
```

We store the graph for later display together with the singlet energy.

```
> P2:=plot(Etriplet,Rh=0..5,E=-0.5..0.5):
> Esinglet:=1/(2*Rh) + (K+A)/(1+S^2):
> P1:=plot(Esinglet,Rh=0..5,E=-0.5..0.5,color=red):
> with(plots):     display({P1,P2});
```

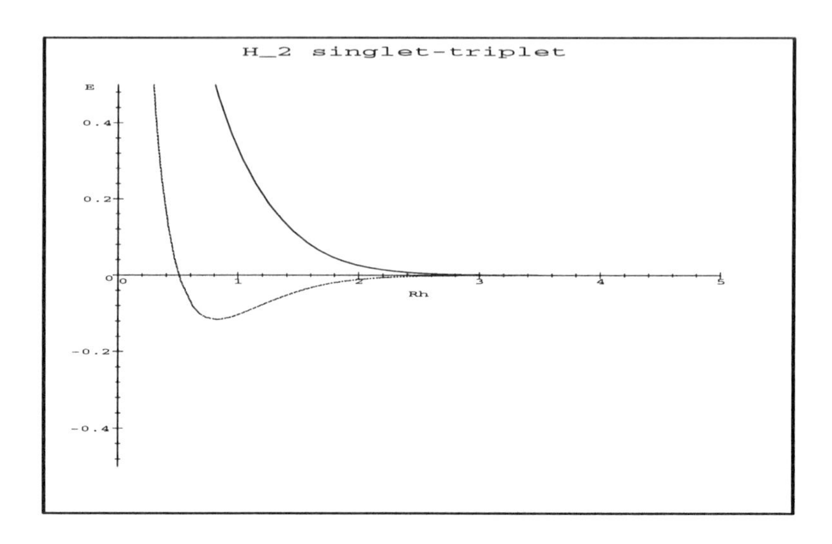

We observe that the triplet case (aligned spins, antibonding) results in a total energy that is always above the dissociation limit $E = 0$. The singlet case on the other hand (antiparallel spins, covalent bond), which represents the ground state in our problem is energetically attractive in the interval $R_1 < R < \infty$. In both cases for small R an increase in the total energy is observed.

We look more closely at the minimum in the electronic energy for the singlet case.

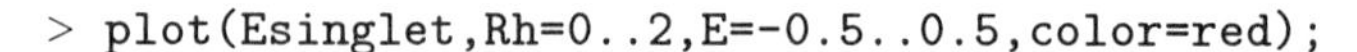

```
> plot(Esinglet,Rh=0..2,E=-0.5..0.5,color=red);
```

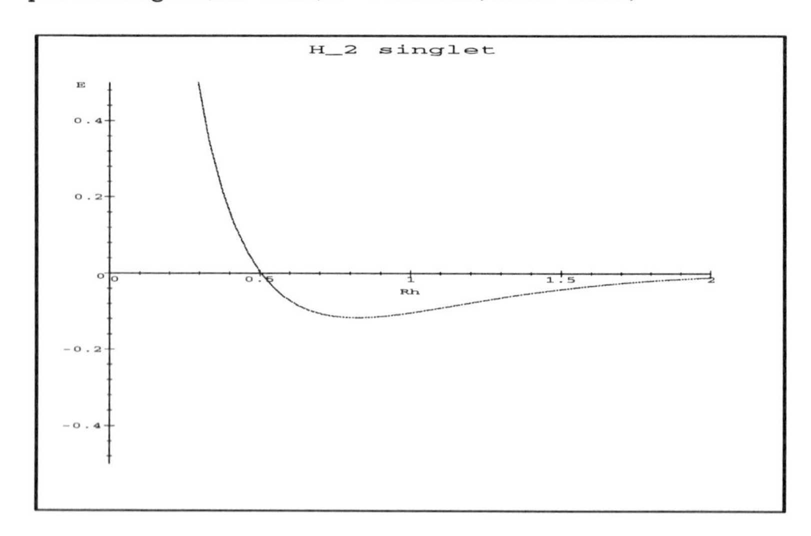

The minimum occurs at $R \approx 1.6$, $E \approx -0.11$ a.u., according to this calculation. The energy is converted to electron volts:

```
> evalf(-0.11*27.12 eV);
```

$$-2.9832 \quad \text{eV}$$

With our simple two-electron wavefunction ansatz we estimate the binding energy to be about 3 eV with a binding radius of 1.6 Bohr radii. What is simple about this wavefunction? It was chosen as an (anti-)symmetrized form of products of hydrogen atom orbitals. This corresponds to the Hartree-Fock ansatz in atoms and one can hope, at best, for the quality of the total energy as achieved in the helium atom.

Quantum chemistry calls for the solution of the nuclear motion in potential energy curves obtained from a solution of the many-electron problem at fixed internuclear separations, particularly for the ground state. This results in rotational-vibrational spectra that represent the possible energy states for the nuclear motion. One can investigate how molecules can store energy, particularly in the rotational motion. This results in an understanding of the specific heat of diatomic gases at low temperatures.

One can also study how electronic excitations are occuring during particle–molecule collisions and how the system can make a transition to a higher level that dissociates, such as the repulsive triplet energy curve calculated above.

Exercise 7.2.1: Find the precise location of the minimum in the singlet-state energy. Consider a local harmonic approximation to this energy $E(R) \approx E_0 + c_2(R - R_0)^2$ and estimate the zero-point energy, as well as the first level spacing for the vibrational modes of the H_2 molecule.

Exercise 7.2.2: Perform the simpler calculation of the symmetric and anti-symmetric ground-state wavefunctions and energies as a function of internuclear distance (correlation diagram) for the hydrogen molecular ion H_2^+.

7.3 Vibrating Molecules

In this section we use a model potential for the singlet state in a symmetric diatomic molecule to calculate the rotational–vibrational excitation spectrum. Details can be found in [Fl71, problem 37]. The potential is given as

$$V(\rho) = -2D(\frac{1}{R} - \frac{1}{2R^2}),$$

where D represents the binding energy, and $\rho = R/a = 1$ is the binding radius.

```
> V:=R->-2*D*(1/R-1/(2*R^2));
```

$$V := R \to -2\,D\left(\frac{1}{R} - \frac{1}{2}\frac{1}{R^2}\right)$$

```
> Vd:=diff(V(R),R);
```

$$Vd := -2\,D\left(-\frac{1}{R^2} + \frac{1}{R^3}\right)$$

```
> sols:=[solve(Vd,R)];      V(1);
```

$$sols := [1] \qquad -D$$

```
> plot(subs(D=1,V(R)),R=0..5,V=-1..1);
```

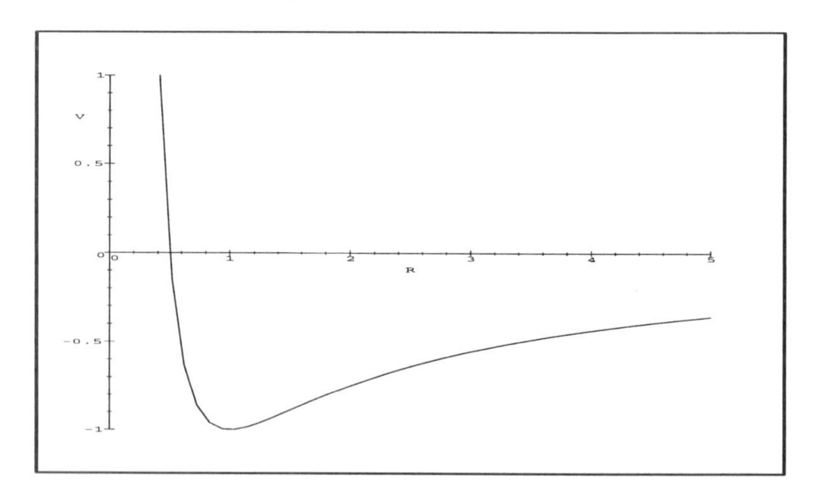

The vibrational levels for the $l = 0$ (s wave) channel are obtained from the solution of the radial Schrödinger equation in this potential for the reduced mass from the two nuclei, $\mu = M_A M_B/(M_A + M_B)$.

For a symmetric molecule: $\mu = M/2$; for hydrogen $\mu = 918m_e$ ($= 918$ in a.u.). Rotational excitations follow from adding the corresponding centrifugal term $l(l+1)/(2\mu R^2)$ to the radial Hamiltonian for the nuclear motion. Note that this term has a small coefficient compared to the potential shown above, as the reduced mass appearing in the denominator is large. In fact, one could approximate the centrifugal term by a constant in which the dynamical variable R is replaced by the binding radius R_0. In this case the moment of inertia of the molecule for rotations about its center, $I_0 = \mu R_0^2$, appears in the denominator.

While the original, exact problem can be solved in closed form using the hypergeometric function (see e.g., [F171]), we justify our approximation on the basis that the centrifugal term represents a correction only. Since the radial functions are centered on R_0, they have a position expectation value of R_0. Therefore we expect the higher l levels to be a perturbation of the $l = 0$ levels of the form

$$\Delta E = \frac{l(l+1)}{2\mu}\langle\Psi(R)|\frac{1}{R^2}|\Psi(R)\rangle \approx \frac{l(l+1)}{2I_0},$$

where $I_0 = \mu R_0^2$. Thus, we solve only the s wave equation. We set $x = \rho = R/a$, $\lambda = E(2\mu a^2/\hbar^2)$, $g^2 = 2\mu a^2 D/\hbar^2$. This is an eigenvalue problem in λ (i.e., the energy E), that Maple can't solve as is. The Schrödinger equation defined below could be solved numerically in Maple for selected λ values such that the wavefunction satisfies the correct asymptotic behaviour.

```
> SE:=diff(chi(x),x$2) + (lambda+2*g^2/x-g^2/x^2)*chi(x) = 0;
```

$$\mathrm{SE} := \left(\frac{\partial^2}{\partial x^2}\chi(x)\right) + \left(\lambda + 2\frac{g^2}{x} - \frac{g^2}{x^2}\right)\chi(x) = 0$$

Alternatively we can approximate the potential by a harmonic oscillator to get around the numerical problem, or apply the variational method. We define the potential for subsequent Taylor expansion around the binding radius $x = R/a = 1$.

```
> Ueff:=x->2*g^2/x-g^2/x^2;
```

$$\mathrm{Ueff} := x \to 2\frac{g^2}{x} - \frac{g^2}{x^2}$$

```
> Ueff4:=y->subs(x=y,convert(taylor(Ueff(x),x=1,5),polynom)):
```

To replace the potential by its Taylor expansion is not unproblematic, however. To show this graphically in arbitrary units we substitute a strength constant $g = 1$.

```
> plot({subs(g=1,-Ueff(x)),subs(g=1,-Ueff4(x))},x=0..5,
> U=-1..1);
```

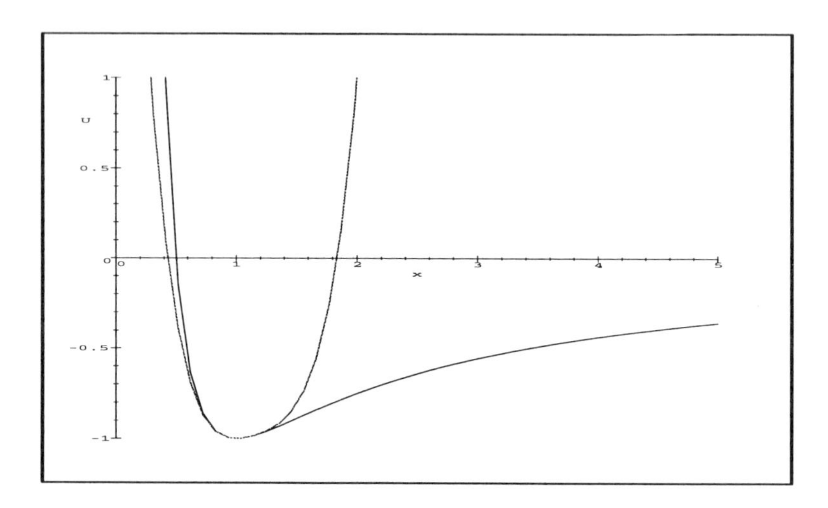

This comparison shows that the Taylor expansion (harmonic approximation) will work only for the lowest energy levels. Of course, the approximate potential knows nothing about dissociation. We estimate the zero-point fluctuation energy and the vibrational level spacing using harmonic wavefunctions.

```
> assume(alpha>0);
```

```
> Hop:=arg->int(arg*(diff(arg,x,x)+Ueff(x)*arg),x=0..infinity)
> /int(arg^2,x=0..infinity);
```

$$\mathrm{Hop} := arg \to \frac{\displaystyle\int_0^\infty arg \left(\left(\frac{\partial^2}{\partial x^2}\, arg\right) + \mathrm{Ueff}(\,x\,)\, arg\right) dx}{\displaystyle\int_0^\infty arg^2\, dx}$$

```
> Hop(exp(-alpha^2*(x-1)^2/2));
```

$$\frac{\displaystyle\int_0^\infty \%1 \left(-\alpha^{\sim 2}\,\%1 + \alpha^{\sim 4}\,(\,x-1\,)^2\,\%1 + \left(2\,\frac{g^2}{x} - \frac{g^2}{x^2}\right)\%1\right) dx}{\displaystyle\frac{1}{2}\,\frac{\sqrt{\pi}}{\alpha^\sim} + \frac{1}{2}\,\frac{\sqrt{\pi}\,\mathrm{erf}(\,\alpha^\sim\,)}{\alpha^\sim}}$$

$$\%1 := \mathrm{e}^{\left(-1/2\,\alpha^{\sim 2}\,(x-1)^2\right)}$$

```
> Hop4:=arg->int(arg*(diff(arg,x,x)+Ueff4(x)*arg),x=0..
> infinity)/int(arg^2,x=0..infinity):
```

```
> E04:=Hop4(exp(-alpha^2*(x-1)^2/2)):
```

We choose a value of the interaction strength parameter g that leads to approximately a minimum in $-U_{\mathrm{eff}}$ that corresponds to a classical binding energy of $D = 0.11$ in atomic units, and a binding radius of 1.6 ($g^2 = 29181.6^2 D \approx 517$). We obtain a maximum as we display the QM binding energy times a factor, i.e., λ defined above:

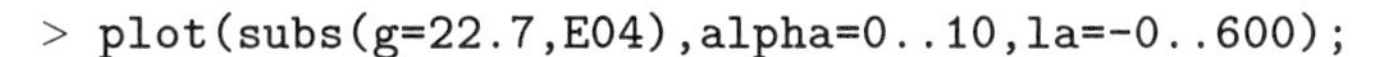

```
> plot(subs(g=22.7,E04),alpha=0..10,la=-0..600);
```

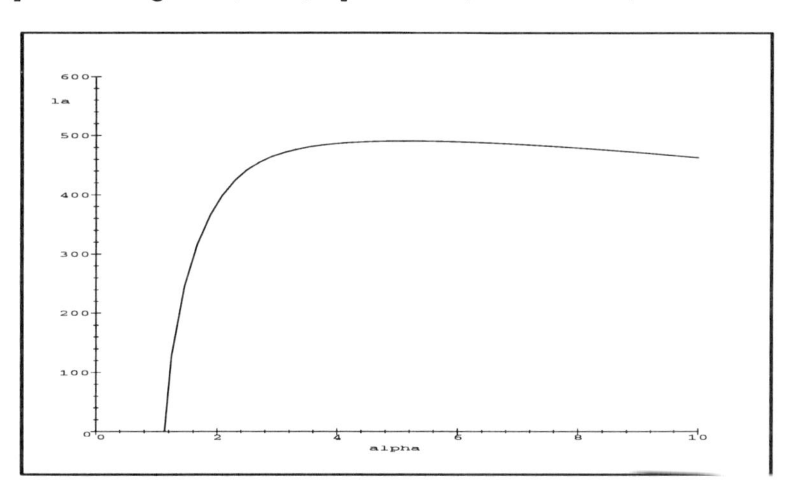

We determine the location of the stationary point to obtain numerically the best value of the wavefunction parameter α for given potential strength $g \approx$ 22.7.

```
> diff(E04,alpha):     fsolve(subs(g=22.7,"),alpha=0..30);
```

$$5.128245667$$

```
> subs(g=22.7,alpha=",E04);
```

$$\left(47.83969567\sqrt{\pi} + 47.83969568\sqrt{\pi}\,\mathrm{erf}(5.128245667) + 48.05268790\,e^{(-26.29890362)}\right) / \left[.09749922925\sqrt{\pi} + .09749922925\sqrt{\pi}\,\mathrm{erf}(5.128245667)\right]$$

```
> E0appr:=evalf(");
```

$$\mathrm{E0appr} := 490.6674243$$

The number above represents the magnitude of the approximate ground-state energy for the nuclear motion, while the next line provides the magnitude of the classical potential at the minimum point.

```
> subs(g=22.7,Ueff4(1));
```

$$515.29$$

Thus, the zero-point energy in the nuclear vibration problem is obtained from

```
> "-E0appr;
                              24.6225757
```

To convert the above answer for the zero-point fluctuations of the vibrating nuclei into eV we need to convert from λ to an energy

```
> "/(2*918*1.6^2)*27.12 eV;
                          .1420726641  eV
```

The latter number shows by how much the dissociation energy estimated on the basis of the minimum in the classical potential is modified due to the zero-point fluctuations. The vibrational excitation energy is given by two times this value, i.e., 0.28 eV. This corresponds to a temperature of 3200°K, according to $E = kT$ and $k = 0.8625 \times 10^{-4}$ eVK^{-1}.

Note that rotational energies have a typical energy spacing that is less by two to three orders of magnitude, while electronic excitations (e.g., singlet–triplet splitting) are larger by one to two orders of magnitude.

We graph the exact and approximate harmonic potentials for the given parameter choice as well as the ground-state energy.

```
> plot({subs(g=22.7,-Ueff(x)),subs(g=22.7,-Ueff4(x)),
> -515.3+24.62},x=0..2,Um=-600..0);
```

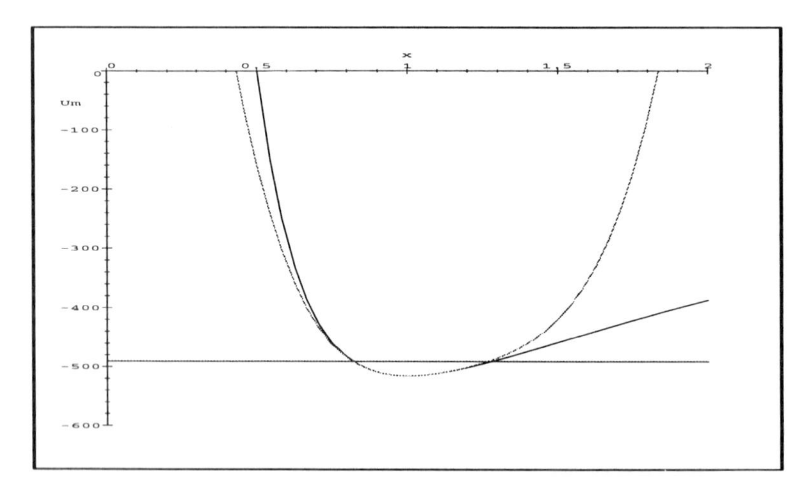

We note that the zero-point fluctuations correspond to $(1/2)\hbar\omega$ in the harmonic approximation. Observe the location of the classical turning points and the estimated location for the first excited level at $(3/2)\hbar\omega$. The harmonic approximation breaks down quickly for increasing n. Note that the classical minimum corresponds in the hydrogen molecule to a few electron volts. The spacing between neighbouring l states (rotational band) is given in atomic units approximately as:

$$2\frac{(l+1)}{2I_0^2} \approx \frac{(l+1)}{3000}\,.$$

In molecular hydrogen H_2, for low l this is of the order of 0.01 eV.

This shows that at room temperature (and below) the kinetic motion of gas molecules (thermal fluctuations) can be converted in collisions to rotational motion. Vibrational modes, on the other hand can only be reached at very high temperatures. At low enough temperatures (50°K and below) the rotational degrees of freedom become inaccessible and the specific heat of the gas changes to a lower value, since the gas loses this energy storage mechanism. Classically the rotational motion is unquantized and this observed phenomenon cannot be explained.

The exact answer to the ro–vib problem with a proper treatment of angular momentum is given by the expression below (n labels the vibrational, l the rotational quantum numbers, units with $\hbar = m_e = 1$ are used, and R_0 denotes the binding radius) [Fl71]:

```
> Eex:=(n,l)->-g^4/(2*mu*R0^2)/(n+1/2+sqrt((l+1/2)^2+g^2))^2;
```

$$\mathrm{Eex} := (n,l) \to -\frac{1}{2}\frac{g^4}{\mu\, R0^2 \left(n+\frac{1}{2}+\mathrm{sqrt}\left(\left(l+\frac{1}{2}\right)^2+g^2\right)\right)^2}$$

This result can be expanded in $1/g^2$ as g^2 is large (proportional to the reduced mass μ).

```
> Eex1:=subs(g=1/eta,Eex(n,l));
```

$$\mathrm{Eex1} := -\frac{1}{2}\frac{1}{\eta^4\,\mu\, R0^2\left(n+\frac{1}{2}+\frac{1}{2}\sqrt{4\,l^2+4\,l+1+4\,\frac{1}{\eta^2}}\right)^2}$$

```
> taylor(g^4*eta^4*Eex1,eta=0);
```

$$-\frac{1}{2}\frac{g^4}{\mu\, R0^2}\eta^2 - \frac{1}{2}\frac{g^4\,(-2\,n-1)}{\mu\, R0^2}\eta^3 -$$
$$\frac{1}{2}\frac{g^4\left(-l^2-l-\frac{1}{4}-\frac{3}{2}(-2\,n-1)\left(n+\frac{1}{2}\right)\right)}{\mu\, R0^2}\eta^4 - \frac{1}{2}g^4$$
$$\left(-\frac{3}{2}(-2\,n-1)\left(\frac{1}{2}l^2+\frac{1}{2}l+\frac{1}{8}\right)-\frac{3}{2}\left(-l^2-l-\frac{1}{4}\right)\left(n+\frac{1}{2}\right)\right.$$
$$\left.+2\,(-2\,n-1)\left(n+\frac{1}{2}\right)^2\right)/(\mu\, R0^2)\eta^5 + \mathrm{O}\left(\eta^6\right)$$

```
> sort(convert(",polynom));
```

$$-\frac{1}{2}\left(-\frac{3}{2}(-2n-1)\left(\frac{1}{2}l^2+\frac{1}{2}l+\frac{1}{8}\right)-\frac{3}{2}\left(-l^2-l-\frac{1}{4}\right)\left(n+\frac{1}{2}\right)\right.$$
$$\left.+2(-2n-1)\left(n+\frac{1}{2}\right)^2\right)g^4\eta^5/(RO^2\mu)$$
$$-\frac{1}{2}\frac{\left(-l^2-l-\frac{1}{4}-\frac{3}{2}(-2n-1)\left(n+\frac{1}{2}\right)\right)g^4\eta^4}{RO^2\mu}$$
$$-\frac{1}{2}\frac{(-2n-1)g^4\eta^3}{RO^2\mu}-\frac{1}{2}\frac{g^4\eta^2}{RO^2\mu}$$

```
> subs(eta=1/g,");
```

$$-\frac{1}{2}\left(-\frac{3}{2}(-2n-1)\left(\frac{1}{2}l^2+\frac{1}{2}l+\frac{1}{8}\right)-\frac{3}{2}\left(-l^2-l-\frac{1}{4}\right)\left(n+\frac{1}{2}\right)\right.$$
$$\left.+2(-2n-1)\left(n+\frac{1}{2}\right)^2\right)/(g\,RO^2\mu)$$
$$-\frac{1}{2}\frac{-l^2-l-\frac{1}{4}-\frac{3}{2}(-2n-1)\left(n+\frac{1}{2}\right)}{\mu\,RO^2}-\frac{1}{2}\frac{(-2n-1)g}{RO^2\mu}$$
$$-\frac{1}{2}\frac{g^2}{RO^2\mu}$$

The last two terms correspond to the classical minimum and the harmonic approximation to the vibrational motion, respectively. We can now go back to the original expansion and improve our understanding of the individual terms:

```
> taylor(g^4*eta^4*Eex1,eta=0,6);
```

$$-\frac{1}{2}\frac{g^4}{\mu\,RO^2}\eta^2-\frac{1}{2}\frac{g^4(-2n-1)}{\mu\,RO^2}\eta^3$$
$$-\frac{1}{2}\frac{g^4\left(-l^2-l-\frac{1}{4}-\frac{3}{2}(-2n-1)\left(n+\frac{1}{2}\right)\right)}{\mu\,RO^2}\eta^4-\frac{1}{2}g^4$$
$$\left(-\frac{3}{2}(-2n-1)\left(\frac{1}{2}l^2+\frac{1}{2}l+\frac{1}{8}\right)-\frac{3}{2}\left(-l^2-l-\frac{1}{4}\right)\left(n+\frac{1}{2}\right)\right.$$
$$\left.+2(-2n-1)\left(n+\frac{1}{2}\right)^2\right)/(\mu\,RO^2)\eta^5+O\left(\eta^6\right)$$

The first term, independent of the quantum numbers, must be the classical potential energy minimum. The second term carries the $(n+1/2)$ dependence characteristic for the harmonic oscillator. The third term contains a pure rotational energy-level dependence $(l+1/2)^2$, as well as anharmonic corrections to the vibrational motion (non-equidistant in the dependence of the energy on n).

Finally, there are couplings between the rotational and vibrational motions caused by the anharmonicity: a mixed dependence on n and l indicates that this coupling occurs. We conclude this section with the remark that a similar rotation–vibration problem occurs in the nuclear many-particle problem [RS80]. The careful treatment of the vibration problem with a potential that is correct to large distances permits us also to connect to the problem of dissociation and nuclear fission.

Exercise 7.3.1: Obtain the estimates for rotational and vibrational excitation energies for the deuterium molecule D_2 in electron volts as well as temperature. The results should be in the vicinity of $T_{\text{rot}} = 44^\circ$ K, and $T_{\text{vib}} = 4500^\circ$ K. Note that the vibrational excitation energies are related to electronic excitation energies (molcculer dissociation) by the factor $\sqrt{m_e/\mu}$, where μ is the reduced mass of the two nuclei involved.

Exercise 7.3.2: A more accurate potential than the simple power-law type discussed above is the so-called Morse potential [Mo29, BM89]:

$$V(R) = A[\exp(-2a(R - R_0)) - 2\exp(-a(R - R_0))]\,.$$

The energy spectrum for this potential can be calculated in closed form. For the oxygen molecule O_2 one can choose the parameters as $A = 5.2$ eV, $R_0 = 0.12$ nm, and $1/a = 27\ \text{nm}^{-1}$. Find the rotational and vibrational excitation energies from an approximate treatment.

7.4 Angular-Momentum Coupling

This section deals with the coupling of angular momenta, and in particular with the calculation of Clebsch-Gordan (CG) coefficients. We consider a composite system of two particles with individual angular momenta with respect to a common origin. The question is how to represent the total angular momentum eigenfunction $Y(JM, j_1, j_2)$ in terms of the products of individual angular momentum eigenfunctions $Y(j_i, m_i)$ (spherical harmonics). The answer can be written as a superposition $(JM) = (j_1 m_1) \otimes (j_2 M_2)$:

$$Y(JM, j_1, j_2) = \sum_{m_1=-j_1}^{j_1} \sum_{m_2=-j_2}^{j_2} C(j_1, j_2, m_1, m_2; J, M) Y(j_1, m_1) Y(j_2, m_2) ,$$

where all three functions Y depend on both particles' polar and azimuthal angles θ_i and ϕ_i. For an introduction to the subject we refer to the intermediate texts on QM cited in Sect. 1.1, as well as to monographs dedicated to angular momentum [Ro57, BS94, Za87, DG90, BV65, BL81, Th94].

We pick some simple examples, e.g., where one angular momentum equals 1/2, or 1. The possible total angular momentum values are $J = |j_1 - j_2|, .., (j_1 + j_2)$, while their projections onto the z axis are denoted by $M = -J, ..., J$.

The CG coefficients can be derived by constructing the matrix representation of $\hat{J}^2$ in the uncoupled basis (in which the CGs are the expansion coefficients) and diagonalizing it. The eigenvectors – once normalized – contain the CG coefficients.

The operator $\hat{J}^2$ is calculated from the angular momentum operators for the two particles as

$$\hat{j}_1^2 + \hat{j}_2^2 + 2\hat{j}_1\hat{j}_2 = \hat{j}_1^2 + \hat{j}_2^2 + 2\hat{j}_{1z}\hat{j}_{2z} + \hat{j}_{1+}\hat{j}_{2-} + \hat{j}_{1-}\hat{j}_{2+} ,$$

with everything diagonal in the uncoupled basis, except the j-plus and j-minus operators that have off-diagonal elements as given by:

$$\langle j, m+1|j_+|j, m\rangle = -\hbar\sqrt{(j-m)(j+m+1)} = \langle j, m|j_-|j, m+1\rangle .$$

We choose the two angular momenta that we wish to couple:

```
> with(linalg):       j1:=3/2;       j2:=1;
```

$$j1 := \frac{3}{2} \qquad j2 := 1$$

We set up an array to hold the matrix representation of the $\hat{J}^2$ operator:

```
> J2:=matrix(2*j2+1,2*j2+1,0):
```

We pick the case of $J = j_1 + j_2$ here, although more important for the calculation is the selection of the z projection of J done below. It is chosen in this example as the most negatively possible M plus $m_2 = j_2$. Any consistent choice for $J = |j_1 - j_2| ... (j_1 + j_2)$, and appropriate projection $M = -J, ..., J$ is allowed.

```
> J:=j1+j2;     M:=-J+j2;
```

$$J := \frac{5}{2} \qquad M := \frac{-3}{2}$$

The matrix representation of $\hat{J}^2$ is set up with the idea in mind that M is fixed, and the coefficients are requested for the expansion of the coupled state

$$(J, M) = \sum_{m_2 = m_2(\min)}^{m_2(\max)} \mathtt{CG}(m_2) \times (j_1 m_1) \otimes (j_2 m_2) \, .$$

One needs only to sum over m_2 as m_1 is fixed by the constraint $m_1 + m_2 = M$. Due to the predefined values the program calculates the CGs for the coupling $(3/2) \otimes (1) = (5/2)$ and we look at the projection $M = -3/2$.

```
> for m2 from -j2 to j2 do

> i1:=m2+j2+1;

> for m2p from -j2 to j2 do

> i1p:=m2p+j2+1;

> m1:=M-m2;

> m1p:=M-m2p;

> if m2=m2p then J2[i1,i1p]:=j1*(j1+1)+j2*(j2+1)+2*m1*m2;

> else

> if m1=m1p-1 and m2=m2p+1 then J2[i1,i1p]:=sqrt((j1-m1)
> *(j1+m1+1)*(j2-m2+1)*(j2+m2)); fi;

> if m1=m1p+1 and m2=m2p-1 then J2[i1,i1p]:=sqrt((j2-m2)
> *(j2+m2+1)*(j1-m1+1)*(j1+m1)); fi;

> fi;

> od;

> od;
```

$$i1 := 1 \qquad i1 := 2 \qquad i1 := 3$$

```
> evalm(J2);
```

$$\begin{bmatrix} \frac{27}{4} & \sqrt{6} & 0 \\ \sqrt{6} & \frac{23}{4} & 0 \\ 0 & 0 & \frac{3}{4} \end{bmatrix}$$

```
> vec1:=eigenvects(J2);
```

$$\mathrm{vec1} := \left[\frac{35}{4}, 1, \left\{\left[\frac{1}{2}\sqrt{6}\,1\,0\right]\right\}\right], \left[\frac{15}{4}, 1, \left\{\left[-\frac{1}{3}\sqrt{6}\,1\,0\right]\right\}\right], \left[\frac{3}{4}, 1, \{[0\,0\,1]\}\right]$$

We find apparently a set of CGs: one for the J^2 that we selected: $J(J+1) = \frac{35}{4}$, i.e., $J = \frac{5}{2}$ with values that still require normalization with the Euclidean norm, as well as two others for different J.

We can lift the required term upon inspection of the result, but note that the order in which eigenvectors are returned by **eigenvects** is (almost) random and can vary from one Maple session to another (depending on the computational history in the session).

```
> op(vec1[1][3]);
```

$$\left[\frac{1}{2}\sqrt{6}\ 1\ 0\right]$$

```
> map(combine,normalize("));
```

$$\left[\frac{1}{5}\sqrt{15}\ \frac{1}{5}\sqrt{10}\ 0\right]$$

Note that while we have specified the total M, J was never used in the calculation. This is why the other combination leading to the same projection M but different J was also calculated.

For a comparison we list some of the CGs from a tabulation of the coefficients for $j_1 \otimes j_2(=1)$ for arbitrary M (our program is missing this generalization). First, we have to unassign some variables:

```
> j1:='j1':     M:='M':
```

For $m_2 = 1$:

```
> sqrt((j1+M)*(j1+M+1)/((2*j1+1)*(2*j1+2)));
```

$$\sqrt{\frac{(j1 + M)(j1 + M + 1)}{(2\,j1 + 1)(2\,j1 + 2)}}$$

For $m_2 = 0$:

```
> sqrt((j1-M+1)*(j1+M+1)/((2*j1+1)*(j1+1)));
```

$$\sqrt{\frac{(j1 - M + 1)(j1 + M + 1)}{(2j1 + 1)(j1 + 1)}}$$

For $m_2 = -1$:

```
> sqrt((j1-M)*(j1-M+1)/((2*j1+1)*(2*j1+2)));
```

$$\sqrt{\frac{(j1 - M)(j1 - M + 1)}{(2j1 + 1)(2j1 + 2)}}$$

We note that for a comparison one has to figure out in which order the matrix was assembled from the various m_2 contributions, $m_2 = -j_2, ..., j_2$. This identifies the coefficients uniquely.

The coupling of angular momenta arises in various contexts in QM. For (effective) one-electron atoms the coupling of orbital angular momentum and spin is important, particularly in the context of a relativistic treatment of spin-$\frac{1}{2}$ fermions due to the presence of a spin–orbit coupling in the Hamiltonian. One can show in that case[Gr94] that the Dirac Hamiltonian commutes with the total angular momentum squared operator $\hat{j}^2$, where $\hat{j} = \hat{l} \oplus \hat{s}$, but not with $\hat{l}^2$. Thus, the relativistic single-electron eigenstates are labeled by j and the z projection m_j. The connection with the Schrödinger representation (that can be supplemented with an $\mathbf{l} \cdot \mathbf{s}$ interaction term) can be established by coupling the $(\hat{l}^2, \hat{s}^2)$ eigenstates.

In the non-relativistic many-electron (many-particle) problem the coupling of multiple angular momenta is important, as the total angular momenta $\hat{J}^2$ and $\hat{J}_z$ are conserved, while the individual particle angular momenta are not, due to the presence of particle–particle interactions in the Hamiltonian. Nevertheless, since independent particle models make use of a single-particle basis, a representation in the uncoupled basis is required. For the coupling of more than two angular momenta one defines additional coupling coefficients, named $(3j)$, $(6j)$ symbols, etc. The reader is referred to the special literature on the subject cited in the beginning of the section.

Exercise 7.4.1: Calculate some of the CG coefficients given at the end of the section in functional form, i.e., for $J = j_1 \otimes 1$.

7.5 Two-Particle Correlations

This section presents an implementation of a symbolic Lanczos algorithm to the problem of two coupled anharmonic oscillators (AHOs). The objective of the exercise is to show on a model problem how a two-particle interaction calls for a two-particle wavefunction that is more sophisticated than the Hartree-Fock ansatz of an antisymmetrized product (Slater determinant) of single-particle orbitals.

This section requires as a prerequisite the study of the Lanczos method itself, which is a powerful tridiagonalization technique for Hamiltonians that depend on many degrees of freedom. The method was introduced in Sect. 2.4 for a single-particle problem. The program given in this section represents a modification of the one discussed before. It allows the use of a non-Gaussian basis and shows an improvement in performance by tabulating integrals beforehand with subsequent look-up as required [Ka94]. The problem from Sect. 2.4 is extended to two coupled oscillators with an obvious extension to N coupled oscillators.

Our aim is to study correlations in a two-particle system. The Lanczos algorithm generates a correlated wavefunction from an independent-particle ansatz that is used as a seed for the iteration. We note that the problems in Sect. 7.1 and 7.2 were solved approximately by such an ansatz.

We may need to increase the numerical accuracy if many iterations are considered. For a start, however, the default accuracy will suffice.

```
> Digits:=10:      assume(n,integer);    N:=5:
```

The variable N denotes the maximum number of Lanczos iterations and determines the size of the tridiagonal Hamiltonian matrix. We choose a modest number to keep the computing time within reasonable limits. In principle, one has to perform a convergence analysis on the number of iterations.

We choose a maximum size for the table of integrals. More Lanczos iterations induce higher powers of x and y in the wavefunctions (Lanczos vectors) and may call for an increase of the maximum.

```
> npmax:=128:      with(linalg):
```

```
> Hdiag:=vector(N,0):    Hodg:=vector(N,0):
```

Now we define the anharmonic oscillator potential for the non-interacting single-particle problem.

```
> Vop:=(arg)->1/2*arg^2+2*arg^4+1/2*arg^6:
```

The complete two-particle interaction contains a coupling term of type cxy. Note that such an interaction can be diagonalized exactly for simple harmonic oscillators in the single-particle problem, but not for the AHO potential considered here. We choose a large value for the two-particle interaction coupling constant $c = 5$ to strongly emphasize its effect.

```
> Vop2:=(arg1,arg2)->Vop(arg1)+Vop(arg2)+5*arg1*arg2:
```

```
> Hop2:=(a,x,y)->-1/2*diff(a,x$2)-1/2*diff(a,y$2)+Vop2(x,y)*a:
```

We pick a harmonic oscillator ground-state wavefunction for each particle to form a simple product state as a seed for the Lanczos iteration.

```
> alpha:=2.;      psi0:=x->exp(-alpha*x^2);
```

$$\alpha := 2. \qquad \psi 0 := x \to e^{(-\alpha x^2)}$$

The entry of α as a floating-point number will force some calculations to be of numerical form. I suggest you experiment with this, i.e., use alternatively an exact number (entered without the period). In the end the algorithm performs a numerical diagonalization of the Hamiltonian matrix, i.e., one can choose at which point to make the conversion to numerical evaluation.

Now define the two-particle wavefunction that seeds the Lanczos iteration.

```
> psi[1]:=(x,y)->psi0(x)*psi0(y):
```

The following procedure pulls from an expression supplied with the variables $x, y, z, \ldots$ the overall factor, as well as the powers for each expression. It is required to perform the required integrals in the matrix elements by look-up.

```
> cextr:=proc() global npmax;
> for i from 1 to nargs-1 do
> cof:=0;
> for n from 0 to npmax while cof=0 do
> cof:=coeff(args[1],args[i+1],n);      od;
> ncof[i]:=n-1;      od;
> fact:=args[1];
> for i from 1 to nargs-1 do
> fact:=fact/args[i+1]^ncof[i];      od;
> RETURN(fact,seq(ncof[nu], nu=1..nargs-1));      end;
```

Now we assemble the 2D integral that is required for the matrix elements of the Hamiltonian. The procedure for the inner product in the two-dimensional space makes use of 1D integrals to be stored in the list `Intt` and uses the power extraction procedure `cextr` defined above.

```
> Dotpro2:=proc();
> arg:=expand(simplify(args[1]/psi[1](x,y)^2));
> sum1:=0;
```

```
> for i from 1 to nops(arg) do
> expr:=op(i,arg);
> cextri:=cextr(expr,x,y);
> sum1:=sum1+cextri[1]*Intt[cextri[2]]*Intt[cextri[3]];
> od:     end;
```

Now we set up a table of integrals. It may be advantageous to use a numerical evaluation, i.e., to replace the `int()` command by an `evalf(Int())` command.

```
> for i from 0 to npmax do
> Intt[i]:=int(x^i*psi0(x)^2,x=-infinity..infinity);    od:
```

The table has been assembled. We try a simple value:

```
> Intt[0];
```

$$.50000000\sqrt{\pi}$$

We prepare for the Lanczos iteration. Note that the program is identical to the single-particle problem dealt with in Sect. 2.4, except that the inner product is no longer done by internal Maple integration.

```
> N1p:=sqrt(Dotpro2(psi[1](x,y)^2));
```

$$N1p := .50000000\sqrt{\pi}$$

```
> phi[1]:=unapply(psi[1](x,y)/(N1p),x,y);
```

$$\phi_1 := (x,y) \to 2.0000000\,\frac{e^{(-2.\,x^2)}\,e^{(-2.\,y^2)}}{\sqrt{\pi}}$$

```
> psi[2]:=unapply(simplify(Hop2(phi[1](x,y),x,y)),x,y);
```

$$\begin{aligned}\psi_2 := (x,y) \to\ & .50000000\,10^{-9}(.90270333\,10^{10} - .16925688\,10^{11}\,x^2 \\ & - .16925688\,10^{11}\,y^2 + .45135167\,10^{10}\,x^4 + .11283792\,10^{10}\,x^6 \\ & + .45135167\,10^{10}\,y^4 + .11283792\,10^{10}\,y^6 + .11283792\,10^{11}\,x\,y) \\ & e^{(-2.\,x^2-2.\,y^2)}\end{aligned}$$

```
> Hd:=Dotpro2(phi[1](x,y)*psi[2](x,y));    Hdiag[1]:=Hd;
```

$$Hd := .74541710\,\pi \qquad Hdiag_1 := .74541710\,\pi$$

```
> xi[2]:=unapply(simplify(psi[2](x,y)-Hd*phi[1](x,y)),x,y);
```

$$\xi_2 := (x, y) \to 1.8710819\, e^{(-2.\,x^2-2.\,y^2)} - 8.4628438\, e^{(-2.\,x^2-2.\,y^2)}\, x^2 - 8.4628438\, e^{(-2.\,x^2-2.\,y^2)}\, y^2 + 2.2567583\, e^{(-2.\,x^2-2.\,y^2)}\, x^4 + .56418958\, e^{(-2.\,x^2-2.\,y^2)}\, x^6 + 2.2567583\, e^{(-2.\,x^2-2.\,y^2)}\, y^4 + .56418958\, e^{(-2.\,x^2-2.\,y^2)}\, y^6 + 5.6418958\, e^{(-2.\,x^2-2.\,y^2)}\, x\, y$$

```
> N2:=sqrt(Dotpro2(xi[2](x,y)^2));      Hodg[2]:=N2;
```

$$N2 := .88977029\sqrt{\pi} \qquad \mathrm{Hodg}_2 := .88977029\sqrt{\pi}$$

```
> phi[2]:=unapply(simplify(xi[2](x,y)/N2),x,y);
```

$$\phi_2 := (x, y) \to .10000000\, 10^{-8}(.11864241\, 10^{10} - .53661584\, 10^{10}\, x^2 - .53661584\, 10^{10}\, y^2 + .14309756\, 10^{10}\, x^4 + .3577439\, 10^{9}\, x^6 + .14309756\, 10^{10}\, y^4 + .3577439\, 10^{9}\, y^6 + .35774389\, 10^{10}\, x\, y)\, e^{(-2.\,x^2-2.\,y^2)}$$

```
> for n from 2 to N-1 do
> psi[n+1]:=unapply(simplify(Hop2(phi[n](x,y),x,y)),x,y);
> Hd:=Dotpro2(phi[n](x,y)*psi[n+1](x,y));
> Hdiag[n]:=Hd;
> xi[n+1]:=unapply(simplify(psi[n+1](x,y)-Hd*phi[n](x,y)-
> Hodg[n]*phi[n-1](x,y)),x,y);
> N2:=sqrt(Dotpro2(xi[n+1](x,y)^2));
> Hodg[n+1]:=N2;
> phi[n+1]:=unapply(simplify(xi[n+1](x,y)/N2),x,y);
> od:
```

The completion of the loop takes a few minutes. Now we complete the matrix by calculation of the last diagonal element:

```
> psi[N+1]:=unapply(simplify(Hop2(phi[N](x,y),x,y)),x,y):
> Hd:=Dotpro2(phi[N](x,y)*psi[N+1](x,y)):
> Hdiag[N]:=Hd:      Hmat:=matrix(N,N,0):
```

The tridiagonal Hamiltonian matrix is assembled from the calculated entries.

```
> for i from 1 to N do
> for j from 1 to N do
> if i=j then Hmat[i,j]:=Hdiag[i] fi;
> if i=j-1 then Hmat[i,j]:=Hodg[j] fi;
> if i=j+1 then Hmat[i,j]:=Hodg[i] fi;
> od:    od:
```

For a simple display we convert to floating-point precision.

```
> evalm(map(evalf,Hmat));
```

$$\begin{bmatrix} 2.34180, & 1.57708, & 0, & 0, & 0 \\ 1.57708, & 7.79927, & 3.89866, & 0, & 0 \\ 0, & 3.898659494, & 14.92104693, & 11.03953111, & 0 \\ 0, & 0, & 11.03953, & 79.16932, & 59.75714 \\ 0, & 0, & 0, & 59.75714, & 173.91037 \end{bmatrix}$$

Next we diagonalize the matrix numerically, including a calculation of the eigenvectors. Note that subsequent calls to **Eigenvals** may return wrong eigenvectors if the array containing the vectors is not reset as done in the next line.

```
> vecs:='vecs':     evalf(Eigenvals(Hmat,vecs),6);
```

$$[1.80238\ 6.10709\ 14.4155\ 52.8984\ 202.918\ ]$$

The ground-state energy is lower than the sum of two single-particle energies corresponding to uncorrelated motion in the x and y direction, which for the potential without xy coupling term would contribute a value of 1 respectively, i.e., for $V(x,y) = V(x) + V(y)$ we have: $E_2^{(0)} = 1 + 1$.

We try to understand why the energy is lowered, given that we added to the Hamiltonian a seemingly positive interaction $5xy$. The answer can be obtained from an inspection of the classical potential: if $\{x > 0$ and $y < 0\}$, and also if $\{x < 0$ and $y > 0\}$ this interaction is attractive. This is demonstrated below in the contour plot of the potential.

The product wavefunction $\psi_0(x)\psi_0(y)$ can sample this interaction, but it has no means to push electron density towards those regions where the interaction is attractive, i.e., it will also sample the repulsive regions. For the product wavefunction the expectation value $\langle xy \rangle$ vanishes exactly due to symmetry.

The correlated wavefunction lowers its energy by avoiding the region near the line $x = y$ (for positive and negative x), while accumulating energy density around $y = -x$.

```
> with (plots):    nu:='nu':
```

We assemble the ground state from the solution of the linear algebra eigenvalue problem. The eigenvector array **vecs** contains the expansion coefficients in its columns. A superposition of the stored Lanczos state vectors (basis functions **phi**) is formed. We have unassigned the dummy variable **nu** to remind the reader that Maple can be stubborn in using such a variable in an internal loop construct, if it has already a pre-set value.

```
> phi0:=sum(vecs[nu,1]*phi[nu](x,y),nu=1..N):
```

First we graph the potential to show the regions of lower classical potential along $y = -x$.

```
> contourplot(Vop2(x,y),x=-1.5..1.5,y=-1.5..1.5,color=red,
> scaling=CONSTRAINED,axes=BOXED);
```

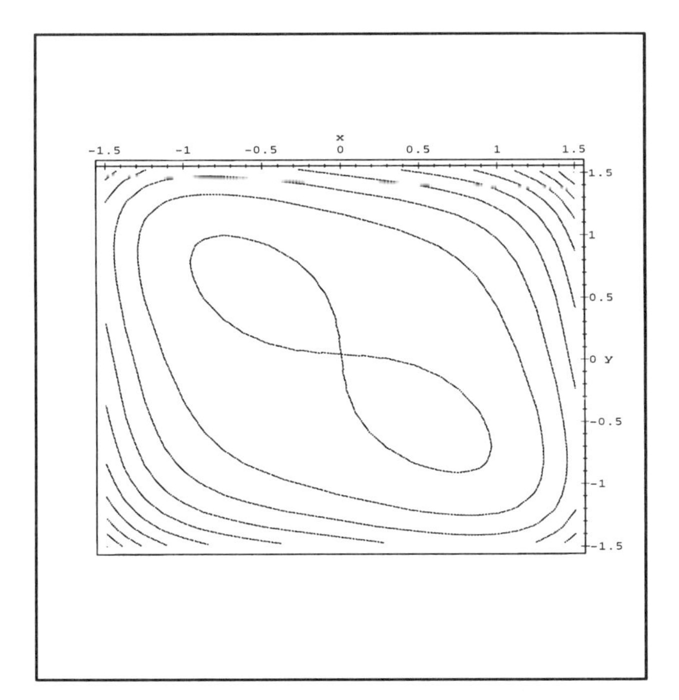

Next we plot the Lanczos start vector which is a product state of the same functions of x and y separately. Note that the scaling on the graphs can distort the $x - y$ symmetry that exists in the problem. We suppress the graph produced by the next command, but emphasize that the start vector is symmetric in x and y as it represents an IPM wavefunction of product type.

```
> contourplot(phi[1](x,y),x=-1.5..1.5,y=-1.5..1.5,color=blue,
> scaling=CONSTRAINED,axes=BOXED):
```

We graph the ground state after Lanczos diagonalization (from a few iterations, it may not be fully converged yet!):

```
> contourplot(phi0,x=-1.5..1.5,y=-1.5..1.5,color=black,
> scaling=CONSTRAINED,axes=BOXED);
```

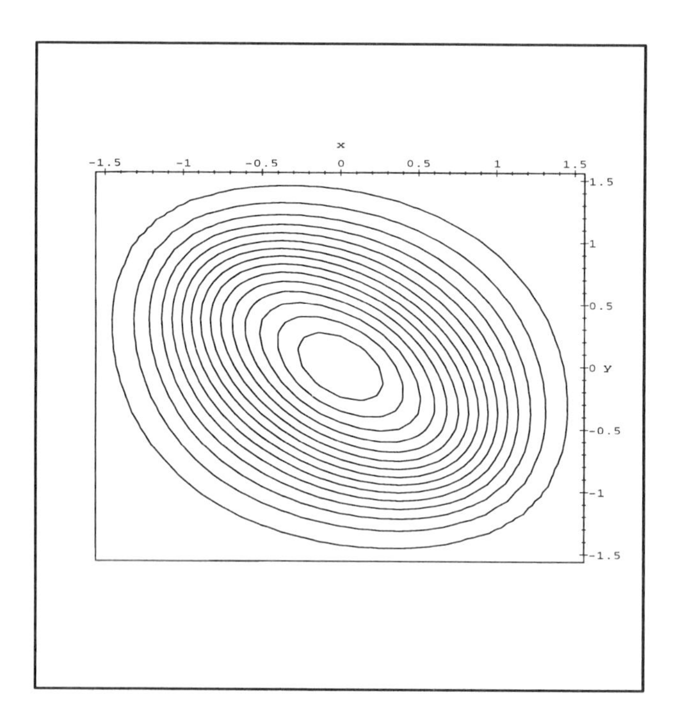

The wavefunction shows a clear deformation such that more probability density is concentrated around $y = -x$ and the region along $y = x$ is avoided. How much deformation occurs depends on the following interplay. The two-particle interaction favours density along $y = -x$. The kinetic energy, however, 'dislikes' curvature (large positive values result). The potential single-particle energy $V(x) + V(y)$ should not acquire a large value either! This limits the role played by the coupling term in this example.

In atomic physics, wavefunction correlations are caused by the mutual Coulomb repulsion. Improvements over the single-particle Hartree-Fock ansatz come from the flexibility for the electrons to avoid each other.

Exercise 7.5.1: Solve the problem of two coupled harmonic oscillators $V(x_1, x_2) = \frac{1}{2}m\omega^2[x_1^2 + x_2^2] + cx_1x_2$ exactly by transforming to new coordinates x_1+x_2 and x_1-x_2. Why is a product ansatz exact in these coordinates?

8. Special Functions

8.1 Hermite Polynomials

We derive the solution of the Hermite differential equation that occurs in the 1D harmonic oscillator problem. The idea of how to generate recursion relations 'by hand' in a symbolic computing environment for polynomials that satisfy an ordinary differential equation (ODE) can be found in the pioneering book by J. Feagin [Fe94]. There are various options to attack the problem in Maple. Sometimes one can find solutions in terms of built-in special functions, such as the Bessel functions, or the hypergeometric function, that are known to Maple. These can be found by a straight call to **dsolve**. More often, one is, however, unsuccessful with such a direct attempt and the expression of the solution in terms of the hypergeometric function is not necessarily very useful for a further understanding. It is possible to find solutions in series form from **dsolve**, which is obtained by invoking the **series** option for ODEs that are linear and have polynomial coefficients. The series solution may be helpful in gaining some understanding, and more importantly can be used directly to extract recursion relations for the coefficients of the series. This is explained in detail in the text by A. Heck [He93, chapter 16], and we will not use that method in this chapter, as we wish to emphasize how recursion relations are derived. For practical purposes, however, the use of the **powseries** package and the command for a truncated power series **tpsform** is generally recommended. There are help pages in Maple for the package and the commands mentioned above.

We define the Hermite differential equation

$$\frac{\mathrm{d}^2 H}{\mathrm{d}z^2} - 2z\frac{\mathrm{d}H}{\mathrm{d}z} + 2nH(z) = 0\,,$$

with $n = 0, 1, 2, ...$ and load the orthogonal polynomial package for comparison purposes.

```
> with(orthopoly):      assume(n,integer);
> HermiteEq:=diff(Hf(z),z,z)-2*z*diff(Hf(z),z)+2*n*Hf(z)=0;
```

$$\mathrm{HermiteEq} := \left(\frac{\partial^2}{\partial z^2}\,\mathrm{Hf}(\,z\,)\right) - 2\,z\,\left(\frac{\partial}{\partial z}\,\mathrm{Hf}(\,z\,)\right) + 2\,n\tilde{}\;\mathrm{Hf}(\,z\,) = 0$$

We can let Maple's **dsolve** facility find symmetric and anti-symmetric solutions. For $n = 0$ and symmetric conditions $H(0) = 1, H'(0) = 0$ we have

```
> dsolve({subs(n=0,HermiteEq),Hf(0)=1,D(Hf)(0)=0},Hf(z));
```

$$\mathrm{Hf}(z) = 1$$

For $n = 2$ and the same symmetric conditions we find

```
> dsolve({subs(n=2,HermiteEq),Hf(0)=1,D(Hf)(0)=0},Hf(z));
```

$$\mathrm{Hf}(z) = -2z^2 + 1$$

For $n = 1$ we try with anti-symmetric conditions at $z = 0$, i.e., $H(0) = 0, H'(0) = 1$:

```
> dsolve({subs(n=1,HermiteEq),Hf(0)=0,D(Hf)(0)=1},Hf(z));
```

$$\mathrm{Hf}(z) = z$$

Now we try the case $n = 3$ with anti-symmetric conditions:

```
> sol3:=dsolve({subs(n=3,HermiteEq),Hf(0)=0,D(Hf)(0)=1},Hf(z));
```

$$\mathrm{sol3} :=$$

Maple's failure to generate a solution is disappointing.

Let's try to generate the series solution by hand (rather than specifying `series` as a `dsolve` option). In principle, we should write an infinite series.

```
> psi(z):=sum(c[n,k]*z^k,k=0..infinity);
```

$$\psi(z) := \sum_{k=0}^{\infty} c_{n\tilde{},k}\, z^k$$

```
> subs(Hf(z)=psi(z),HermiteEq);
```

$$\frac{\partial^2}{\partial z^2}\left(\sum_{k=0}^{\infty} c_{n\tilde{},k}\, z^k\right) - 2z\left(\frac{\partial}{\partial z}\left(\sum_{k=0}^{\infty} c_{n\tilde{},k}\, z^k\right)\right) + 2n\tilde{}\left(\sum_{k=0}^{\infty} c_{n\tilde{},k}\, z^k\right) = 0$$

```
> HEq1:=simplify(");
```

$$\mathrm{HEq1} := \sum_{k=0}^{\infty}\left(c_{n\tilde{},k}\, z^{(k-2)}\, k^2 - c_{n\tilde{},k}\, z^{(k-2)}\, k\right) - 2z\left(\sum_{k=0}^{\infty} c_{n\tilde{},k}\, z^{(k-1)}\, k\right)$$
$$+ 2n\tilde{}\left(\sum_{k=0}^{\infty} c_{n\tilde{},k}\, z^k\right) = 0$$

The trick is now to combine the sums and to isolate the powers of z such as to equate the individual coefficients of z^k to zero.

```
> HEq2:=collect(",z);
```

$$\mathrm{HEq2} := \sum_{k=0}^{\infty} \left(c_{n\tilde{},k}\, z^{(k-2)}\, k^2 - c_{n\tilde{},k}\, z^{(k-2)}\, k \right) - 2\,z \left(\sum_{k=0}^{\infty} c_{n\tilde{},k}\, z^{(k-1)}\, k \right) + 2\,n\tilde{} \left(\sum_{k=0}^{\infty} c_{n\tilde{},k}\, z^k \right) = 0$$

```
> coeff(HEq2,z,k):     simplify(");
```

$$\mathrm{coeff}\left(\sum_{k=0}^{\infty} \left(c_{n\tilde{},k}\, z^{(k-2)}\, k^2 - c_{n\tilde{},k}\, z^{(k-2)}\, k \right) - 2\,z \left(\sum_{k=0}^{\infty} c_{n\tilde{},k}\, z^{(k-1)}\, k \right) + 2\,n\tilde{} \left(\sum_{k=0}^{\infty} c_{n\tilde{},k}\, z^k \right) = 0, z, k \right)$$

Maple can't work this out on its own without additional help. Let's try the terms one-by-one and learn how to decompose expressions:

```
> op(1,HermiteEq);          op(1,op(1,HermiteEq));
```

$$\left(\frac{\partial^2}{\partial z^2} \mathrm{Hf}(\,z\,) \right) - 2\,z \left(\frac{\partial}{\partial z} \mathrm{Hf}(\,z\,) \right) + 2\,n\tilde{}\ \mathrm{Hf}(\,z\,) \qquad \frac{\partial^2}{\partial z^2} \mathrm{Hf}(\,z\,)$$

```
> Term1:=subs(Hf(z)=psi(z),op(1,op(1,HermiteEq)));
```

$$\mathrm{Term1} := \frac{\partial^2}{\partial z^2} \left(\sum_{k=0}^{\infty} c_{n\tilde{},k}\, z^k \right)$$

```
> Term1:=simplify(Term1);
```

$$\mathrm{Term1} := \sum_{k=0}^{\infty} \left(c_{n\tilde{},k}\, z^{(k-2)}\, k^2 - c_{n\tilde{},k}\, z^{(k-2)}\, k \right)$$

How can we shift the sum? Obviously $k = 0$ doesn't contribute. In fact, $k = 0$ and $k = 1$ are meaningless as the differentiation rules that Maple employed are valid only from $k = 2$ onwards.

We wish to drop the first two terms and then to relabel such that the polynomial starts properly with `kp` $= k' = 0$. How can we accomplish this? We practice some more anatomy of expressions in Maple:

```
> op(1,Term1);         op(2,Term1);
```

$$c_{n\tilde{},k}\, z^{(k-2)}\, k^2 - c_{n\tilde{},k}\, z^{(k-2)}\, k \qquad k = 0..\infty$$

```
> nops(op(2,Term1)); op(1,op(2,Term1)); op(2,op(2,Term1));
```

$$2 \qquad k \qquad 0..\infty$$

```
> Term1p:=sum(op(1,Term1),k=1..infinity);
```

$$\mathrm{Term1p} := \sum_{k=1}^{\infty} \left(c_{n^{\sim},k}\, z^{(k-2)}\, k^2 - c_{n^{\sim},k}\, z^{(k-2)}\, k \right)$$

Now that we know how to isolate the pieces, we can perform the index shift by a substitution.

```
> Term1p:=sum(subs(k=k+2,op(1,Term1)),k=0..infinity);
```

$$\mathrm{Term1p} := \sum_{k=0}^{\infty} \left(c_{n^{\sim},k+2}\, z^k\, (k+2)^2 - c_{n^{\sim},k+2}\, z^k\, (k+2) \right)$$

We perform the corresponding shift in the first derivative term.

```
> Term2:=subs(Hf(z)=psi(z),op(2,op(1,HermiteEq)));
```

$$\mathrm{Term2} := -2\, z \left(\frac{\partial}{\partial z} \left(\sum_{k=0}^{\infty} c_{n^{\sim},k}\, z^k \right) \right)$$

```
> Term2:=simplify(Term2);
```

$$\mathrm{Term2} := -2\, z \left(\sum_{k=0}^{\infty} c_{n^{\sim},k}\, z^{(k-1)}\, k \right)$$

```
> nops(Term2);  op(1,Term2);  op(2,Term2);  op(3,Term2);
```

$$3 \qquad -2 \qquad z \qquad \sum_{k=0}^{\infty} c_{n^{\sim},k}\, z^{(k-1)}\, k$$

```
> Term2p:=-2*z*sum(subs(k=k+1,op(1,op(3,Term2))),
> k=0..infinity);
```

$$\mathrm{Term2p} := -2\, z \left(\sum_{k=0}^{\infty} c_{n^{\sim},k+1}\, z^k\, (k+1) \right)$$

```
> Term1p+Term2+subs(Hf(z)=psi(z),op(3,op(1,HermiteEq)));
```

$$\sum_{k=0}^{\infty} \left(c_{n^{\sim},k+2}\, z^k\, (k+2)^2 - c_{n^{\sim},k+2}\, z^k\, (k+2) \right) - 2\, z \left(\sum_{k=0}^{\infty} c_{n^{\sim},k}\, z^{(k-1)}\, k \right)$$

$$+ 2\,n\tilde{}\left(\sum_{k=0}^{\infty} c_{n\tilde{},k}\, z^k\right)$$

We can now collect the left-hand side of the Hermite equation with appropriately running dummy indices k.

```
> HeLHS:=simplify(");
```

$$\mathrm{HeLHS} := \sum_{k=0}^{\infty} \left(c_{n\tilde{},k+2}\, z^k\, k^2 + 3\, c_{n\tilde{},k+2}\, z^k\, k + 2\, c_{n\tilde{},k+2}\, z^k\right)$$
$$- 2\,z\left(\sum_{k=0}^{\infty} c_{n\tilde{},k}\, z^{(k-1)}\, k\right) + 2\,n\tilde{}\left(\sum_{k=0}^{\infty} c_{n\tilde{},k}\, z^k\right)$$

We study how to dissect this expression. Our objective is to collect the expression and to obtain a comparison of coefficients of powers z^k.

```
> nops(");          op(1,HeLHS);
```

$$3 \qquad \sum_{k=0}^{\infty} \left(c_{n\tilde{},k+2}\, z^k\, k^2 + 3\, c_{n\tilde{},k+2}\, z^k\, k + 2\, c_{n\tilde{},k+2}\, z^k\right)$$

```
> op(1,op(1,HeLHS));
```

$$c_{n\tilde{},k+2}\, z^k\, k^2 + 3\, c_{n\tilde{},k+2}\, z^k\, k + 2\, c_{n\tilde{},k+2}\, z^k$$

```
> op(1,op(3,op(2,HeLHS)));    op(1,op(3,op(3,HeLHS)));
```

$$c_{n\tilde{},k}\, z^{(k-1)}\, k \qquad c_{n\tilde{},k}\, z^k$$

We are ready to extract the coefficient of z^k:

```
> op(1,op(1,HeLHS))-2*z*op(1,op(3,op(2,HeLHS)))+
> 2*n*op(1,op(3,op(3,HeLHS)));
```

$$c_{n\tilde{},k+2}\, z^k\, k^2 + 3\, c_{n\tilde{},k+2}\, z^k\, k + 2\, c_{n\tilde{},k+2}\, z^k - 2\,z\, c_{n\tilde{},k}\, z^{(k-1)}\, k + 2\,n\tilde{}\, c_{n\tilde{},k}\, z^k$$

The recursion relation is obtained by first removing z^k

```
> RecRel:=simplify("/z^k);
```

$$\mathrm{RecRel} := c_{n\tilde{},k+2}\, k^2 + 3\, c_{n\tilde{},k+2}\, k + 2\, c_{n\tilde{},k+2} - 2\, c_{n\tilde{},k}\, k + 2\,n\tilde{}\, c_{n\tilde{},k}$$

and then solving for $c_{n,k+2}$. We remember that the right-hand side (RHS) of the equation equals zero, and thus:

```
> RRrhs:=simplify(solve(RecRel,c[n,k+2]));
```

$$\text{RRrhs} := 2\,\frac{c_{n\tilde{\ },k}\,(\,k - n\tilde{\ }\,)}{k^2 + 3\,k + 2}$$

We see that the recursion relation connects only every second coefficient. This forms the basis for the generation of polynomials with purely even or purely odd terms (symmetric and anti-symmetric polynomials). It is convenient to shift the index k back by two to determine $c_{n,k}$ in terms of $c_{n,k-2}$.

```
> simplify(subs(k=k-2,RRrhs));
```

$$2\,\frac{c_{n\tilde{\ },k-2}\,(\,k - 2 - n\tilde{\ }\,)}{k\,(\,k-1\,)}$$

Now we can make the following argument:

The termination of the series solution is required in order to obtain a normalizable wavefunction. One can show that without the truncation to a finite polynomial the infinite series blows up at infinity faster than $\exp(-z^2/2)$ attenuates. A termination of the recursion (as derived heuristically in Sect. 1.2) is possible, if for given, chosen n at a certain value of k, namely for $k = n+2$, the recursion stops due to $c_{n,k} = 0$. All subsequent coefficients also vanish. To calculate the recursion, a choice of start value $c_{n,0}$ is required (resulting in even-power terms only), or alternatively a non-vanishing $c_{n,1}$ to generate odd-order terms. One can see that as the power series terminates at $k = n$, an even or odd polynomial is generated. The start value $c_{n,0}$ or $c_{n,1}$, respectively, normalizes the polynomial.

We now define the recursion to generate a Hermite polynomial:

```
> Herm:=proc() global c;

> n:=args[1]; z:=args[2];

> if type(n,even)=true then i0:=0 else if type(n,odd)=true
> then i0:=1 else RETURN (`n not integer in Herm`,n) fi; fi;

> c[n,i0]:=1; sumH:=c[n,i0]*z^i0;

> for k from i0+2 by 2 to n do:

> c[n,k]:=2*c[n,k-2]*(k-2-n)/(k*(k-1));

> sumH:=sumH+c[n,k]*z^k;

> od;     sumH;     end:
```

We try the procedure and list for comparison the result from the built-in Hermite polynomial `H(n,x)`.

```
> Herm(5,x);       H(5,x);
```

$$x - \frac{4}{3}\,x^3 + \frac{4}{15}\,x^5 \qquad\qquad 32\,x^5 - 160\,x^3 + 120\,x$$

```
> Herm(6,y);        H(6,y);
```

$$1 - 6\,y^2 + 4\,y^4 - \frac{8}{15}\,y^6 \qquad 64\,y^6 - 480\,y^4 + 720\,y^2 - 120$$

```
> simplify(H(6,y)/Herm(6,y));
```

$$-120$$

By convention the Hermite polynomials are normalized such that the leading power carries a factor of 2^n. We could fix our procedure by dividing the final polynomial by $c_{n,n}$ and multiplying by 2^n.

We wish to investigate the asymptotic behaviour of the series generated by the recursion. The ratio of two neighbouring terms for large k is

```
> ratio:=2*(k-2-n)/(k*(k-1));
```

$$\mathrm{ratio} := 2\,\frac{k - 2 - n\tilde{}}{k\,(k-1)}$$

```
> series(ratio,k=infinity);
```

$$2\,\frac{1}{k} + \frac{-2 - 2\,n\tilde{}}{k^2} + \frac{-2 - 2\,n\tilde{}}{k^3} + \frac{-2 - 2\,n\tilde{}}{k^4} + \frac{-2 - 2\,n\tilde{}}{k^5} + \mathrm{O}\left(\frac{1}{k^6}\right)$$

Consider now the Taylor coefficient of $\exp(z^2)$:

```
> ratioT:=n->simplify((subs(z=0,diff(exp(z^2),z$n))/n!)/
> (subs(z=0,diff(exp(z^2),z$(n-2))/(n-2)!)));
```

$$\mathrm{ratioT} := n \to \mathrm{simplify}\left(\frac{\mathrm{subs}(\,z = 0, \mathrm{diff}(\,\mathrm{e}^{(\,z^2\,)}, z\,\$\,n\,)\,)}{n!\,\mathrm{subs}\left(z = 0, \dfrac{\mathrm{diff}(\,\mathrm{e}^{(\,z^2\,)}, z\,\$\,(\,n-2\,)\,)}{(\,n-2\,)!}\right)}\right)$$

```
> ratioT(10);    ratioT(20);    ratioT(50);
```

$$\frac{1}{5} \qquad \frac{1}{10} \qquad \frac{1}{25}$$

Apparently the Taylor series for $\exp(z^2)$ in leading order has precisely the behaviour

$$\frac{a_{k+2}}{a_k} = \frac{2}{k}\,.$$

It has only even powers as the function is symmetric in z.

We have completed the argument that the only way to obtain normalizable eigenfunctions for the harmonic oscillator is by truncating the series ansatz such that the recursion terminates and produces a finite-order polynomial. We proceed with a demonstration of some properties of the Hermite polynomials.

– (1) The Rodrigues formula to generate the Hermite polynomials:

```
> HRod:=(n,z)->expand((-1)^n*exp(z^2)*diff(exp(-z^2),z$n));
```

$$\mathrm{HRod} := (n, z) \rightarrow \mathrm{expand}\left((-1)^n \, e^{(z^2)} \, \mathrm{diff}(e^{(-z^2)}, z \, \$ \, n)\right)$$

```
> HRod(3,x);        H(3,x);
```

$$-12\,x + 8\,x^3 \qquad -12\,x + 8\,x^3$$

```
> HRod(17,x)-H(17,x);
```

$$0$$

The Rodrigues formula can be used to obtain the normalization integral for arbitrary n

$$\int_{-\infty}^{\infty} \exp(-z^2) H_n(z)^2 dz = 2^n n! \sqrt{\pi} \, .$$

Exercise 8.1.1: Use n-fold integration by parts on Rodrigues' formula to derive the above result for some particular n values.

– (2) The recurrence relation of the Hermite polynomials:

$$H_{n+1}(z) - 2zH_n(z) + 2nH_{n-1}(z) = 0 \, .$$

```
> Rrel:=(n,z)->simplify(H(n+1,z)-2*z*H(n,z)+2*n*H(n-1,z));
```

$$\mathrm{Rrel} := (n, z) \rightarrow \mathrm{simplify}(\mathrm{H}(n+1, z) - 2\,z\,\mathrm{H}(n, z) + 2\,n\,\mathrm{H}(n-1, z))$$

```
> Rrel(1,x);        Rrel(14,y);
```

$$0 \qquad 0$$

Recurrence relations play an important role in the numerical computation of orthogonal polynomials.

– (3) The generating function for the Hermite polynomials can be checked out numerically after truncation of the infinite series on the RHS

$$\exp(-t^2 + 2tz) = \sum_{n=0}^{\infty} H_n(z) \frac{t^n}{n!} \, .$$

```
> GenF:=(t,z,nmax)->sum(H(n,z)*t^n/n!,n=0..nmax);
```

$$\mathrm{GenF} := (t, z, nmax) \rightarrow \sum_{n=0}^{nmax} \frac{\mathrm{H}(n, z)\, t^n}{n!}$$

```
> fun1:=(t,z)->exp(-t^2+2*t*z);
```

$$\text{fun1} := (t, z) \to e^{(-t^2+2tz)}$$

We try for a small value $t = 1/2$ and truncate at $n_{\max} = 4$:

```
> plot({fun1(1/2,z),GenF(1/2,z,4)},z=-5..5);
```

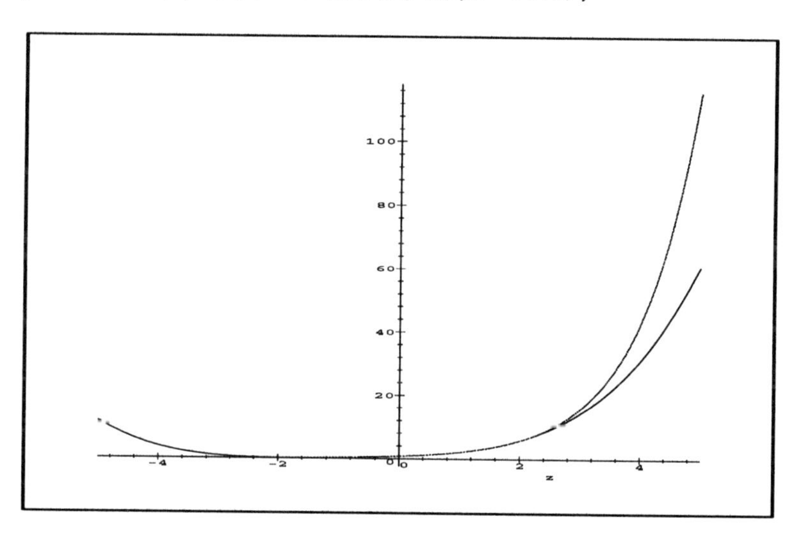

Exercise 8.1.2: Demonstrate the convergence properties of the generating functional for large z and t as a function of $n_{\max}$.

We conclude the discussion of properties by emphasizing that the orthogonality relation for the Hermite polynomials requires the $\exp(-z^2)$ factor as a weight in the definition of the inner product.

The Hermite polynomials together with the $\exp(-z^2/2)$ factor form the eigenfunctions of the harmonic oscillator Schrödinger equation, i.e, of a Sturm-Liouville problem. Any reasonably behaved function on the interval $-\infty < x < \infty$ can be expanded in terms of these functions. The expansion coefficients are obtained from the projection onto the basis functions, *viz.*,

$$f(x) = \sum_{n=0}^{\infty} c_n \exp(-x^2/2) H_n(x)\,,$$

with

$$c_n = \int_{-\infty}^{\infty} f(z) \exp(-z^2/2) H_n(z) \mathrm{d}z\,.$$

We try this on some examples. Note that $f(x)$ is not required to have particular symmetry properties. We have a discrete alternative to the continuous Fourier representation for functions on the infinite interval.

In practice we have to truncate the expansion at some finite order (and usually pay a price for this). In a computer algebra system we can make use of the expansion as long as the inner product can be calculated exactly. Otherwise one is limited to numerical techniques. We use an inner product for real functions.

```
> InProd:=proc();
> f1:=args[1]; f2:=args[2]; weight:=1; x:=args[3];
> int(f1*f2*weight,x=-infinity..infinity);
> end:
```

Now we define normalized basis functions

```
> Hermbas:=proc();
> n:=args[1]; x:=args[2];
> exp(-x^2/2)*H(n,x)/sqrt(InProd(exp(-y^2/2)*H(n,y),
> exp(-y^2/2)*H(n,y),y));
> end:
```

We are ready to combine the prescription for the expansion coefficients c_n to define a procedure to calculate the expansion of a function defined on the interval $-\infty < x < \infty$ in terms of the Hermite basis. The procedure is written such that it expects a Maple function (mapping) as a first argument.

```
> Hermfun:=proc();
> func:=args[1]; z:=args[2]; nmax:=args[3]; sumf:=0;
> for n from 0 to nmax do
> sumf:=sumf+InProd(func(x),Hermbas(n,x),x)*Hermbas(n,z);
> od;
> end:
```

We expand $\exp(-t^2/2)$ using a single term only.

```
> Hermfun(t->exp(-t^2),z,0);
```

$$\frac{1}{3}\sqrt{3}\,\sqrt{2}\,e^{(-1/2\,z^2)}$$

```
> plot({",exp(-z^2)},z=-5..5);
```

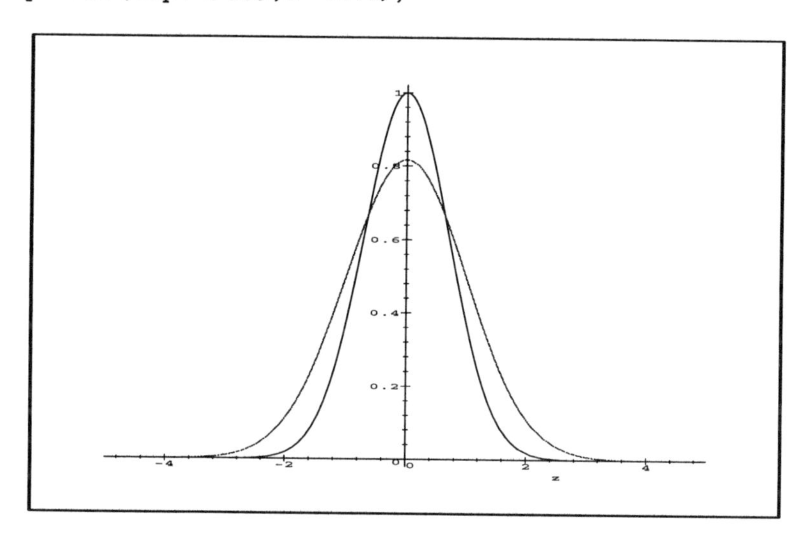

We attempt to expand a Gaussian with one oscillator constant in terms of the complete set of harmonic oscillator eigenfunctions for another oscillator constant. The function is symmetric and only even powers appear. As $n_{\max}$ is increased we expect the expansion to converge.

```
> Hermfun(t->exp(-t^2),z,3);
```

$$\frac{1}{3}\sqrt{3}\sqrt{2}\,\mathrm{e}^{(-1/2\,z^2)} - \frac{1}{36}\sqrt{3}\,\mathrm{e}^{(-1/2\,z^2)}\,(4\,z^2-2)\sqrt{2}$$

```
> plot({",exp(-z^2)},z=-5..5);
```

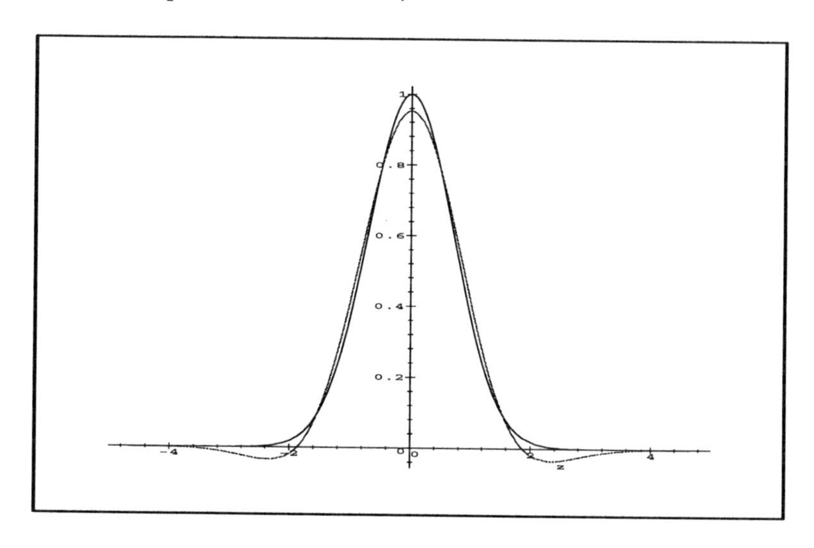

The expansion appears to converge. We increase $n_{\max}$ to 10 and observe the resulting error.

```
> Hermfun(t->exp(-t^2),z,10):
```

The error in the polynomial representation is small now and changes sign many times:

```
> plot("-exp(-z^2),z=-5..5,title=`Error for nmax=10`);
```

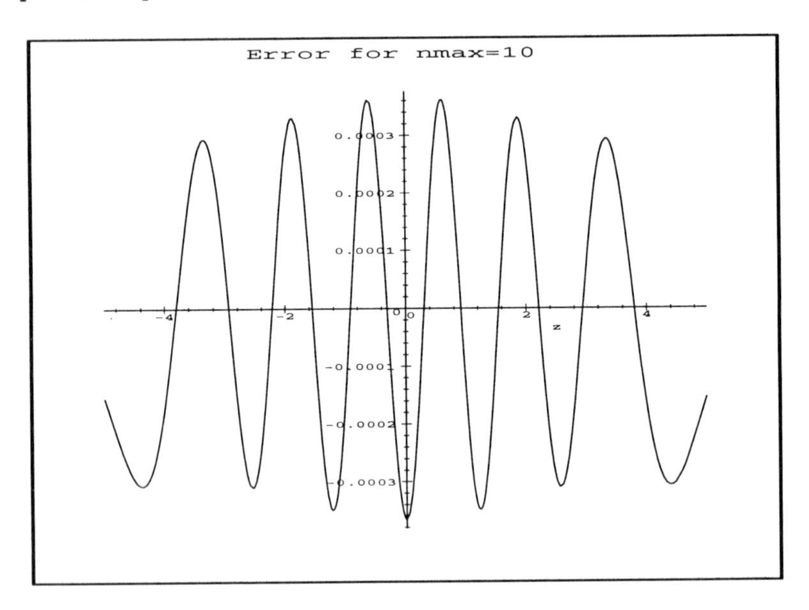

We see that we can achieve convergence on a finite interval. I leave it to the reader to consider what difficulties the method can run into at very large x values. In order to warn about potential problems in basis function expansions, and to encourage a critical curiosity, we work through an example of a function that grows rapidly asymptotically. We try the expansion for a non-symmetric function, such as e^{-t}:

```
> Hermfun(t->exp(-t),z,0);
```

$$e^{(1/2)}\sqrt{2}\,e^{(-1/2\,z^2)}$$

```
> plot({",exp(-z)},z=-5..5,y=-1..4):
```

It does not appear to be meaningful to approximate this function starting with a Gaussian. Nevertheless, we can try to push to some finite order, e.g., $n_{\max} = 3$:

```
> Hermfun(t->exp(-t),z,3);
```

$$e^{(1/2)}\sqrt{2}\,e^{(-1/2\,z^2)} - 2\,e^{(1/2)}\,e^{(-1/2\,z^2)}\,z\,\sqrt{2} + \frac{3}{4}\,e^{(1/2)}\,e^{(-1/2\,z^2)}$$

$$\times\ (4z^2-2)\sqrt{2}-\frac{5}{12}\,e^{(1/2)}\sqrt{2}\,e^{(-1/2z^2)}\,(8z^3-12z)$$

```
> plot({",exp(-z)},z=-5..5,y=-1..4);
```

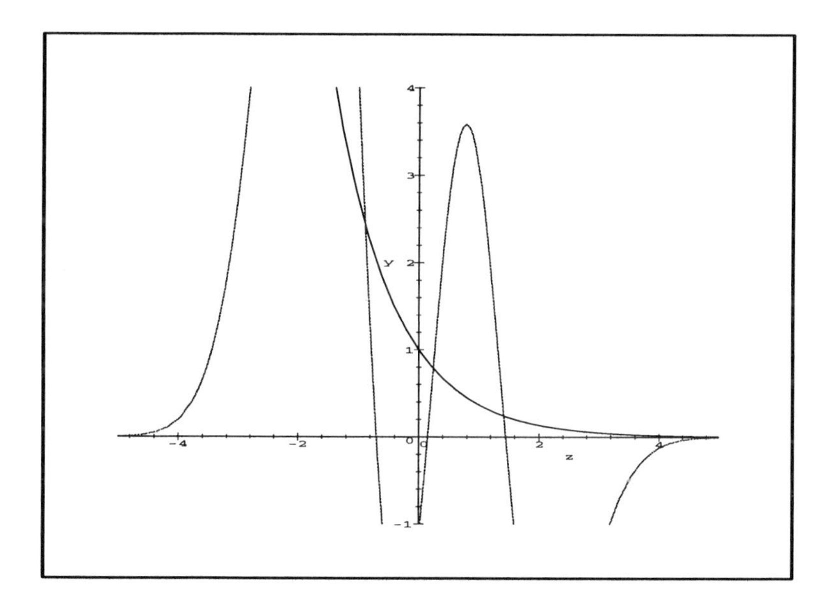

This looks pretty hopeless! We assume that it is the divergence of $\exp(-z)$ for negative z that causes the problem. Therefore, we look at the expansion for an exponential function cut off at $z=0$ to a zero value by means of the step function. This, of course, introduces a discontinuity at $z=0$.

```
> Hermfun(t->Heaviside(t)*exp(-t),z,1):
> plot({",Heaviside(z)*exp(-z)},z=-5..5,-1..1):
```

We expand up to $n=3$ and observe that there is no disaster occuring this time, but that the expansion has the same problems as the Fourier–series expansion of functions with discontinuities: overshooting occurs, and it is not clear whether an increase in the order of the expansion fixes the problems near the discontinuity.

```
> Hermfun(t->Heaviside(t)*exp(-t),z,3):
> plot({",Heaviside(z)*exp(-z)},z=-5..5,-1..1);
```

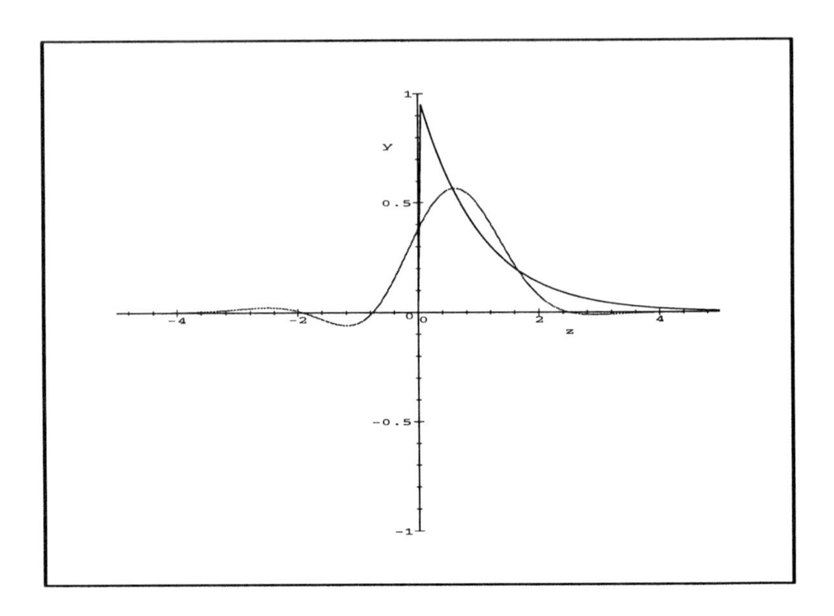

This is clearly a tough example (discontinuity at $z = 0$), which shows what is meant by the requirement that the function $f(x)$ should be well-behaved. The increase in $n_{\max}$ lets the expansion struggle with another problem apart from the discontinuity, namely that the Gaussian attenuation factor in the basis is too strong for the function $\exp(-x)$. Thus, we learn that basis-state expansions in function space for the interval $-\infty < x < \infty$ work only if the basis is properly adjusted for the problem.

Exercise 8.1.3: Expand functions of your choice in Hermite polynomials times a Gaussian over the interval $\infty < x < \infty$. Try this for a two-centre Gaussian wavefunction and vary the distance between the two Gaussians.

8.2 Laguerre Polynomials

Laguerre polynomials are relevant for Schrödinger problems in the radial coordinate $0 \leq r < \infty$, such as the spherical harmonic oscillator and the hydrogen atom problem. We consider the hydrogen atom problem. After the separation of the angular part of the wavefunction using spherical harmonics, a radial equation with labels n (principal quantum number) and l (magnitude of angular momentum) results for $u_{nl}(r) = rR_{nl}(r)$. We use atomic (Bohr) units, and **eps** $= \epsilon = 2E$ represents twice the eigenenergy.

```
> SErad:=diff(u(r),r$2)+(eps + 2*Z/r-l*(l+1)/r^2)*u(r)=0;
```

$$\mathrm{SErad} := \left(\frac{\partial^2}{\partial r^2}\mathrm{u}(r)\right) + \left(\mathrm{eps} + 2\,\frac{Z}{r} - \frac{l\,(l+1)}{r^2}\right)\mathrm{u}(r) = 0$$

The analysis of the asymptotic behaviour for $r \to \infty$ shows that $u(r)$ falls as an exponential with coefficient $\alpha = \sqrt{-\epsilon}$. We expect a negative energy eigenvalue $E = \epsilon/2$ in analogy to the classical Kepler problem (see, e.g., [Sy71, Go80, MT70]); $E > 0$ corresponds to scattering solutions between an electron and a proton. The exponentially growing solution is not normalizable. Thus, we write an equation for $f_{nl}(r) = u_{nl}(r)\exp(\alpha r)$.

```
> expand(exp(sqrt(-eps)*r)*subs(u(r)=f(r)*exp(-sqrt(-eps)*r),
> SErad));
```

$$\left(\frac{\partial^2}{\partial r^2}\mathrm{f}(r)\right) - 2\left(\frac{\partial}{\partial r}\mathrm{f}(r)\right)\sqrt{-\mathrm{eps}} + 2\,\frac{\mathrm{f}(r)\,Z}{r} - \frac{\mathrm{f}(r)\,l^2}{r^2} - \frac{\mathrm{f}(r)\,l}{r^2} = 0$$

```
> LaEq:=subs(sqrt(-Enl)=alpha,");
```

$$\mathrm{LaEq} := \left(\frac{\partial^2}{\partial r^2}\mathrm{f}(r)\right) - 2\left(\frac{\partial}{\partial r}\mathrm{f}(r)\right)\alpha + 2\,\frac{\mathrm{f}(r)\,Z}{r} - \frac{\mathrm{f}(r)\,l^2}{r^2} - \frac{\mathrm{f}(r)\,l}{r^2} = 0$$

Maple can find the series solution easily. The boundary condition f(0) = 0 is required for $R_{nl}(r) = u_{nl}(r)/r$ to remain finite at $r = 0$. Note that there are two independent parts in the solution that are pre-multiplied by the integration constants $\_C1$ and $\_C2$.

```
> dsolve({LaEq,f(0)=0},f(r),series);
```

$$\mathrm{f}(r) = \_C1\, r^{(-l)}\left(1 + \frac{l\,\alpha + Z}{l}\,r + 2\,\frac{(l\,\alpha - \alpha + Z)(l\,\alpha + Z)}{l\,(4l-2)}\,r^2\right.$$
$$+4\,\frac{(l\,\alpha - 2\,\alpha + Z)(l\,\alpha - \alpha + Z)(l\,\alpha + Z)}{l\,(4l-2)(6l-6)}\,r^3 + 8$$
$$(l\,\alpha - 3\,\alpha + Z)(l\,\alpha - 2\,\alpha + Z)(l\,\alpha - \alpha + Z)(l\,\alpha + Z)$$
$$/(l\,(4l-2)(6l-6)(8l-12))r^4 + 16$$
$$(l\,\alpha - 4\,\alpha + Z)(l\,\alpha - 3\,\alpha + Z)(l\,\alpha - 2\,\alpha + Z)$$
$$(l\,\alpha - \alpha + Z)(l\,\alpha + Z)/(l\,(4l-2)(6l-6)(8l-12)$$

$$(10\,l-20))r^5+\mathrm{O}(r^6)\Big)\ +\ \_C2\,r^{(l+1)}\Big(1+$$

$$2\,\frac{-(l+1)\,\alpha+Z}{l^2-(l+1)^2-1}\,r+4\,\frac{\%2\,(-(l+1)\,\alpha+Z)}{(l^2-(l+1)^2-1)\ \%1}\,r^2+$$

$$8\,\frac{(-(l+1)\,\alpha-2\,\alpha+Z)\,\%2\,(-(l+1)\,\alpha+Z)}{(l^2-(l+1)^2-1)\ \%1\ (l^2-4\,l-(l+1)^2-11)}\,r^3+$$

$$16(-(l+1)\,\alpha-3\,\alpha+Z)\,(-(l+1)\,\alpha-2\,\alpha+Z)\,\%2$$

$$(-(l+1)\,\alpha+Z)\Big/\Big((l^2-(l+1)^2-1)\ \%1$$

$$(l^2-4\,l-(l+1)^2-11)\ (l^2-6\,l-(l+1)^2-19)\Big)\,r^4+$$

$$32(-(l+1)\,\alpha-4\,\alpha+Z)\,(-(l+1)\,\alpha-3\,\alpha+Z)$$

$$(-(l+1)\,\alpha-2\,\alpha+Z)\,\%2\,(-(l+1)\,\alpha+Z)\Big/$$

$$\Big((l^2-(l+1)^2-1)\ \%1\ (l^2-4\,l-(l+1)^2-11)$$

$$(l^2-6\,l-(l+1)^2-19)\ (l^2-8\,l-(l+1)^2-29)\Big)\,r^5+$$

$$\mathrm{O}(r^6)\Big)$$

$$\%1 := l^2-2\,l-(l+1)^2-5$$
$$\%2 := -(l+1)\,\alpha-\alpha+Z$$

While this result exposes the structure of the solution, as well as the possibility to terminate the series by a choice of α at any order n (all subsequent terms vanish), there is still a combination of two independent solutions present. The first of these carries a common factor of r^{-l}, while the second series solution has an overall factor of r^{l+1}. We are interested in a derivation of the recursion relation. This can be performed automatically by Maple as indicated in the introduction to the previous section. Nevertheless, we derive the recursion manually.

To satisfy the requirement of a finite solution at $r = 0$ the leading behaviour for $u_{nl}(r)$ has to be r^{l+1}. We substitute a single term of the series expansion into the equation:

```
> LaEq1:=simplify(expand(subs(f(r)=r^(l+1)*a[n,k]*r^k,LaEq)));
```

$$\begin{aligned}\mathrm{LaEq1} := {} & 2\,r^{(l+k-1)}\,l\,a_{n,k}\,k+r^{(l+k-1)}\,a_{n,k}\,k+r^{(l+k-1)}\,a_{n,k}\,k^2\\ & -2\,\alpha\,r^{(l+k)}\,l\,a_{n,k}-2\,\alpha\,r^{(l+k)}\,a_{n,k}-2\,\alpha\,r^{(l+k)}\,a_{n,k}\,k\\ & +2\,r^{(l+k)}\,a_{n,k}\,Z=0\end{aligned}$$

Now we divide out r^{l+k-1} and simplify.

```
> assume(l,integer);    assume(k,integer);
```

```
> LaEq1a:=simplify(expand(lhs(LaEq1))/r^(l+k-1));
```

$$LaEq1a := 2\,l^{\sim}\,a_{n,k^{\sim}}\,k^{\sim} + a_{n,k^{\sim}}\,k^{\sim} + a_{n,k^{\sim}}\,k^{\sim 2} - 2\,r\,\alpha\,l^{\sim}\,a_{n,k^{\sim}} - 2\,r\,\alpha\,a_{n,k^{\sim}} - 2\,r\,\alpha\,a_{n,k^{\sim}}\,k^{\sim} + 2\,r\,a_{n,k^{\sim}}\,Z$$

To isolate the coefficient of r^{l+k-1} we set $r = 0$:

```
> cof1:=simplify(subs(r=0,LaEq1a));
```

$$cof1 := 2\,l^{\sim}\,a_{n,k^{\sim}}\,k^{\sim} + a_{n,k^{\sim}}\,k^{\sim} + a_{n,k^{\sim}}\,k^{\sim 2}$$

To facilitate the comparison of coefficients for r^{k+l} we shift the index k by 1 and collect the coefficient of $a_{n,k+1}$:

```
> cof1s:=collect(subs(k=k+1,cof1),a[n,k+1]);
```

$$cof1s := \left(2\,l^{\sim}\,(\,k^{\sim} + 1\,) + k^{\sim} + 1 + (\,k^{\sim} + 1\,)^{2}\right)\,a_{n,k^{\sim}+1}$$

We obtain the coefficient of r^{l+k} by dividing out r^{l+k}:

```
> LaEq1b:=simplify(expand(lhs(LaEq1))/r^(l+k));
```

$$LaEq1b := \frac{a_{n,k^{\sim}}\,(\,2\,l^{\sim}\,k^{\sim} + k^{\sim} + k^{\sim 2} - 2\,r\,\alpha\,l^{\sim} - 2\,r\,\alpha - 2\,r\,\alpha\,k^{\sim} + 2\,r\,Z\,)}{r}$$

and, then, we take the limit $r \to \infty$, which makes the r^{l+k-1} contribution to the equation disappear:

```
> cof2:=limit(LaEq1b,r=infinity);
```

$$cof2 := -2\,a_{n,k^{\sim}}\,(\,\alpha + \alpha\,k^{\sim} - Z + \alpha\,l^{\sim}\,)$$

Now we equate the overall coefficient of r^{k+l} to zero:

```
> RecRel:=cof1s+cof2=0;
```

$$RecRel := \left(2\,l^{\sim}\,(\,k^{\sim} + 1\,) + k^{\sim} + 1 + (\,k^{\sim} + 1\,)^{2}\right)\,a_{n,k^{\sim}+1} - 2\,a_{n,k^{\sim}}\,(\,\alpha + \alpha\,k^{\sim} - Z + \alpha\,l^{\sim}\,) = 0$$

We can read off the eigenvalue condition for α from the requirement to terminate the recursion relation at some given power $k = n$. Thus we have:

```
> assume(n_r,integer);     alpha_nr:=Z/(n_r+l+1);
```

$$alpha\_nr := \frac{Z}{n\_r^{\sim} + l^{\sim} + 1}$$

Here n_r is a radial quantum number and can take on values $n_r = 0, 1, 2, \ldots$ determining the number of nodes and the order of the polynomial. It is possible to combine the denominator in the above expression to a principal quantum number n and the energy then follows from **eps** $= \epsilon = -\alpha_{\mathrm{n}}^2$ and is independent of l. If we take into account the factor of 2 that arose in the conversion of the radial SE to the eigenvalue problem in **eps**, and remember that in atomic units $\hbar = m_e = e = 1$, the hydrogen-like bound-state energy values become

```
> E_nl:=n->-Z^2/(2*n^2);
```

$$E\_nl := n \to -\frac{1}{2}\frac{Z^2}{n^2}$$

Exercise 8.2.1: Understand where the scaling behaviour Z^2 comes from, given that the potential is $-Z/r$ by considering the (kinetic and) potential energy expectation values.

We are ready now to generate the recursion:

```
> ankp1:=collect(solve(RecRel,a[n,k+1]),a[n,k]);
```

$$\text{ankp1} := -\frac{(-2\,\alpha - 2\,\alpha\,k^{\sim} + 2\,Z - 2\,\alpha\,l^{\sim})\,a_{n,k^{\sim}}}{2\,l^{\sim}\,k^{\sim} + 2\,l^{\sim} + 3\,k^{\sim} + 2 + k^{\sim 2}}$$

Exercise 8.2.2: Demonstrate how for given $n_r = 0, 1, 2, ...$ the recursion terminates due to the choice of α.

Now we prepare the ground for the implementation of a procedure that generates the polynomials from the recursion relation. We use the **unapply** feature to let Maple generate a mapping from the derived expression that permits the nuclear charge Z and the quantum numbers to be passed. This mapping is made known to the procedure through a **global** declaration.

```
> RHSRR0:=subs(alpha=Z/nu,ankp1);
```

$$\text{RHSRR0} := -\frac{\left(-2\,\frac{Z}{\nu} - 2\,\frac{Z\,k^{\sim}}{\nu} + 2\,Z - 2\,\frac{Z\,l^{\sim}}{\nu}\right)\,a_{n,k^{\sim}}}{2\,l^{\sim}\,k^{\sim} + 2\,l^{\sim} + 3\,k^{\sim} + 2 + k^{\sim 2}}$$

```
> RHSRR:=unapply(RHSRR0,Z,nu,k,l,a[n,k]):
> Poly:=proc() global a,RHSRR;
> n1:=args[1]; l1:=args[2]; Z:=args[3]; r:=args[4];
> if type(n1,integer)=true then
> a[n1,0]:=1; alpha:=Z/(n1); sumP:=a[n1,0];
> for k from 0 to n1-1 do:
> a[n1,k+1]:=RHSRR(Z,n1,k,l1,a[n1,k]);
> sumP:=sumP+a[n1,k+1]*r^(k+1);
```

```
> od;
> sumP*r^(l1+1);
> else
> RETURN(`Call Poly with integer as first argument`,n1);
> fi;        end:
```

For $n = 1, l = 0, Z = 1$ and variable r we obtain

```
> Poly(1,0,1,r);
```

$$r$$

The procedure is protected against non-integer arguments in n:

```
> Poly(1.5,0,1,r);
```

Call Poly with integer as first argument, 1.5

but it does not check whether n and l are in the allowed range:

```
> Poly(1,1,Z,r);
```

$$\left(1 + \frac{1}{2} r Z\right) r^2$$

No check was made in the program that $n = 1$, $l = 1$ is impossible as $n = n_r + l + 1$ and $n_r = 0, 1, 2, \ldots$ are required! For $n = 2$ and $l = 1$ and $n = 2$ and $l = 0$ respectively it returns the polynomial parts to the radial 2p and 2s wavefunctions for a hydrogen-like atom:

```
> Poly(2,1,Z,r);     Poly(2,0,Z,r);
```

$$r^2 \qquad \left(1 - \frac{1}{2} r Z\right) r$$

A 3s-state radial function without the exponential factor is given by:

```
> Poly(3,0,Z,r);
```

$$\left(1 - \frac{2}{3} r Z + \frac{2}{27} Z^2 r^2\right) r$$

The polynomials derived for the hydrogen atom are the Laguerre polynomials with arguments $L_{n+l}^{2l+1}(r)$ times r. We compare them now with the built-in generalized (or associated) Laguerre polynomials $L(n, a, x)$, which when the second argument is omitted reduce to the ordinary ones, i.e., $L(n, x)$.

```
> with(orthopoly):          L(2,x);
```

$$1 - 2x + \frac{1}{2} x^2$$

The connection with the polynomials required for the radial solution to the hydrogen atom problem is as follows:

```
> wL:=(n,l,r)->L(n-l-1,2*l+1,r);
```

$$wL := (n, l, r) \to L(n - l - 1, 2l + 1, r)$$

```
> wL(1,0,r);      wL(2,0,r);      wL(3,0,r);
```

$$1 \qquad 2 - r \qquad 3 - 3r + \frac{1}{2}r^2$$

Clearly, we need a modification of the argument r in order to obtain the functions derived above for arbitrary Z. In atomic units the radial wavefunction (without the factor r and not yet normalized) is given as:

```
> Rnl:=(Z,n,l,r)->r^l*exp(-Z*r/n)*wL(n,l,2*Z*r/n):
> Rnl(Z,3,0,r);
```

$$e^{(-1/3\, r\, Z)}\left(3 - 2\, r\, Z + \frac{2}{9}r^2 Z^2\right)$$

We can plot the radial functions within a symmetry sector l:
1) The 1s, 2s, 3s states:

```
> plot({Rnl(1,1,0,r),Rnl(1,2,0,r),Rnl(1,3,0,r)},r=0..20);
```

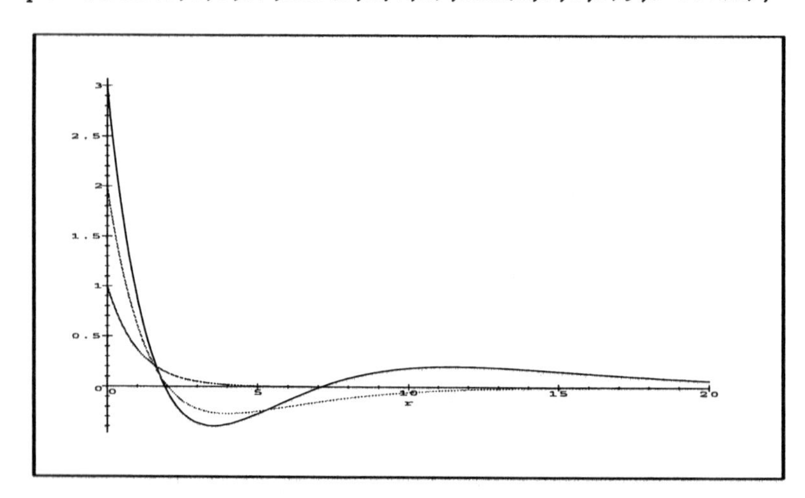

The normalized functions are given by the following mapping:

```
> NRnl:=(Z,n,l,r)->simplify(Rnl(Z,n,l,r)/sqrt(int((Rnl(Z,n,l,r)
> *r)^2,r=0..infinity))):
```

As an example we evaluate the hydrogenic ($Z = 1$) 3s state:

```
> NRnl(1,3,0,r);
```

$$\frac{2}{243}\left(27-18\,r+2\,r^2\right)\sqrt{3}\,e^{(-1/3\,r)}$$

A graph of the radial occupation probabilities for 1s, 2s, and 3s follows below.

```
> plot({(r*NRnl(1,1,0,r))^2,(r*NRnl(1,2,0,r))^2,
> (r*NRnl(1,3,0,r))^2},r=0..20);
```

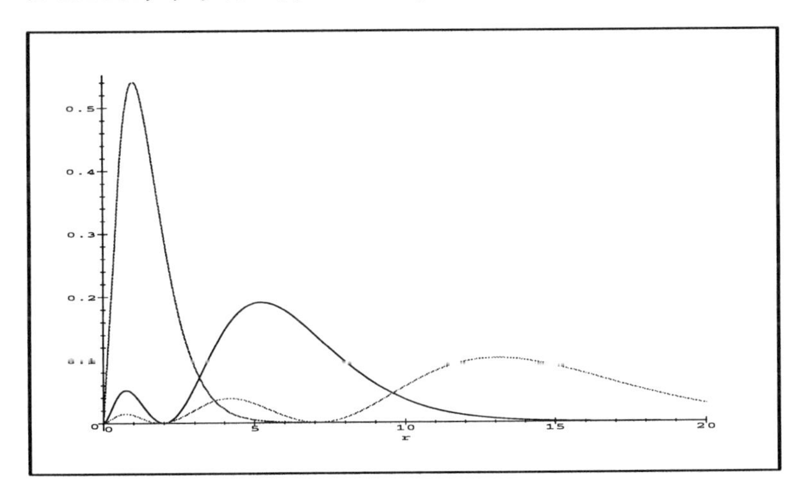

In the $l = 1$ sector we graph the occupation probabilities for the 2p, 3p, and 4p states.

```
> plot({(r*NRnl(1,2,1,r))^2,(r*NRnl(1,3,1,r))^2,
> (r*NRnl(1,4,1,r))^2},r=0..40);
```

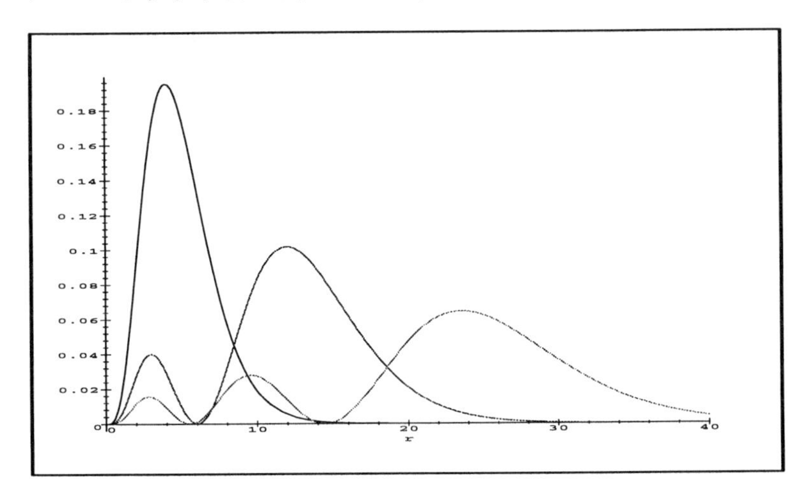

The Laguerre polynomials can also be calculated in the following way:

```
> Lag:=(n,xi)->expand(exp(xi)*diff(xi^n*exp(-xi),xi$n)):
```

While calling this function for $n = 0$

```
> Lag(0,z);
  Error, (in Lag)
  wrong number (or type) of parameters in function diff
```

we find that Maple doesn't recognize that zero-fold differentiation is equivalent to returning the function itself. We should avoid this $n = 0$ case in the procedure with an `if-then-else-fi` construct. For $n > 1$, however, the polynomials are obtained as follows:

```
> Lag(1,z);    Lag(3,z);
```

$$1 - z \qquad 6 - 18\,z + 9\,z^2 - z^3$$

We compare the latter result with the built-in function's answer:

```
> L(3,z);
```

$$1 - 3\,z + \frac{3}{2}\,z^2 - \frac{1}{6}\,z^3$$

The normalization is slightly different! One has to watch out for this when comparing different authors' textbooks. The associated Laguerre polynomials are obtained by an additional m-fold differentiation.

```
> ALag:=(n,m,xi)->diff(Lag(n,xi),xi$m):
```

```
> wLalt:=(n,l,r)->ALag(n+l,2*l+1,z):
```

We verify the difference in the normalization on the example of $n = 3$:

```
> wLalt(3,0,z);    wL(3,0,z);
```

$$-18 + 18\,z - 3\,z^2 \qquad 3 - 3\,z + \frac{1}{2}\,z^2$$

Note that the associated Laguerre polynomials are not identical with the built-in generalized Laguerre polynomials $L(n, a, x)$ in which a is not restricted to be of integer type. However, the hydrogenic eigenfunctions can be constructed from either set. For the built-in functions one uses arguments $n - l - 1$ and $2l + 1$, while for the ones generated above the arguments are $n + l$ and $2l + 1$. Both definitions can be found in the literature.

These polynomials represent special cases of the hypergeometric function $F(n, d, z)$. Maple has reserved by default `F` for generating Fermat numbers and one has to perform a library call `readlib(hypergeom)` to load the hypergeometric function. The help facility does know `hypergeom`, the default name for the this very powerful function.

```
> readlib(hypergeom):
```

The hypergeometric differential equation is defined as

```
> HGde:=r*diff(f(r),r$2)+(2*l+2-r)*diff(f(r),r)-(l+1-kappa)
> *f(r)=0;
```

$$\mathrm{HGde} := r\left(\frac{\partial^2}{\partial r^2}\,\mathrm{f}(r)\right) + (2\,l\tilde{} + 2 - r)\left(\frac{\partial}{\partial r}\,\mathrm{f}(r)\right) - (l\tilde{} + 1 - \kappa)\,\mathrm{f}(r) = 0$$

We now engage in an ambitious task, namely we attempt a verification of the solution to the hypergeometric differential equation for arbitrary l and k.

```
> simplify(subs(f(r)=hypergeom([l+1-kappa],[2*l+2],r),HGde));
```

$$\frac{1}{2}(l\tilde{} + 1 - \kappa)(r\,\%3\,l\tilde{} + 2\,r\,\%3 - r\,\%3\,\kappa + 4\,\%2\,l\tilde{}^{2} + 10\,\%2\,l\tilde{} + 6\,\%2$$
$$- 2\,\%2\,r\,l\tilde{} - 3\,\%2\,r - 4\,\%1\,l\tilde{}^{2} - 10\,\%1\,l\tilde{} - 6\,\%1)\,/(2\,l\tilde{} + 3)\,(l\tilde{} + 1)$$
$$= 0$$
$$\%1 := \mathrm{hypergeom}([\,l\tilde{} + 1 - \kappa\,], [\,2\,l\tilde{} + 2\,], r)$$
$$\%2 := \mathrm{hypergeom}([\,l\tilde{} + 2 - \kappa\,], [\,2\,l\tilde{} + 3\,], r)$$
$$\%3 := \mathrm{hypergeom}([\,l\tilde{} + 3 - \kappa\,], [\,2\,l\tilde{} + 4\,], r)$$

Now we substitute $\kappa = n + l + 1$:

```
> subs(kappa=n+l+1,"):
```

This does not simplify, however, and we hope for some help from the assumption that n is restricted to integer values.

```
> assume(n,integer);   simplify(");
```

$$-\frac{1}{2}n\tilde{}\,(r\,\mathrm{hypergeom}([\,2 - n\tilde{}\,], [\,2\,l\tilde{} + 4\,], r)$$
$$- r\,\mathrm{hypergeom}([\,2 - n\tilde{}\,], [\,2\,l\tilde{} + 4\,], r)\,n\tilde{} + 4\,\%1\,l\tilde{}^{2} + 10\,\%1\,l\tilde{} + 6\,\%1$$
$$- 2\,\%1\,r\,l\tilde{} - 3\,\%1\,r - 4\,\mathrm{hypergeom}([\,-n\tilde{}\,], [\,2\,l\tilde{} + 2\,], r)\,l\tilde{}^{2}$$
$$- 10\,\mathrm{hypergeom}([\,-n\tilde{}\,], [\,2\,l\tilde{} + 2\,], r)\,l\tilde{}$$
$$- 6\,\mathrm{hypergeom}([\,-n\tilde{}\,], [\,2\,l\tilde{} + 2\,], r))\,/((2\,l\tilde{} + 3)\,(l\tilde{} + 1)) = 0$$
$$\%1 := \mathrm{hypergeom}([\,1 - n\tilde{}\,], [\,2\,l\tilde{} + 3\,], r)$$

We give up on the general case, but try an explicit example:

```
> subs(n=2,l=0,");
```

$$\frac{1}{3}\,r\,\mathrm{hypergeom}([\,0\,], [\,4\,], r) - 2\,\mathrm{hypergeom}([\,-1\,], [\,3\,], r)$$
$$+ \mathrm{hypergeom}([\,-1\,], [\,3\,], r)\,r + 2\,\mathrm{hypergeom}([\,-2\,], [\,2\,], r) = 0$$

```
> simplify(");
```

$$0 = 0$$

Some good news at last. We define now the hydrogenic bound-state wavefunctions using the hypergeometric function.

```
> HydF:=(n,l,r)->simplify(hypergeom([l+1-n],[2*l+2],2*r/n)):
```
For explicit parameter choices simple polyonomials emerge as answers:

```
> HydF(1,0,r);        HydF(2,0,r);        HydF(3,0,r);
```

$$1 \qquad 1-\frac{1}{2}r \qquad 1-\frac{2}{3}r+\frac{2}{27}r^2$$

The help facility for the hypergeometric function explains under what circumstances the infinite series for this function terminates as a finite-order polynomial. The hypergeometric differential equation is general enough to encompass a wide class of equations that define orthogonal polynomials.

Exercise 8.2.3: Verify the orthonormality relation for Laguerre polynomials

$$\int_0^\infty \mathrm{e}^{-x} L_m(x) L_n(x) \mathrm{d}x = \delta_{m,n} \, .$$

Exercise 8.2.4: Investigate the associate Laguerre equation

$$x f''(x) + (k+1-x) f'(x) + n f(x) = 0 \, ,$$

i.e., verify that it is satisfied by $f(x) = L_n^k(x)$.

Exercise 8.2.5: The Schrödinger equation for the 3D harmonic oscillator is given by

$$-\frac{\hbar^2}{2\mu}\nabla^2\psi + \frac{1}{2}\mu\omega^2 r^2\psi = E\psi \, .$$

The equation can be solved in Cartesian coordinates and the eigenfunctions are given as products of Hermite polynomials (cf. Sect. 8.1). Using the parameter $\beta = \mu\omega/\hbar$ find the solution in spherical polar coordinates. Verify that the radial equation in an angular momentum sector labeled by l is solved by $\mathrm{e}^{-\beta r^2/2} r^l L_{(n-l-1)/2}^{l+1/2}(\beta r^2)$. For a solution in terms of the hypergeometric function cf. [Fl71, problems 35 and 36].

Exercise 8.2.6: Verify the normalization condition for the associated Laguerre polynomials

$$\int_0^\infty \mathrm{e}^{-x} x^{k+1} L_n^k(x) L_n^k(x) \mathrm{d}x = \frac{(n+k)!}{n!}(2n+k+1) \, .$$

(cf. [Ar85, p. 728]).

8.3 Legendre Polynomials

In this section we consider the Legendre polynomials that satisfy the differential equation

$$(1-x^2)\frac{\mathrm{d}^2 f}{\mathrm{d}x^2} - 2x\frac{\mathrm{d}f}{\mathrm{d}x} + l(l+1)f = 0\,,$$

with $f(x)$ defined on the interval $-1 \leq x \leq 1$.

```
> assume(l,integer);
> LeE:=(1-x^2)*diff(f(x),x$2)-2*x*diff(f(x),x)+l*(l+1)*f(x)=0;
```

$$\mathrm{LeE} := (1-x^2)\left(\frac{\partial^2}{\partial x^2}\mathrm{f}(x)\right) - 2\,x\left(\frac{\partial}{\partial x}\mathrm{f}(x)\right) + l\tilde{}\,(l\tilde{}+1)\,\mathrm{f}(x) = 0$$

```
> dsolve(LeE,f(x),series);
```

$$\mathrm{f}(x) = \mathrm{f}(0) + \mathrm{D}(\mathrm{f})(0)\,x - \frac{1}{2}\,l\tilde{}\,(l\tilde{}+1)\,\mathrm{f}(0)\,x^2 + \left(-\frac{1}{6}\,l\tilde{}^2\,\mathrm{D}(\mathrm{f})(0) - \frac{1}{6}\,l\tilde{}\,\mathrm{D}(\mathrm{f})(0) + \frac{1}{3}\,\mathrm{D}(\mathrm{f})(0)\right)x^3 + \frac{1}{24}\,(-6 + l\tilde{}^2 + l\tilde{}\,)\,l\tilde{}\,(l\tilde{}+1)\,\mathrm{f}(0)\,x^4 + \frac{1}{120}\,(-12 + l\tilde{}^2 + l\tilde{}\,)\,\mathrm{D}(\mathrm{f})(0)\,(l\tilde{}^2 + l\tilde{} - 2)\,x^5 + \mathrm{O}(x^6)$$

We see that the even-order terms depend on $f(0)$, while the odd-order terms depend on the first derivative at $x = 0$. We also observe that the coefficient at a given even (or odd) power of x appears also as a factor in the next higher even (or odd) powers of x. This is important as it permits to terminate the recursion at a given order. We develop the recursion relation that defines the series. An individual term is given by

```
> LeE1:=simplify(expand(subs(f(x)=c[k]*x^k,LeE)));
```

$$\mathrm{LeE1} := c_k\,x^{(k-2)}\,k^2 - c_k\,x^{(k-2)}\,k - c_k\,x^k\,k^2 - c_k\,x^k\,k + l\tilde{}^2\,c_k\,x^k + l\tilde{}\,c_k\,x^k = 0$$

For the series solution to be valid for any x, the individual coefficients of x^k have to vanish separately. To collect the powers of x we shift the terms proportional to x^{k-2} by two.

```
> LeE1a:=simplify(lhs(LeE1)/x^(k-2));
```

$$\mathrm{LeE1a} := c_k\,k^2 - c_k\,k - x^2\,c_k\,k^2 - x^2\,c_k\,k + x^2\,l\tilde{}^2\,c_k + x^2\,l\tilde{}\,c_k$$

We extract the terms that have no x dependence after the division by x^{k-2}:

```
> LeE1a:=subs(x=0,LeE1a);
```

$$\text{LeE1a} := c_k\,k^2 - c_k\,k$$

We have isolated the coefficient of the power x^{k-2}. The coefficient of x^k is obtained after division by x^k and truncation of terms proportional to x^{-2}:

```
> LeE1b:=limit(simplify(lhs(LeE1)/x^k),x=infinity);
```

$$\text{LeE1b} := c_k\,(-k^2 - k + l\tilde{}^{\,2} + l\tilde{}\,)$$

We apply the shift to the x^{k-2} term:

```
> LeE1as:=subs(k=k+2,LeE1a);
```

$$\text{LeE1as} := c_{k+2}\,(k+2)^2 - c_{k+2}\,(k+2)$$

and the recursion relation reads now

```
> RecRel:=LeE1as+LeE1b=0;
```

$$\text{RecRel} := c_{k+2}\,(k+2)^2 - c_{k+2}\,(k+2) + c_k\,(-k^2 - k + l\tilde{}^{\,2} + l\tilde{}\,) = 0$$

```
> RR0:=factor(solve(RecRel,c[k+2]));
```

$$\text{RR0} := -\frac{c_k\,(l\tilde{} + 1 + k)\,(l\tilde{} - k)}{(k+2)\,(k+1)}$$

For given integer l the recursion terminates at $k = l$ resulting in a polynomial of order l. If l in the differential equation is not an integer, then the solution represents an infinite series. We define a function that contains the recursion relation in order to pass it into the procedure that assembles Legendre polynomials generated from the initial conditions $c_0 = 1, c_1 = 0$ or $c_0 = 0, c_1 = 1$, respectively.

```
> RR1:=unapply(RR0,l,k,c[k]);
```

$$\text{RR1} := (\,y1, y2, y3\,) \to -\frac{y3\,(\,y1 + 1 + y2\,)\,(\,y1 - y2\,)}{(\,y2 + 2\,)\,(\,y2 + 1\,)}$$

```
> LegP:=proc() global RR1;
> l:=args[1]; x:=args[2];
> if type(l,integer)=false then RETURN(`Integer value l please
> in LegP: `,l) fi;
> if type(l,even)=true then
```

```
> if l=0 then RETURN(1); fi;
> c[0]:=1; sumP:=1;
> for k from 0 by 2 to l-2 do:
> c[k+2]:=RR1(l,k,c[k]);
> sumP:=sumP+c[k+2]*x^(k+2);
> od;
> else
> if l=1 then RETURN(x); fi;
> c[1]:=1; sumP:=x;
> for k from 1 by 2 to l-2 do:
> c[k+2]:=RR1(l,k,c[k]);
> sumP:=sumP+c[k+2]*x^(k+2);
> od;
> fi;    end:
```

The procedure is protected from non-integer arguments l

```
> LegP(2.5,z);
```

Integer value l please in LegP : , 2.5

and tested here for $l = 3$:

```
> LegP(3,z);
```

$$z - \frac{5}{3} z^3$$

We wish to compare with the built-in Legendre polynomials.

```
> with(orthopoly):    P(3,z);
```

$$\frac{5}{2} z^3 - \frac{3}{2} z$$

We see that by convention the Legendre polynomials are seeded not always by $c_{0/1} = 1$, but differently, depending on the order l. We can simply normalize the Legendre polynomials. Making use of the property that they are orthogonal on the interval $[-1, 1]$ it makes sense to normalize them on this interval. By convention one chooses the normalization integral to give $2/(2l + 1)$ as follows:

```
> int(P(3,x)*P(3,x),x=-1..1);   int(P(2,x)*P(2,x),x=-1..1);
```

$$\frac{2}{7} \qquad \frac{2}{5}$$

We redefine our polynomials **LegP** into conventionally normalized ones:

```
> LegPN:=(n,x)->simplify(LegP(n,x)/sqrt(int(LegP(n,x)^2,
> x=-1..1)*(2*n+1)/2)):
```

```
> LegPN(3,z);
```

$$\frac{3}{2}z - \frac{5}{2}z^3$$

We verify that they agree now with the built-in polynomials for $l = 8$ as an example.

```
> LegPN(8,x)-P(8,x);
```

$$0$$

We include the graphs of a few even-order and odd-order cases to demonstrate how they span function space on $-1 \leq x \leq 1$.

```
> plot({P(0,x),P(2,x),P(4,x)},x=-1..1);
```

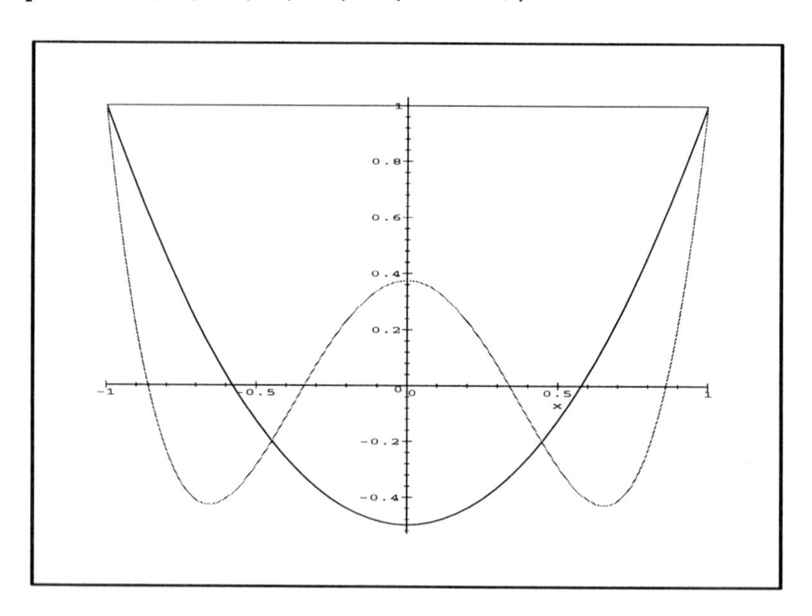

For odd-order l the polynomials are anti-symmetric:

```
> plot({P(1,x),P(3,x),P(5,x)},x=-1..1);
```

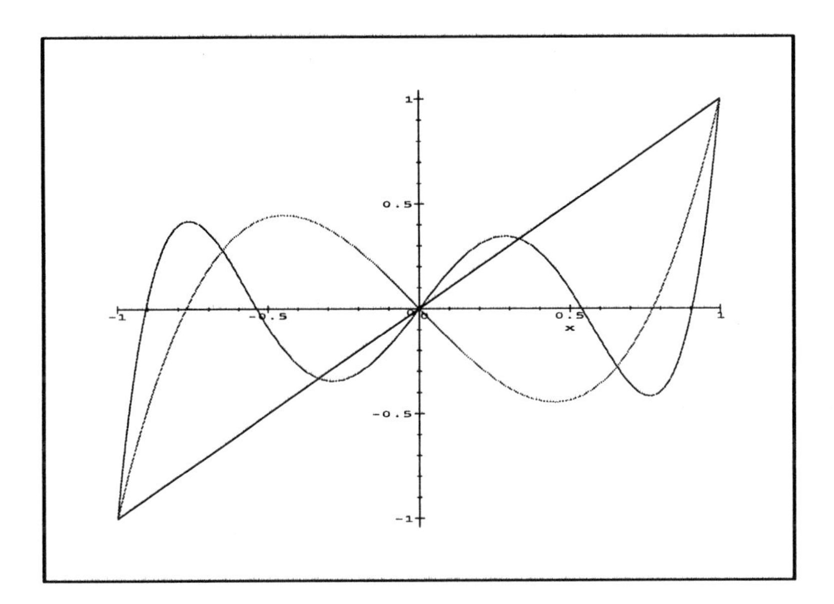

One can calculate the Legendre polynomials also from Rodrigues' formula

```
> LP:=(n,x)->expand(diff((x^2-1)^n,x$n)/(2^n*n!));
```

$$LP := (n, x) \to \text{expand}\left(\frac{\text{diff}\left((x^2-1)^n, x \$ n\right)}{2^n\, n!}\right)$$

```
> LP(3,x);
```

$$\frac{5}{2}x^3 - \frac{3}{2}x$$

Again, we verify our results for $l = 8$:

```
> LP(8,x)-P(8,x);
```

$$0$$

Now we present the following solved problem: Represent the polynomial $P(x) = x^4 + 2x^3 + 2x^2 - x - 3$ by Legendre polynomials.

This can be done by linear algebra. We first try, however, expansion techniques in function space. We work strictly on the interval $[-1, 1]$ and define an inner product for real functions:

```
> InProd:=(f,g,x)->int(f*g,x=-1..1):
```

```
> Px:=x^4+2*x^3+2*x^2-x-3:
```

We calculate the expansion coefficients by projection of Px onto the basis functions, and sum only up to $n = 4$ (all higher terms vanish). We have to keep in mind that the polynomials are not normalized to unity and insert the appropriate correction in the expansion, i.e.,

$$P(x) = \sum_{l=0}^{4} a_l P_l(x) \frac{2l+1}{2},$$

where

$$a_l = \int_{-1}^{1} P_l(x) P(x) \mathrm{d}x .$$

The results from the loop executed below are printed to show how $P(x)$ is built up in a Legendre polynomial representation.

```
> sumP:=0:        for i from 0 to 4 do:
> a[i]:=InProd(Px,P(i,x),x);
> sumP:=sumP+a[i]*P(i,x)*(2*i+1)/2;     od;
```

$$a_0 := \frac{-64}{15} \qquad \text{sumP} := \frac{-32}{15}$$

$$a_1 := \frac{2}{15} \qquad \text{sumP} := -\frac{32}{15} + \frac{1}{5}x$$

$$a_2 := \frac{16}{21} \qquad \text{sumP} := -\frac{108}{35} + \frac{1}{5}x + \frac{20}{7}x^2$$

$$a_3 := \frac{8}{35} \qquad \text{sumP} := -\frac{108}{35} - x + \frac{20}{7}x^2 + 2x^3$$

$$a_4 := \frac{16}{315} \qquad \text{sumP} := x^4 + 2x^3 + 2x^2 - x - 3$$

```
> sumP-Px;
```

$$0$$

We graph the two identical expressions, and emphasize that Legendre polynomials represent one particular convenient set of polynomials to span functions on the interval $-1 \le x \le 1$, which occurs most often in the context of expansions in terms of the polar angle after the substitution $x = \cos\theta$. Orthogonal polynomials can be constructed by Gram-Schmidt orthogonalization of the monomials after a choice of inner product for some interval $a \le x \le b$. However, they are normally classified according to the differential equation which they satisfy.

```
> plot({sumP,Px},x=-2..2,y=-10..10);
```

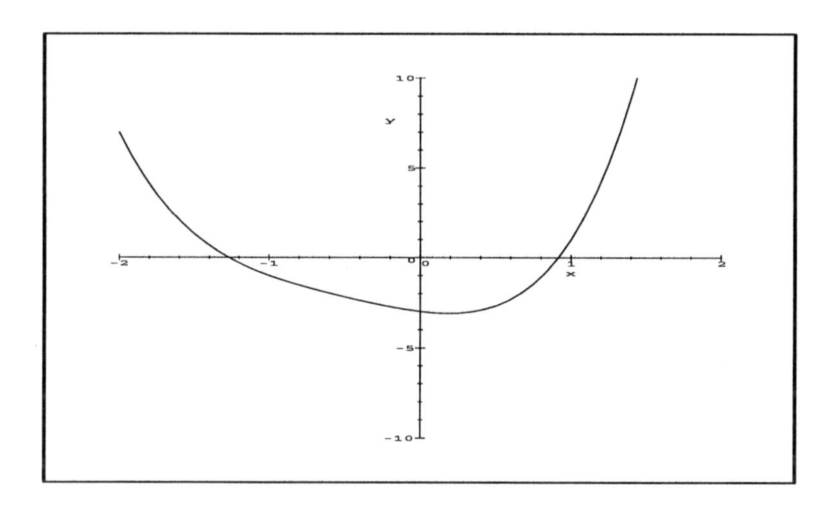

The function (a polynomial in our case) is represented exactly in terms of Legendre polynomials and the expansion holds also outside the interval $[-1, 1]$. For completeness we also show the calculation by hand using linear algebra [Ay52, p. 226]:

```
> P4=P(4,x);
```

$$P4 = \frac{35}{8}x^4 - \frac{15}{4}x^2 + \frac{3}{8}$$

We solve $P_4(x)$ for the x^4 term

```
> x4:=solve(",x^4);
```

$$x4 := \frac{8}{35}P4 + \frac{6}{7}x^2 - \frac{3}{35}$$

and substitute the x^4 term in $P(x)$ to express it in terms of $P_4(x)$ and lower powers x^i.

```
> Px1:=subs(x^4=x4,Px);
```

$$Px1 := \frac{8}{35}P4 + \frac{20}{7}x^2 - \frac{108}{35} + 2x^3 - x$$

```
> P3=P(3,x);
```

$$P3 = \frac{5}{2}x^3 - \frac{3}{2}x$$

Now we repeat the procedure by solving $P_3(x)$ for x^3

```
> x3:=solve(",x^3);
```

$$x3 := \frac{2}{5}\,P3 + \frac{3}{5}\,x$$

and substituting x^3 in `Px1` in terms of $P_3(x)$ and lower powers x^i.

```
> Px2:=subs(x^3=x3,Px1);
```

$$Px2 := \frac{8}{35}\,P4 + \frac{20}{7}\,x^2 - \frac{108}{35} + \frac{4}{5}\,P3 + \frac{1}{5}\,x$$

```
> P2=P(2,x);
```

$$P2 = \frac{3}{2}\,x^2 - \frac{1}{2}$$

Now x^2 is substituted in terms of $P_2(x)$ and lower powers x^i,

```
> x2:=solve(",x^2);
```

$$x2 := \frac{2}{3}\,P2 + \frac{1}{3}$$

```
> Px3:=subs(x^2=x2,Px2);
```

$$Px3 := \frac{8}{35}\,P4 + \frac{40}{21}\,P2 - \frac{32}{15} + \frac{4}{5}\,P3 + \frac{1}{5}\,x$$

and finally x itself is replaced by $P_1(x)$.

```
> Px4:=subs(x=P1,Px3);
```

$$Px4 := \frac{8}{35}\,P4 + \frac{40}{21}\,P2 - \frac{32}{15} + \frac{4}{5}\,P3 + \frac{1}{5}\,P1$$

We save ourselves the substitution $1 =$ `P(0,x)` and note that the original polynomial is exactly represented by the expansion coefficients for the Legendre polynomials

$$\{a_k\} = \{-\frac{32}{15}, \frac{1}{5}, \frac{40}{21}, \frac{4}{5}, \frac{8}{35}\}.$$

```
> subs(P4=P(4,x),P3=P(3,x),P2=P(2,x),P1=P(1,x),Px4)-Px;
```

$$0$$

In the original monomial basis representation the polynomial $P(x)$ is defined as

$$\{b_k\} = \{-3, -1, 2, 2, 1\}.$$

Exercise 8.3.1: Use Rodrigues' formula and integration by parts to prove the orthogonality of the Legendre polynomials on the interval $[-1, 1]$.

8.4 Coulomb Waves

Physicists denote as Coulomb functions the continuum ($E > 0$) solutions to the hydrogen atom problem, which describe proton–electron (or charged particle) scattering. These solutions are analogous to the plane wave scattering solutions to the free particle problem in 3D, except that they take into account the presence of a charge Z at the origin. One can consider an attractive or repulsive Coulomb potential at the origin. Due to the spherical symmetry of the Hamiltonian one splits the angular part from the wavefunction in the usual way, i.e., one performs a partial wave decomposition.

The radial equations in each angular momentum symmetry sector labeled by l have two independent solutions. We begin with the regular solution for $l = 0$ (s wave):

```
> assume(k>0);  SE:=-diff(u(r),r$2)/2-(Z/r+k^2/2)*u(r)=0;
```

$$\mathrm{SE} := -\frac{1}{2}\left(\frac{\partial^2}{\partial r^2}\mathrm{u}(r)\right) - \left(\frac{Z}{r} + \frac{1}{2}k\tilde{~}^2\right)\mathrm{u}(r) = 0$$

First we solve for $Z = 1$ and some fixed wavenumber, e.g., $k = 1$. We attempt a series solution as otherwise no answer is found by Maple.

```
> sl1:=dsolve({subs(Z=1,k=1,SE),u(0)=0,D(u)(0)=1},u(r),series);
```

$$\mathrm{sl1} := \mathrm{u}(r) = r - r^2 + \frac{1}{6}r^3 + \frac{1}{18}r^4 - \frac{1}{72}r^5 + \mathrm{O}(r^6)$$

Now we lift the restriction of a numerical k value:

```
> sl2:=dsolve({subs(Z=1,SE),u(0)=0,D(u)(0)=1},u(r),series);
```

$$sl2 := \mathrm{u}(r) = r - r^2 + \left(\frac{1}{3} - \frac{1}{6}k\tilde{~}^{-2}\right)r^3 + \left(\frac{1}{9}k\tilde{~}^{-2} - \frac{1}{18}\right)r^4$$
$$+ \left(-\frac{1}{36}k\tilde{~}^{-2} + \frac{1}{120}k\tilde{~}^{-4} + \frac{1}{180}\right)r^5 + \mathrm{O}(r^6)$$

Finally we generalize to arbitrary nuclear charge Z:

```
> sl3:=dsolve({SE,u(0)=0,D(u)(0)=1},u(r),series);
```

$$\mathrm{sl3} := \mathrm{u}(r) = r - Z\,r^2 + \left(-\frac{1}{6}k\tilde{~}^{-2} + \frac{1}{3}Z^2\right)r^3 + \left(\frac{1}{9}k\tilde{~}^{-2}Z - \frac{1}{18}Z^3\right)r^4$$
$$+ \left(\frac{1}{120}k\tilde{~}^{-4} - \frac{1}{36}k\tilde{~}^{-2}Z^2 + \frac{1}{180}Z^4\right)r^5 + \mathrm{O}(r^6)$$

```
> rhs(sl3);
```

$$r - Z\,r^2 + \left(-\frac{1}{6}k\tilde{~}^{-2} + \frac{1}{3}Z^2\right)r^3 + \left(\frac{1}{9}k\tilde{~}^{-2}Z - \frac{1}{18}Z^3\right)r^4$$

$$+\left(\frac{1}{120}\,k^{\sim 4}-\frac{1}{36}\,k^{\sim 2}\,Z^2+\frac{1}{180}\,Z^4\right)r^5+O(r^6)$$

Now we show that the result is given in terms of the Kummer function (confluent hypergeometric function). The Kummer function is obtained from a restriction of the general hypergeometric function.

```
> readlib(hypergeom):   F11:=(a,b,z)->hypergeom([a],[b],z):
```

```
> F11(1+I*Z/k,2,2*I*k*r)*exp(-I*k*r)*r;
```

$$\mathrm{hypergeom}\left(\left[1+\frac{I\,Z}{k^{\sim}}\right],[2],2\,I\,k^{\sim}\,r\right)\mathrm{e}^{(-I\,k^{\sim}\,r)}\,r$$

```
> myres:=simplify(evalc(series(",r=0)));
```

$$\mathrm{myres} := r - Z\,r^2+\left(-\frac{1}{6}\,k^{\sim 2}+\frac{1}{3}\,Z^2\right)r^3+\left(\frac{1}{9}\,k^{\sim 2}\,Z-\frac{1}{18}\,Z^3\right)r^4$$
$$+\left(\frac{1}{120}\,k^{\sim 4}-\frac{1}{36}\,k^{\sim 2}\,Z^2+\frac{1}{180}\,Z^4\right)r^5+O(r^6)$$

```
> simplify(convert(myres-rhs(sol3),polynom));
```

$$0$$

We have shown that (at least up to some finite order) the result of solving the Schrödinger equation for the Coulomb problem at positive energies is given by the answer in terms of the confluent hypergeometric function. However, up to release 3 in Maple the hypergeometric function does not evaluate numerically for complex argument:

```
> CW:=(Z,k,r)->simplify(evalc(F11(1+I*Z/k,2,2*I*k*r)
> *exp(-I*k*r)*r)):
```

```
> evalf(CW(1,1,5));
```

$$1.418310928\,\mathrm{hypergeom}([1+I],[2],10.\,I)$$
$$+\,4.794621374\,I\,\mathrm{hypergeom}([1+I],[2],10.\,I)$$

One can get some idea about the solution from the series expansion. However, one needs to include sufficiently many terms in the expansion. In the two examples shown below we request 50 terms:

```
> plot(convert(series(CW(1,1,r),r=0,50),polynom),
> r=0..20,-1..1,title=`Z=1, k=1`);
```

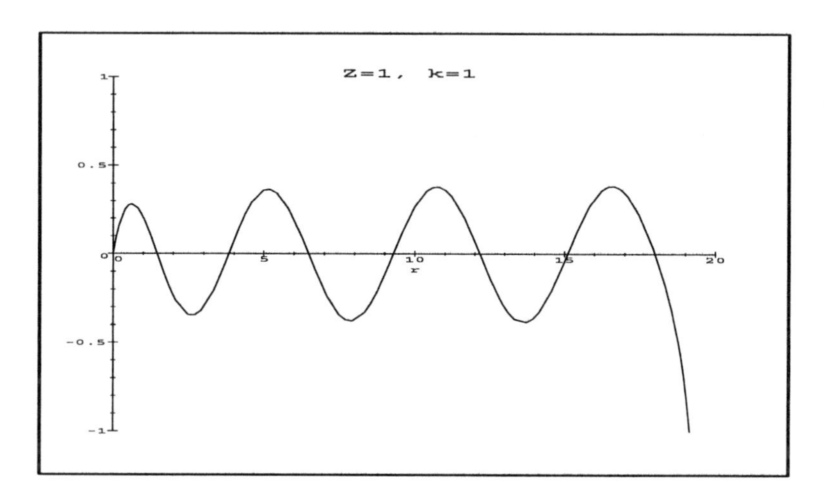


```
> plot(convert(series(CW(1,1/2,r),r=0,50),polynom),
> r=0..20,-1..1,title='Z=1, k=1/2');
```

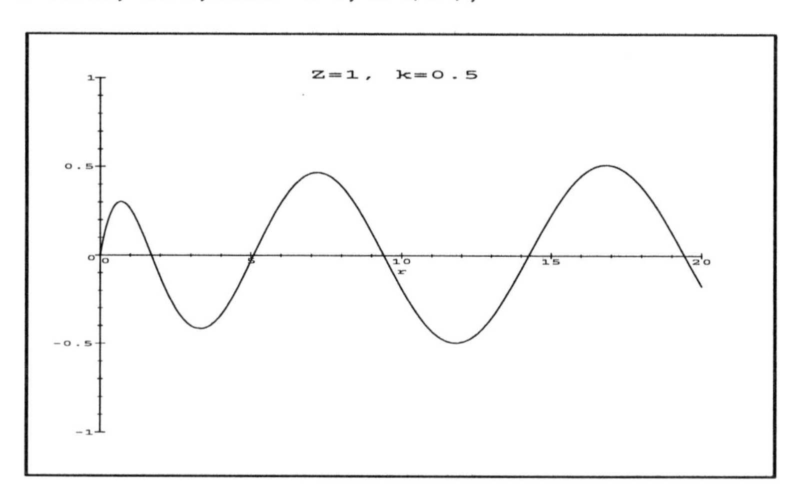

It is apparent how the wavelength of the scattering wave decreases rapidly as the wave moves towards $r = 0$ and how the particle's kinetic energy increases due to the Coulomb singularity in the potential. For smaller wavenumber k the effect is more pronounced.

Now we show that as the restriction $u(0) = 0$ is lifted the solution can acquire an irregular piece at the origin. The additional term displays a $\ln r$ behaviour, and is linearly independent from the regular solution.

```
> sl4:=dsolve({SE},u(r),series);
```

$$\text{sl4} := \mathrm{u}(r) = \_C1\, r \left(1 - Z\, r + \left(-\frac{1}{6} k^{\sim 2} + \frac{1}{3} Z^2\right) r^2 + \frac{1}{36} Z\,(-2\,Z^2 + 4\,k^{\sim 2})\, r^3 + \left(\frac{1}{120} k^{\sim 4} - \frac{1}{36} k^{\sim 2} Z^2 + \frac{1}{180} Z^4\right) r^4 - \frac{1}{43200} Z\,(16\,Z^4 - 160\,k^{\sim 2} Z^2 + 184\,k^{\sim 4})\, r^5 + \mathrm{O}(r^6)\right)$$

$$+ \_C2 \left(\ln(r) \left(-2\,Z\,r + 2\,Z^2 r^2 + \frac{1}{3} Z\,(-2\,Z^2 + k^{\sim 2})\, r^3 + \left(\frac{1}{9} Z^4 - \frac{2}{9} k^{\sim 2} Z^2\right) r^4 - \frac{1}{1440} Z\,(16\,Z^4 - 80\,k^{\sim 2} Z^2 + 24\,k^{\sim 4})\, r^5 + \mathrm{O}(r^6)\right) + \left(1 + \left(-\frac{1}{2} k^{\sim 2} - 3\,Z^2\right) r^2 + \left(\frac{2}{3} k^{\sim 2} Z - \frac{7}{9} Z\,(-2\,Z^2 + k^{\sim 2})\right) r^3 + \left(\frac{43}{108} k^{\sim 2} Z^2 + \frac{1}{24} k^{\sim 4} - \frac{35}{108} Z^4\right) r^4 + \left[-\frac{1}{1440} Z\,(-64\,k^{\sim 2} Z^2 + 68\,k^{\sim 4}) + \frac{101}{43200} Z\,(16\,Z^4 - 80\,k^{\sim 2} Z^2 + 24\,k^{\sim 4})\right] r^5 + \mathrm{O}(r^6)\right)\right)$$

For specific parameter values we obtain:

```
> subs(_C1=1,_C2=1,k=1,Z=1,rhs(sl4));
```

$$r \left(1 - r + \frac{1}{6} r^2 + \frac{1}{18} r^3 - \frac{1}{72} r^4 - \frac{1}{1080} r^5 + \mathrm{O}(r^6)\right) + \ln(r) \left(-2\,r + 2\,r^2 - \frac{1}{3} r^3 - \frac{1}{9} r^4 + \frac{1}{36} r^5 + \mathrm{O}(r^6)\right) + \left(1 - \frac{7}{2} r^2 + \frac{13}{9} r^3 + \frac{25}{216} r^4 - \frac{13}{135} r^5 + \mathrm{O}(r^6)\right)$$

```
> series(",r=0):      res:=convert(",polynom);
```

$$\text{res} := 1 + \left(2 \ln\left(\frac{1}{r}\right) + 1\right) r + \left(-\frac{9}{2} - 2 \ln\left(\frac{1}{r}\right)\right) r^2 + \left(\frac{29}{18} + \frac{1}{3} \ln\left(\frac{1}{r}\right)\right) r^3 + \left(\frac{37}{216} + \frac{1}{9} \ln\left(\frac{1}{r}\right)\right) r^4 + \left(-\frac{119}{1080} - \frac{1}{36} \ln\left(\frac{1}{r}\right)\right) r^5$$

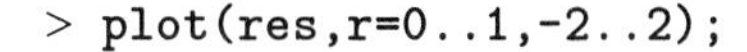

```
> plot(res,r=0..1,-2..2);
```

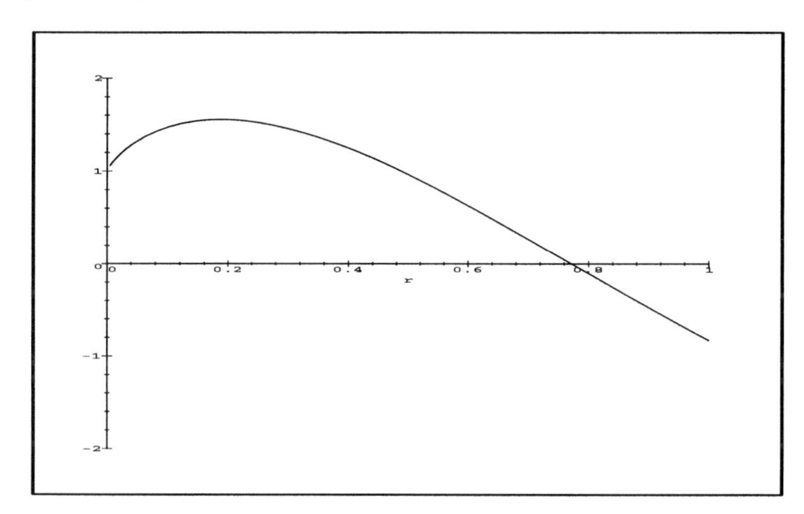

We note that the solutions starts with an infinite slope due to the $\ln(r)$ contribution. Once divided by r the wavefunction diverges at the origin and is, therefore, irregular.

```
> diff(res,r):     limit(",r=0);
```

$$\infty$$

The regular and irregular solutions are required to specify a general solution to a scattering problem with a Coulomb-like long-range potential. The irregular solution conventionally is defined in a complicated way [AS65].

The bound-state solutions to the Coulomb problem can also be understood from the point of view of the confluent hypergeometric function. They correspond to the special case when the series for the Kummer function terminates at finite order and becomes a Laguerre polynomial (cf. Sect. 8.2).

We provide the regular Coulomb wave for arbitrary partial wave number L:

```
> CWL:=(Z,k,L,r)->simplify(evalc(F11(1+L+I*Z/k,2+2*L,2*I*k*r)
> *exp(-I*k*r)*r^(L+1))):
```

We choose $Z = 1$, $k = 1$ ($E = 1/2$), and $L = 1$:

```
> psi1:=CWL(1,1,1,r);
```

$$\psi 1 := \mathrm{hypergeom}(\,[2+I\,],[\,4\,],2\,I\,r\,)\,r^2\cos(\,r\,) \\ - I\,\mathrm{hypergeom}(\,[2+I\,],[\,4\,],2\,I\,r\,)\,r^2\sin(\,r\,)$$

```
> psi1a:=convert(series(psi1,r=0,8),polynom);
```

$$\psi 1a := r^2 - \frac{1}{2}r^3 + \frac{1}{36}r^5 - \frac{1}{504}r^6 - \frac{1}{1680}r^7$$

```
> psi1a:=convert(series(psi1,r=0,28),polynom):
```

We graph the radial probability density shifted by the total scattering energy together with the effective potential:

```
> plot({-1/r+1/r^2,1/2+psi1a^2},r=0..8,V=-0.5..1.5,
> title='Radial probability, Z=1, k=1');
```

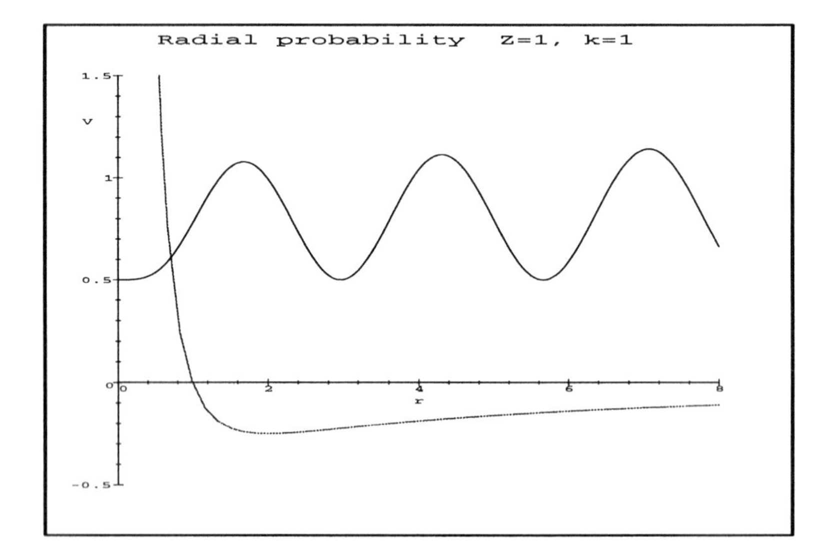

Note that the radial probability density acquires a constant 'amplitude' at large distances. Now we increase the charge to $Z = 4$, and demonstrate the change. The potential is more attractive, as the relationship between Coulomb potential $-Z/r$ and the centrifugal potential $l(l+1)/(2r^2)$ changes:

```
> psi1:=CWL(4,1,1,r);
```

$$\psi 1 := \text{hypergeom}([2+4I],[4],2Ir)\,r^2\cos(r) - I\,\text{hypergeom}([2+4I],[4],2Ir)\,r^2\sin(r)$$

```
> psi1b:=convert(series(psi1,r=0,28),polynom):
```

```
> plot({-4/r+1/r^2,1/2+100*psi1b^2},r=0..8,V=-4..2,
> title='Radial probability, Z=4, k=1');
```

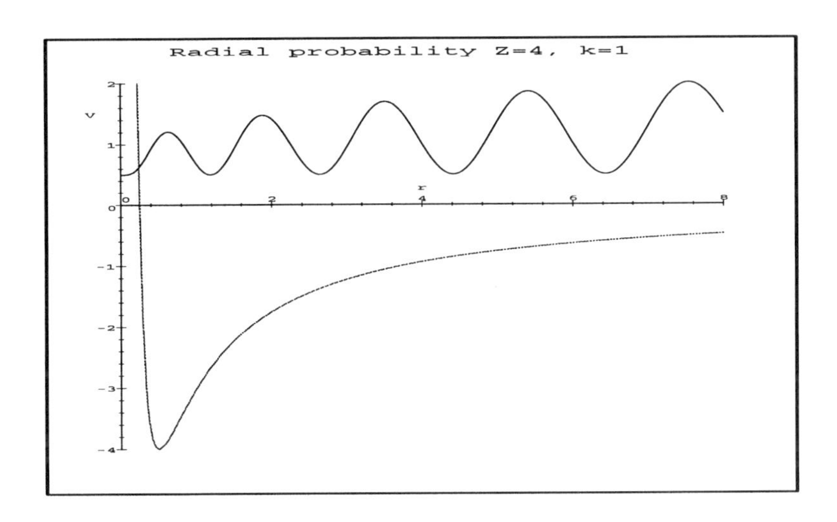

Now we increase the energy to $E = 2 = k^2/2$ and keep the charge as $Z = 4$:

```
> psi1:=CWL(4,2,1,r);
```

$$\psi 1 := \mathrm{hypergeom}([2+2I],[4],4Ir)\,r^2\cos(2r)$$
$$- I\,\mathrm{hypergeom}([2+2I],[4],4Ir)\,r^2\sin(2r)$$

```
> psi1c:=convert(series(psi1,r=0,48),polynom):
```

```
> plot({-4/r+1/r^2,2+200*psi1c^2},r=0..8,V=-4..4,
> numpoints=500,title=`Radial probability, Z=4, k=2`);
```

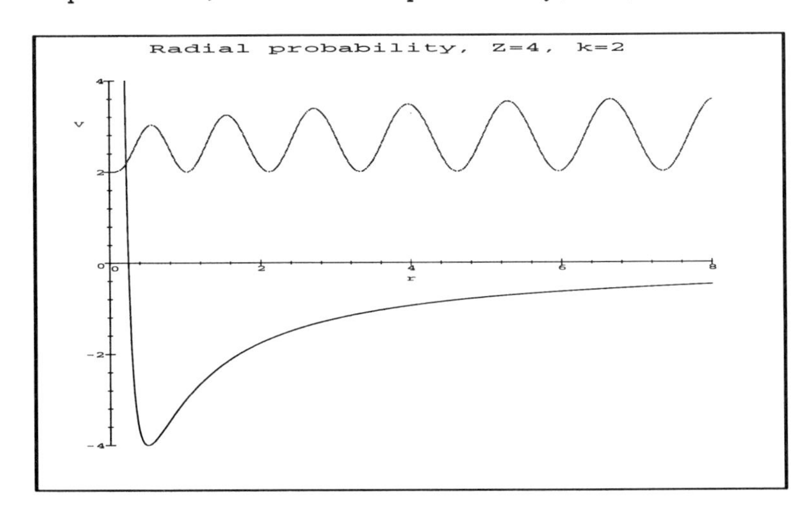

The local free-particle wavenumber $k_0(r)$ defined roughly as the separation between two nodes of the wavefunction (or the wavelength) changes little as a function of r as k is large. What happens for small k? We consider $k = 0.1$:

```
> psi1:=CWL(4,1/10,1,r);
```

$$\psi 1 := \text{hypergeom}\left([2+40\,I],[4],\frac{1}{5}\,I\,r\right)\,r^2\cos\left(\frac{1}{10}\,r\right)$$
$$-\,I\,\text{hypergeom}\left([2+40\,I],[4],\frac{1}{5}\,I\,r\right)\,r^2\sin\left(\frac{1}{10}\,r\right)$$

```
> psi1d:=convert(series(psi1,r=0,28),polynom):
> plot({-4/r+1/r^2,180*psi1d^2},r=0..8,V=-4..4,
> numpoints=500,title=`Radial probability, Z=4, k=0.1`);
```

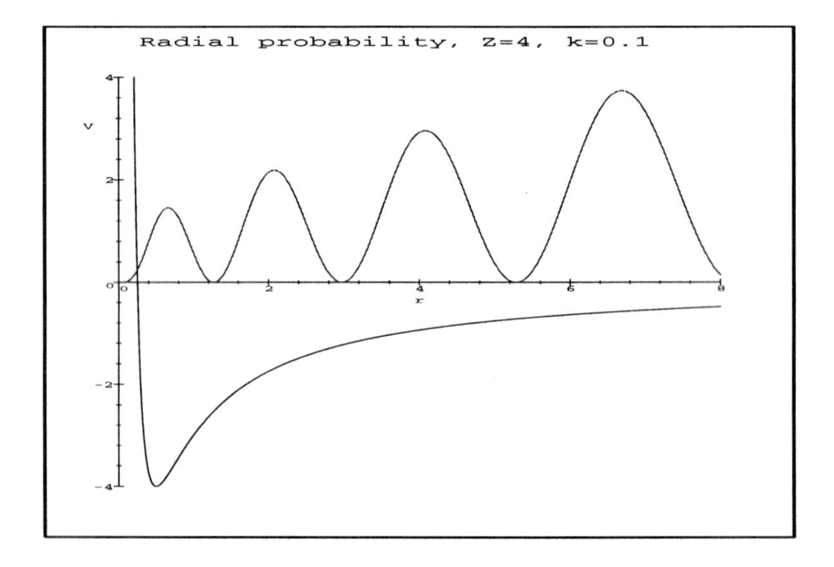

Note that if one graphs the magnitude squared of the wavefunction without the volume element r^2 the picture looks quite different. It is tempting to view this picture as the more appropriate one, as it shows a large probability where the potential is most attractive. However, this is misleading. The radial probability density is more relevant as it takes into account the volume of space involved at a given radial distance. The scattering states are not normalizable in the same sense as the bound states. A constant amplitude A in the asymptotic region corresponds to an incident wave with some amount of flux $|A|^2$ associated with it. The scattering part of the wave is proportional to the amplitude A, and, therefore, A drops out of the calculation.

Note that while the square of the wavefunction is labeled as a probability, this is not meant literally. The continuum state is not normalized in the usual sense. To achieve that, one would have to truncate space at a finite distance L (cf. Sect. 1.1), which is, however, in conflict with the Coulomb problem with its infinite-range potential.

```
> plot({-4/r+1/r^2,140*(psi1d/r)^2},r=0..8,V=-4..4,
> numpoints=500,title='Probability, Z=4, k=0.1');
```

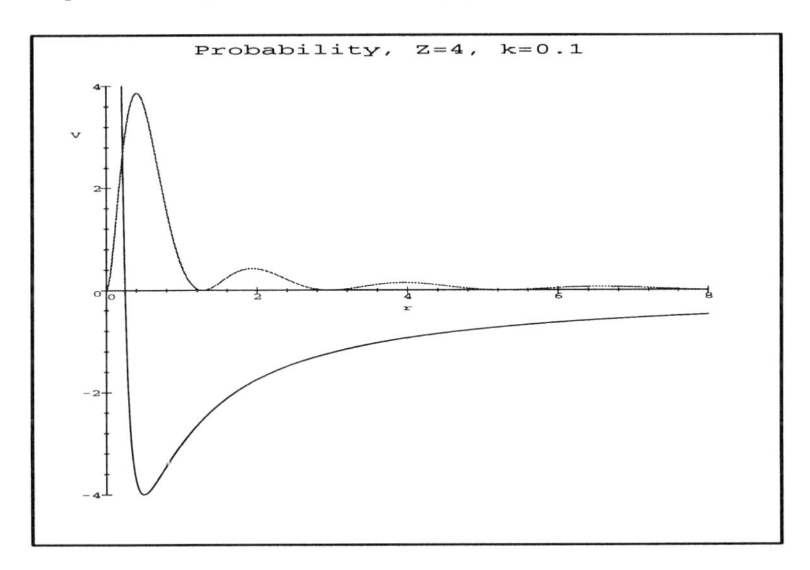

Finally, we show what happens for a repulsive Coulomb potential, i.e., for $Z = -1$. The wave cannot penetrate much into the classically forbidden region that is marked by the crossing point of the baseline for the probability plot (scattering energy) and the classical potential function, i.e., the classical turning point.

```
> psi1:=CWL(-1,1,1,r);
```

$$\psi 1 := \text{hypergeom}(\,[2 - I\,], [\,4\,], 2\,I\,r\,)\,r^2 \cos(\,r\,) \\ - I\,\text{hypergeom}(\,[2 - I\,], [\,4\,], 2\,I\,r\,)\,r^2 \sin(\,r\,)$$

```
> psi1e:=convert(series(psi1,r=0,28),polynom):
```

```
> plot({1/r+1/r^2,0.5+psi1e^2/200},r=0..10,V=0..4,
> numpoints=500,title='Radial probability, Z=-1, k=1');
```

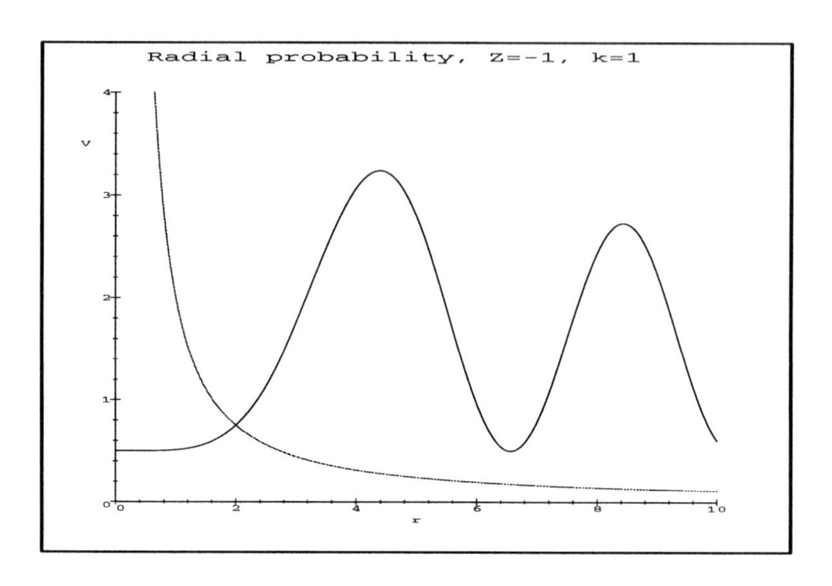

Exercise 8.4.1: Define the Kummer differential equation (also named the confluent hypergeometric equation)

$$xy''(x) + (c - x)y'(x) - ay(x) = 0\,,$$

and find the solution in terms of hypergeometric functions. How does this compare to the conventional definition of the general solution [Ar85, AS65]? Show that the regular solution to the Coulomb problem

$$\frac{\mathrm{d}^2 f}{\mathrm{d}\rho^2} + \left[1 - \frac{2\eta}{\rho} - \frac{L(L+1)}{\rho^2}\right] f(\rho) = 0\,,$$

up to some normalization is given by

$$f_L(\eta, \rho) = \rho^{L+1} \exp(-\mathrm{i}\rho) M(L + 1 - \mathrm{i}\eta, 2L + 2; 2\mathrm{i}\rho)\,,$$

where $M(a, c; x)$ is the regular solution to the Kummer equation.

References

[Ac90] F.S. Acton. *Numerical Methods that Usually Work*, Mathematical Association of America, Washington, 1990.

[Ar85] G. Arfken. *Mathematical Methods for Physicists*, 3rd ed, Academic Press, Orlando, 1985.

[An71] E.E. Anderson. *Modern Physics and Quantum Mechanics*, Saunders, Philadelphia, 1971.

[AS65] M. Abramowitz, I. Stegun. *Handbook of Mathematical Functions with Formulas, Graphs, and Mathematical Tables*, Dover, New York, 1965.

[AS+89] S. Augst, D. Strickland, D.D. Meyerhofer, S.L. Chin, J.H. Eberly. Phys. Rev. Lett. **63**, 2212 (1989).

[Ay52] F. Ayres Jr.. *Schaum's Outline of Theory and Problems of Differential Equations* McGraw-Hill, New York 1952.

[Ba90] L.E. Ballentine. *Quantum Mechanics*, Prentice-Hall, Englewood Cliffs, 1990.

[Ba94] W.E. Baylis. *Theoretical Methods in the Physical Sciences, An Introduction to Problem Solving using Maple V*, Birkhäuser, Boston, 1994.

[BG80] T. Barnes, G.I. Ghandour. Phys. Rev. **D 22**, 924 (1980).

[BD91] S. Brandt, H.D. Dahmen. *Quantum Mechanics on the Personal Computer*, Springer, Berlin, 1991.

[Bl64] D.I. Blokhintsev. *Principles of Quantum Mechanics*, Allyn and Bacon, Boston, 1964.

[BL81] L.C. Biedenharn, J.D. Louck. *Angular Momentum in Quantum Physics*, Addison-Wesley, Reading, 1981.

[BM89] J.J. Brehm, W.J. Mullin. *Introduction to the Structure of Matter*, Wiley, New York, 1989.

[BM94] N.R. Blachmann, M. Mossinghoff. *Maple V Quick Reference*, Brooks/Cole, Pacific Grove, 1994.

[Br70] B.H. Bransden. *Atomic Collision Theory*, Benjamin, New York, 1970.

[BS57] H.A. Bethe, E.E. Salpeter. *Quantum Mechanics of One- and Two-Electron Atoms*, Springer, New York, 1957.

[BS94] D.M. Brink, G.R. Satchler. *Angular Momentum*, 3rd ed, Clarendon, Oxford, 1994.

[BV65] L.C. Biedenharn, H. Van Dam. *Quantum Theory of Angular Momentum*, Academic Press, New York, 1965.

[CD+77] C. Cohen-Tannoudji, B. Diu, F. Laloë. *Quantum Mechanics*, Vols. 1-2, Wiley, New York, 1977.

[CG+93] B.W. Char, K.O. Geddes, G.H. Gonnet, B.L. Leong, M.B. Monagan, S.M. Watt. *First Leaves: A Tutorial Introduction to Maple V*, Springer, New York, 1993.

[CH89] R. Courant, D. Hilbert. *Methods of Mathematical Physics*, Vols. 1-2, Wiley, New York, 1989.

[CO80] E.U. Condon, H. Odabaşi. *Atomic Structure*, Cambridge University Press, Cambridge, 1980.
[CR74] E. Clementi, C. Roetti. At. Data and Nucl. Data Tables **14**, 177 (1974).
[Da69] A.S. Davydov. *Quantum Mechanics*, NEO press, Ann Arbor, 1969.
[DG90] M. Danos, V. Gillet. *Angular Momentum Calculus in Quantum Physics*, World Scientific, Singapore, 1990.
[DH83] J.W. Darewych, M. Horbatsch. Phys. Rev. **A 27**, 2245 (1983).
[DHK92] J.W. Darewych, M. Horbatsch, R. Koniuk. Phys. Rev. **D 45**, 675 (1992).
[EJ+92] W. Ellis, E. Johnson, E. Lodi, D. Schwalbe. *Maple V Flight Manual*, Brooks/Cole, Pacific Grove, 1992.
[ER74] R.M. Eisberg, R. Resnick. *Quantum Physics of Atoms, Molecules, Solids, Nuclei, and Particles*, Wiley, New York, 1974.
[Fe94] J.M. Feagin. *Quantum Methods with Mathematica*, Springer (TELOS), New York, 1994.
[FH65] R.P. Feynman, A.R. Hibbs. *Quantum Mechanics and Path Integrals*, McGraw-Hill, New York, 1965.
[Fl71] S. Flügge. *Practical Quantum Mechanics*, Vols. 1-2 Springer, New York, 1971.
[Ga74] S. Gasiorowicz. *Quantum Physics*, Wiley, New York, 1974.
[Ga79] S. Gasiorowicz. *The Structure of Matter*, Addison-Wesley, Reading, 1979.
[Go80] H. Goldstein. *Classical Mechanics*, 2nd ed, Addison-Wesley, Reading, 1980.
[Gr93] W. Greiner. *Quantum Mechanics: an Introduction*, 2nd ed, Springer, Berlin, 1993.
[Gr94] W. Greiner. *Relativistic Quantum Mechanics: Wave Equations*, 2nd ed, Springer, Berlin, 1994.
[GR80] I.S. Gradshteyn, I.M. Ryzhik. *Table of Integrals, Series and Products*, Academic Press, Orlando, 1980.
[GR86] E.K.U. Gross, E. Runge. *Vielteilchentheorie*, Teubner, Stuttgart 1986.
[GRH91] E.K.U. Gross, E. Runge, O. Heinonen. *Many-Particle Theory*, Adam Hilger IOP Publishing, Bristol, 1991.
[GS+71] A.E.S. Green, D.L. Sellin, G. Darewych. Phys. Rev. **A 3**, 159 (1971).
[GW64] M.L. Goldberger, K.M. Watson. *Collision Theory*, Wiley, New York, 1964.
[HC+82] S. Hagmann, C.L. Cocke, J.R. McDonald, P. Richard, H. Schmidt-Böcking, R. Schuch. Phys. Rev. **A 25**, 1918 (1982).
[He93] A. Heck. *Introduction to Maple*, Springer, New York, 1993.
[HM84] F. Halzen, A.D. Martin. *Quarks and Leptons: An Introductory Course in Modern Particle Physics*, Wiley, New York, 1984.
[Hy30] E. Hylleraas. Z. Physik **65**, 209 (1930).
[Jo79] C.J. Joachain. *Quantum Collision Theory*, North Holland, Amsterdam, 1979.
[Ka93] M. Kaku. *Quantum Field Theory*, Oxford University Press, New York, 1993.
[Ka94] M. Kaluža. Comp. Phys. Comm. **79**, 425 (1994).
[Ko86] S.E. Koonin. *Computational Physics*, Benjamin/Cummings, Menlo Park, 1986.
[LAD87] H.J. Lüdde, H. Ast, R.M. Dreizler. Phys. Lett. **A 125**, 197 (1987).
[LD81] H.J. Lüdde, R.M. Dreizler. J. Phys. **B 14**, 2181 (1981).
[LE62] G.J. Lockwood, E. Everhart. Phys. Rev. **125**, 567 (1962).
[Li92] R.L. Liboff. *Introductory Quantum Mechanics*, 2nd ed, Addison-Wesley, Reading, 1992.
[LL77] L.D. Landau, E.M. Lifshitz. *Quantum Mechanics: Non-relativistic Theory*, Pergamon, Oxford, 1977.

[LP82] L. Lapidus, G.F. Pinder. *Numerical Solution of Partial Differential Equations in Science and Engineering*, Wiley, New York, 1982.

[McG71] J.D. McGervey. *Introduction to Modern Physics*, Academic Press, New York, 1971.

[Mer70] E. Merzbacher. *Quantum Mechanics*, 2nd ed, Wiley, New York, 1970.

[Mes70] A. Messiah. *Quantum Mechanics*, Vols.1-2, North-Holland, Amsterdam, 1970.

[MM71] N.F. Mott, H.S.W. Massey. *The Theory of Atomic Collisions* Clarendon, Oxford, 1971.

[Mo29] P.M. Morse. Phys. Rev. **34**, 57 (1929).

[Mo90] M.A. Morrison. *Understanding Quantum Mechanics*, Prentice-Hall, Englewood Cliffs, 1990.

[MT70] J.B. Marion, S.T. Thornton. *Classical Dynamics*, Harcourt Brace Jovanovich, Orlando, 1970.

[Na90] O. Nachtmann. *Elementary Particle Physics*, Springer, Berlin, 1990.

[Ne82] R. Newton. *Scattering Theory of Waves and Particles*, Springer, New York, 1982.

[PDG92] Particle Data Group. Phys. Rev. **D 45**, 1 (1992).

[Pe78] G. Peach. J. Phys. B 11, 2107 (1978).

[Pe82] G. Peach. Comm. At. Mol. Phys. **16**, 101 (1982).

[PF+92] W.H. Press, B.P. Flannery, S.A. Teukolsky, W.T. Vetterling. *Numerical Recipes: The Art of Scientific Computing*, 2nd ed, Cambridge University Press, Cambridge, 1992.

[Re93] D. Redfern. *The Maple Handbook* Springer, New York, 1993.

[Ro57] M.E. Rose. *Elementary Theory of Angular Momentum*, Wiley, New York, 1957.

[RS80] P. Ring, P. Schuck. *The Nuclear Many-Body Problem* Springer, New York, 1980.

[RT67] L.S. Rodberg, R.M. Thaler. *Introduction to the Quantum Theory of Scattering*, Academic Press, New York, 1967.

[Sa94] J.J. Sakurai. *Modern Quantum Mechanics*, Revised ed, Addison-Wesley, Reading, 1994.

[Sch68] L.I. Schiff. *Quantum Mechanics*, 3rd ed, McGraw-Hill, New York, 1968.

[Si91] A.G. Sitenko. *Scattering Theory*, Springer, Berlin, 1991.

[St85] P.M. Stevenson. Phys Rev. **D 32**, 1389 (1985).

[Sy71] K.R. Symon. *Mechanics*, 3rd ed, Addison-Wesley, Reading, 1971.

[Th92] W.J. Thompson. *Computing for Scientists and Engineers*, Wiley, New York, 1992.

[Th94] W.J. Thompson. *Angular Momentum*, Wiley, New York, 1994.

[TH+87] A. Toepfer, A. Henne, H.J. Lüdde, M. Horbatsch, R.M. Dreizler. Phys. Lett. **A 126**, 11 (1987).

[To92] J.S. Townsend. *A Modern Approach to Quantum Mechanics*, McGraw-Hill, New York, 1992.

[Vi91] P.B. Visscher, Comp. in Phys. **5**, 596-98, (1991).

[VK81] V. Vemuri, W.J. Karplus. *Digital Computer Treatment of Partial Differential Equations*, Prentice-Hall, Englewood Cliffs, 1981.

[WH+89] S.J. Ward, M. Horbatsch, R.P. McEachran, A.D. Stauffer. J. Phys. **B 22**, 3763 (1989).

[XDH92] L. Xiao, J.W. Darewych, M. Horbatsch. Phys. Rev. **A 46**, 4026 (1992).

[Za87] R.N. Zare. *Angular Momentum*, Wiley, New York, 1987.

[Zw92] D. Zwillinger. *Handbook of Integration*, Jones and Bartlett, Boston, 1992.

Index

A. Heck

Introduction to Maple

1993. XIII, 497 pp.
84 figs.
ISBN 3-540-97662-0

Introduction to Maple intends to teach the reader not only **what** can be done by Maple, but also **how** it can be done.

It is not only a readable manual, explaining how to use Maple as a symbolic calculator, it also provides the necessary background to those who want to extend the built-in knowledge of Maple by implementing new algorithms.

The book contains both elementary and more sophisticated examples and many exercises.

1994. 3 1/2" Diskette,
manual with VI, 56 pp.,
9 figs.
ISBN 3-540-14209-6

Waterloo Maple Software

Maple V Student Version

Release 3. Notes

1994. 6 3 1/2" diskettes,
handbook with 70 pp.,
60 notes
ISBN 3-540-14216-9

Waterloo Maple Software

Maple V Student Version

Release 3, DOS/Windows

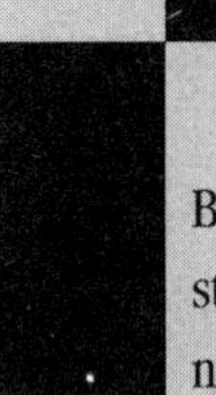

Both **Maple V^{TM} Student Versions** provide the science student or enthusiast with the powerful symbolic and numeric computational capabilities of the Maple V computer algebra system.

An easy-to-implement programming language and comprehensive library of built-in mathematical functions makes solving problems from algebra to differential equations fast and reliable.

Maple's open architecture gives one the freedom to develop a personalized library of functions or to customize the system for almost any application. And most of all, Maple's two and three dimensional graphics puts the power of scientific visualization on your desktop.

Tm.BA95.04.24